METHODS IN MOLECULAR BIOLOGY

Series Editor
**John M. Walker
School of Life and Medical Sciences
University of Hertfordshire
Hatfield, Hertfordshire, AL10 9AB, UK**

For further volumes:
http://www.springer.com/series/7651

Mycotoxigenic Fungi

Methods and Protocols

Edited by

Antonio Moretti

Institute of Sciences of Food Production,
National Research Council, Bari, Italy

Antonia Susca

Institute of Sciences of Food Production,
National Research Council, Bari, Italy

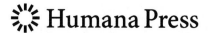

Editors
Antonio Moretti
Institute of Sciences of Food Production
National Research Council
Bari, Italy

Antonia Susca
Institute of Sciences of Food Production
National Research Council
Bari, Italy

ISSN 1064-3745   ISSN 1940-6029   (electronic)
Methods in Molecular Biology
ISBN 978-1-4939-8279-0   ISBN 978-1-4939-6707-0   (eBook)
DOI 10.1007/978-1-4939-6707-0

This Humana Press imprint is published by Springer Nature
The registered company is Springer Science+Business Media LLC
The registered company address is: 233 Spring Street, New York, NY 10013, U.S.A.

Preface

Mycotoxins are toxic fungal metabolites that cause severe health problems in humans and animals after exposure to contaminated food and feed, having a broad range of toxic effects, including carcinogenicity, neurotoxicity, and reproductive and developmental toxicity. The United Nations Commission on Sustainable Development approved in 1996 a work program on indicators of sustainable development that included mycotoxins in food as one of the components related to protection and promotion of human health.

From that program, the concern due to mycotoxin contamination of agro-food crops is in continuous growth worldwide since the level of their occurrence in final products is still high and the consequent impact on human and animal health significant. Moreover, the economic costs for the whole agricultural sector can be enormous, even in developed countries as shown by the losses in the United States alone that can be around $5 billion per annum. Different approaches have been used in mycotoxin research through years. First, implications of mycotoxins in humans were investigated in medicine; later agro-ecological aspects and the fundamental mystery of the biological role for production of secondary metabolites are still analyzed. Regulatory limits, imposed in about 80 countries to minimize human and animal exposure to mycotoxins, also have tremendous economic impact on international trading and must be developed using science-based risk assessments, such as expensive analytical methods used to detect mycotoxins eventually occurring in food and feed. On the other hand, decontamination strategies for mycotoxins in foods and feeds include treatments that could show inappropriate results because nutritional and organoleptic benefits could be deteriorated by the process. Alternatively, programs of mycotoxin prevention and control could be applied through evaluating the contamination of foodstuffs by the related mycotoxin-producing fungi and therefore screening the potential mycotoxin risk associated.

Because mycotoxins are produced within certain groups of fungi, the understanding of their population biology, speciation, phylogeny, and evolution is a key aspect for establishing well-addressed mycotoxin reduction programs. This perspective is of fundamental importance to the correct identification of the mycotoxigenic fungi, since each species/genus can have a species-specific mycotoxin profile which would change the health risks associated with each fungal species. The previous use of comparative morphology has been quickly replaced in the last two decades by comparative DNA analyses that provide a more objective interpretation of data. Advances in molecular biology techniques and the ability to sequence DNA at very low cost contributed to the development of alternative techniques to assess possible occurrence of mycotoxins in foods and feeds based on fungal genetic variability in conserved functional genes or regions of taxonomical interest, or by focusing on the mycotoxigenic genes and their expression. The possibility of using a highly standardized, rapid, and practical PCR-based protocol that can be easily used both by researchers and by nonexperts for practical uses is currently available for some species/mycotoxins and hereby proposed. Further progress in transcriptomics, proteomics, and metabolomics will continue to advance the understanding of fungal secondary metabolism and provide insight into possible actions to reduce mycotoxin contamination of crop plants and the food/feed by-products.

Finally, we do hope that readers will find the chapters of *Mycotoxigenic Fungi: Methods and Protocols* helpful and informative for their own work, and we deeply thank all authors for their enthusiastic and effective work that made the preparation of this book possible.

Bari, Italy   *Antonio Moretti*
                                                                                                            Antonia Susca

Contents

Contributors

MARÍA J. ANDRADE • *Faculty of Veterinary Science, Food Hygiene and Safety, Meat and Meat Products Research Institute, University of Extremadura, Cáceres, Spain*

ALI ATOUI • *Lebanese Atomic Energy Commission-CNRS, Riad El Solh, Beirut, Lebanon; Laboratory of Microbiology, Department of Natural Sciences and Earth, Faculty of Sciences I, Lebanese University, Hadath Campus, Beirut, Lebanon*

ROSA AZNAR • *Department of Biotechnology, Institute of Agrochemistry and Food Technology, IATA-CSIC, Valencia, Spain; Department of Microbiology and Ecology and Spanish Type Culture Collection (CECT), University of Valencia, Valencia, Spain*

SCOTT E. BAKER • *US Department of Energy, Environmental Molecular Sciences Laboratory, Pacific Northwest National Laboratory, Richland, WA, USA*

GIOVANNI A.L. BROGGINI • *Institute for Plant Production Sciences, Agroscope, Wädenswil, Switzerland*

DAREN W. BROWN • *Mycotoxin Prevention and Applied Microbiology Research, US Department of Agriculture, Agricultural Research Service, National Center for Agricultural Utilization Research (USDA–ARS–NCAUR), Peoria, IL, USA*

ROBERT L. BROWN • *Southern Regional Research Center, SDA-ARS New Orleans, LA, USA*

FRANCESCA CARDINALE • *Department of Agricultural, Forest and Food Sciences, University of Turin, Grugliasco, Italy*

JUAN J. CÓRDOBA • *Faculty of Veterinary Science, Food Hygiene and Safety, Meat and Meat Products Research Institute, University of Extremadura, Cáceres, Spain*

ANDRÉ EL KHOURY • *Centre D'Analyses Et De Recherches, Faculté des Sciences, Université Saint-Joseph, Beyrouth, Lebanon*

TAMÁS EMRI • *Faculty of Science and Technology, Department of Biotechnology and Microbiology, University of Debrecen, Debrecen, Hungary*

AHMAD M. FAKHOURY • *Department of Plant Soil and Agriculture Systems, Southern Illinois University, Carbondale, IL, USA*

ANTONIA GALLO • *Institute of Sciences of Food Production (ISPA), National Research Council (CNR), Lecce, Italy*

TERESA GARCÍA • *Facultad de Veterinaria, Departamento de Nutrición, Bromatología y Tecnología de los Alimentos, Universidad Complutense de Madrid, Madrid, Spain*

MARGA VAN GENT • *Biointeractions and Plant Health, Wageningen UR, Wageningen, The Netherlands*

JÉSSICA GIL-SERNA • *Facultad de Ciencias Biologicas, Departamento de Microbiologia, Universidad Complutense de Madrid, Jose Antonio Novais, Madrid, Spain*

ISABEL GONZÁLEZ • *Facultad de Veterinaria, Departamento de Nutrición, Bromatología y Tecnología de los Alimentos, Universidad Complutense de Madrid, Madrid, Spain*

M. TERESA GONZÁLEZ-JAÉN • *Facultad de Ciencias Biologicas, Departamento de Genetica, Universidad Complutense de Madrid, Jose Antonio Novais, Madrid, Spain*

MICHÈLE GUILLOUX-BÉNATIER • *Institut Universitaire de la Vigne et du Vin "Jules Guyot", Université de Bourgogne, Dijon Cedex, France*

MIGUEL JURADO • *Facultad de Ciencias Biologicas, Departamento de Genetica, Universidad Complutense de Madrid, Jose Antonio Novais, Madrid, Spain*

SÁNDOR KOCSUBÉ • *Faculty of Science and Informatics, Department of Microbiology, University of Szeged, Szeged, Hungary*

THEO VAN DER LEE • *Biointeractions and Plant Health, Wageningen UR, Wageningen, The Netherlands*

ANTONIO F. LOGRIECO • *Institute of Sciences of Food Production, National Research Council, Bari, Italy*

INÉS MARÍA LÓPEZ-CALLEJA • *Facultad de Veterinaria, Departamento de Nutrición, Bromatología y Tecnología de los Alimentos, Universidad Complutense de Madrid, Madrid, Spain*

ROSARIO MARTÍN • *Facultad de Veterinaria, Departamento de Nutrición, Bromatología y Tecnología de los Alimentos, Universidad Complutense de Madrid, Madrid, Spain*

PEDRO MARTÍNEZ-CULEBRAS • *Department of Preventive Medicine, Public Health, Food Science and Technology, Bromatology, Toxicology, and Legal Medicine, University of Valencia, Valencia, Spain; Department of Biotechnology, Institute of Agrochemistry and Food Technology (IATA-CSIC), Valencia, Spain*

SALVADOR MIRETE • *Facultad de Ciencias Biologicas, Departamento de Genetica, Universidad Complutense de Madrid, Jose Antonio Novais, Madrid, Spain*

ANTONIO MORETTI • *Institute of Sciences of Food Production, National Research Council, Bari, Italy*

GARY P. MUNKVOLD • *Department of Plant Pathology and Microbiology, Seed Science Center, Iowa State University, Ames, IA, USA*

SIDDAIAH CHANDRA NAYAKA • *DOS in Biotechnology, University of Mysore, Manasagangotri, Mysuru, India*

BELÉN PATIÑO • *Facultad de Ciencias Biologicas, Departamento de Microbiologia, Universidad Complutense de Madrid, Jose Antonio Novais, Madrid, Spain*

ANDREA PATRIARCA • *Laboratorio de Microbiología de Alimentos, Departamento de Química Orgánica, Facultad de Ciencias Exactas y Naturales, Universidad de Buenos Aires, Buenos Aires, Argentina*

MIGUEL ÁNGEL PAVÓN • *Facultad de Veterinaria, Departamento de Nutrición, Bromatología y Tecnología de los Alimentos, Universidad Complutense de Madrid, Madrid, Spain*

GIANCARLO PERRONE • *Institute of Sciences of Food Production (ISPA), National Research Council (CNR), Bari, Italy*

STEPHEN W. PETERSON • *Bacterial Foodborne Pathogens and Mycology Research Unit, National Center for Agricultural Utilization Research, Agricultural Research Service, U.S. Department of Agriculture, Peoria, IL, USA*

VIRGINIA ELENA FERNÁNDEZ PINTO • *Laboratorio de Microbiología de Alimentos, Departamento de Química Orgánica, Facultad de Ciencias Exactas y Naturales, Universidad de Buenos Aires, Buenos Aires, Argentina*

ISTVÁN PÓCSI • *Faculty of Science and Technology, Department of Biotechnology and Microbiology, University of Debrecen, Debrecen, Hungary*

ROBERT H. PROCTOR • *USDA ARS NCAUR, Peoria, IL, USA; United States Department of Agriculture, National Center for Agricultural Utilization Research, Peoria, IL, USA*

PAULA CRISTINA AZEVEDO RODRIGUES • *CIMO/School of Agriculture, The Polytechnic Institute of Bragança, Bragança, Portugal*

ALICIA RODRÍGUEZ • *Faculty of Veterinary Science, Food Hygiene and Safety, Meat and Meat Products Research Institute, University of Extremadura, Cáceres, Spain*

MAR RODRÍGUEZ • *Faculty of Veterinary Science, Food Hygiene and Safety, Meat and Meat Products Research Institute, University of Extremadura, Cáceres, Spain*

SANDRINE ROUSSEAUX • *Institut Universitaire de la Vigne et du Vin "Jules Guyot", Université de Bourgogne, Dijon, France*

VALERIA SCALA • *Department of Environmental Biology, University of Rome "Sapienza", Rome, Italy*

MARÍA VICTORIA SELMA • *Research Group on Quality Safety and Bioactivity of Plant Foods, Department of Food Science and Technology, CEBAS-CSIC, Murcia, Spain*

ALI Y. SROUR • *Department of Plant Soil and Agriculture Systems, Southern Illinois University, Carbondale, IL, USA*

MICHELANGELO STORARI • *Institute for Food Sciences, Agroscope, Bern, Switzerland*

ANTONIA SUSCA • *Institute of Sciences of Food Production, National Research Council, Bari, Italy*

JÁNOS VARGA • *Faculty of Science and Informatics, Department of Microbiology, University of Szeged, Szeged, Hungary*

MARTHA M. VAUGHAN • *United States Department of Agriculture, National Center for Agricultural Utilization Research, Peoria, IL, USA*

COVADONGA VÁZQUEZ • *Facultad de Ciencias Biologicas, Departamento de Microbiologia, Universidad Complutense de Madrid, Jose Antonio Novais, Madrid, Spain*

MUDILI VENKATARAMANA • *Microbiology Division, DRDO-BU-Centre for Life sciences, Bharathiar University Campus, Coimbatore, Tamil Nadu, India*

ELS VERSTAPPEN • *Biointeractions and Plant Health, Wageningen UR, Wageningen, The Netherlands*

IVAN VISENTIN • *Department of Agricultural, Forest and Food Sciences, University of Turin, Grugliasco, Italy*

CEES WAALWIJK • *Biointeractions and Plant Health, Wageningen UR, Wageningen, The Netherlands*

SONGHONG WEI • *College of Plant Protection, Shenyang Agricultural University, Shenyang, Liaoning, China*

TAPANI YLI-MATTILA • *Molecular Plant Biology, Department of Biochemistry, University of Turku, Turku, Finland*

EMRE YÖRÜK • *Department of Molecular Biology and Genetics, Faculty of Arts and Sciences, Istanbul Yeni Yuzyil University, Istanbul, Turkey*

ANNA ZALKA • *Kromat Ltd., Budapest, Hungary*

Part I

Fungal Genera and Species of Major Significance and Their Associated Mycotoxins

Chapter 1

Mycotoxins: An Underhand Food Problem

Antonio Moretti, Antonio F. Logrieco, and Antonia Susca

Abstract

Among the food safety issues, the occurrence of fungal species able to produce toxic metabolites on the agro-food products has acquired a general attention. These compounds, the mycotoxins, generally provided of low molecular weight, are the result of the secondary metabolism of the toxigenic fungi. They may have toxic activity toward the plants, but mostly represent a serious risk for human and animal health worldwide, since they can be accumulated on many final crop products and they have a broad range of toxic biological activities. In particular, mainly cereals are the most sensitive crops to the colonization of toxigenic fungal species which accumulate in the grains the related mycotoxins both in the field, until the harvest stage, and in the storage. According to a Food and Agriculture Organization study, approximately 25 % of the global food and feed output is contaminated by mycotoxins. Therefore, since a large proportion of the world's population consumes, as a staple food, the cereals, the consumption of mycotoxin-contaminated cereals is a main issue for health risk worldwide. Furthermore, mycotoxin contamination can have a huge economic and social impact, especially when mycotoxin occurrence on the food commodities is over the regulation limits established by different national and transnational institutions, implying that contaminated products must be discarded. Finally, the climate change due to the global warming can alter stages and rates of toxigenic fungi development and modify host-resistance and host-pathogen interactions, influencing deeply also the conditions for mycotoxin production that vary for each individual pathogen. New combinations of mycotoxins/host plants/geographical areas are arising to the attention of the scientific community and require new diagnostic tools and deeper knowledge of both biology and genetics of toxigenic fungi. Moreover, to spread awareness and knowledge at international level on both the hazard that mycotoxins represent for consumers and costs for stakeholders is of key importance for developing all possible measures aimed to control such dangerous contaminants worldwide.

Key words *Aspergillus*, *Fusarium*, *Penicillium*, Aflatoxins, Health impact, Economic impact

1 Introduction

> *"Indeed, some authorities now believe that, apart from food security, the single most effective and beneficial change that could be made in human diets around the world would be the elimination of mycotoxins from food."*
> *[Mary Webb]*[1]

---

[1] Mary Webb: New concerns on food-borne mycotoxins, ACIAR Postharvest Newsletter No. 58, 09/ 2001.

Antonio Moretti and Antonia Susca (eds.), *Mycotoxigenic Fungi: Methods and Protocols*, Methods in Molecular Biology, vol. 1542, DOI 10.1007/978-1-4939-6707-0_1, © Springer Science+Business Media LLC 2017

The need of ensuring food safety to consumers is considered a main issue at worldwide level. Problems related to several kinds of food contamination harmful for human and animal health have been increasing in the recent years. Globalization and development of an exchange-based worldwide economy have deeply influenced and enlarged the food market. However, at the same time, the expanded marketing of food products increased the exposure to natural and chemical contaminants. Among the emerging issues in food safety, the increase of plant diseases associated with the occurrence of toxigenic fungal species and their secondary metabolites is of major importance. These fungi can synthesize hundreds of different secondary metabolites, most of whose function is completely unknown. Among these metabolites, the mycotoxins, characterized by low molecular weight, may have toxic activity to several human and animal physiological functions [1]. These pathogenic fungi cause considerable yield losses for crops because mycotoxins can be accumulated in the final crop products and on many products of agro-food interest. Moreover, many of them can also be toxic toward the plants inducing a wide range of symptoms [2]. This contamination can occur both in the field, until the harvest stage, and in the grain storage. According to a Food and Agriculture Organization (FAO) study, approximately 25 % of the global food and feed crop output is affected by mycotoxins [3]. Due to their broad range of biological activities, many of them discovered in the recent decades, the consumption of mycotoxin-contaminated foods became a main issue in food safety worldwide. This is particularly so since a large proportion of the world's population consumes, as a staple food, cereals. The mycotoxin contamination of crops is generally regulated by two main factors: susceptibility of the host plant, on the one hand, and the geographic and climatic conditions, on the other hand. Mycotoxins are produced on the plants before the harvest due to toxigenic fungal contamination in the field and also at the postharvest stage, encompassing stages of the food chain (i.e., storage, processing, and transportation). Moreover, mycotoxins can also be accumulated in animal by-products, due to a carry-over effect, as a consequence of the use of highly contaminated feed. Up to now, the mycotoxins identified show, even in low concentration, carcinogenic, mutagenic, teratogenic, and immuno-, hepato-, nephro-, and neurotoxic properties [4]. Mycotoxins are very stable and are hardly destroyed by processing or boiling of food. They are mainly problematic due to their chronic effects. The farmer operators and crop-processing and livestock-producing industries need rapid methods for detection of both mycotoxigenic fungi and mycotoxin levels in crops in order to reduce the risks for consumers. Additionally, public awareness concerning health risks caused by long-term-exposed mycotoxins is poor or even does not exist. Some mycotoxins are now under regulation in several countries, while the risk related to emerging problems and/or new discovered mycotoxins requires urgent and wide investigations. Main mycotoxin-producing genera are primarily *Aspergillus*, *Fusarium*, and *Penicillium* [5]. However, also the genus *Alternaria* includes several mycotoxigenic species [6]. Most of the species can produce more than a single mycotoxin, but a given mycotoxin can also be produced by species that belong to different genera. Factors that increase the stress status in plants, such as a lack of water and an unbalanced absorption of nutrients, and therefore reduce their immune system, can lead to a higher exposure to mycotoxin contamination. In addition, specific climatic conditions and environmental factors, as temperature and humidity, can influence the growth of mycotoxigenic fungi and eventually stimulate their ability to produce mycotoxins. Finally, mycotoxin contamination can have a huge economic and social impact since their occurrence on the food commodities can be over the regulation limits established by different national and transnational institutions. Therefore, to increase awareness and knowledge, at international level, about the role that mycotoxins can play in food safety is of key importance for developing all possible measures for improving the control of such dangerous contaminants, worldwide.

2   The Impact on Human and Animal Health

Mycotoxins are among the most important food contaminants to control, in order to protect public health around the world. According to Kuiper-Goodman [7], mycotoxins are the most important chronic dietary risk factor, higher than synthetic contaminants, plant toxins, food additives, or pesticide residues. Their associated diseases range from cancers to acute toxicities to developmental effects, including kidney damage, gastrointestinal disturbances, reproductive disorders, or suppression of the immune system. Typically, health effects, associated with mycotoxin exposures, affect populations in low-income nations, where dietary staples are frequently contaminated and control measures are scarce. Although toxigenic fungi can produce hundreds of toxic metabolites, only few of them represent a serious concern for human and animal health worldwide: aflatoxins, produced by species of *Aspergillus* genus; fumonisins, produced mainly by species of *Fusarium*, but also belonging to *Aspergillus* genus; ochratoxin A, produced by species of *Aspergillus* and *Penicillium* genera; patulin, produced by *Penicillium* species; and the mycotoxins produced by *Fusarium* species such as trichothecenes [mainly T-2 and HT-2 toxins (for trichothecenes type A), and deoxynivalenol, nivalenol and related derivatives (for trichothecenes type B)], and zearalenone [5]. Due to their toxicity, a tolerable daily intake (TDI) has been established for the most dangerous mycotoxins that estimates the quantity of a given mycotoxin to which someone can be exposed to daily over a lifetime without it posing a significant risk to health. Aflatoxins are the most toxic mycotoxins and have been shown to be genotoxic, i.e., can damage DNA and cause cancer in animal species, and there is also evidence that they can cause liver cancer in humans [8]. Because aflatoxin contamination is one of the most important risk factors for one of the deadliest cancers worldwide, liver cancer, its eradication in the food supply is critical. It is responsible for up to 172,000 liver cancer cases per year, most of which would result in mortality within several months of diagnosis [9]. Moreover, the link between aflatoxin exposure and childhood stunting is highly worrisome, which can lead to a variety of adverse health conditions that last well beyond childhood. For other agriculturally important mycotoxins such as fumonisins, trichothecenes, and ochratoxin A, proofs that link the exposure to specific human health effects are relatively lower. The role of fumonisins in esophageal cancer is evident, although it may be contributory rather than causal. Trichothecenes have been implicated in acute toxicities and gastrointestinal disorders, and other more long-term adverse effects may be caused by trichothecene exposures. With ochratoxin A, impacts to human populations are limited; however, animal studies suggest possible contributions to toxic effects. The potential for decreased food security, should such foods become less available to a growing human world population, must counterbalance the assessment of human health risks and removal of mycotoxin-contaminated foods from the human food supply. A variety of methods exist by which to mitigate the risks associated with mycotoxins in the diet. Interventions into preharvest, postharvest, dietary, and clinical methods of reducing the risks of mycotoxins to human health, through either direct reduction of mycotoxin levels in crops or reducing their adverse effects in the human body, have been set up [10]. Preharvest interventions include good agricultural practices, breeding, insect pest damage or fungal infection, and biocontrol. Postharvest interventions focus largely on proper sorting, drying, and storage of food crops to reduce the risk of fungal growth and subsequent mycotoxin accumulation. Dietary interventions include the addition of toxin-adsorbing agents into the diet, or increasing dietary diversity where possible. Finally, the mycotoxin exposure in human populations could be added to the effects caused by other factors such as interaction with nutrients or other diet contaminants or environmental conditions. Therefore mycotoxins can also be merely increasing factors for health risks. This is particularly true in vulnerable categories such as young people or pregnant women or populations living in poor/degraded areas. In these situations, mycotoxin exposure may cause even greater damage to human health than previously supposed when evaluated separately. Conversely, reducing mycotoxin exposure in high-risk populations may result in even greater health benefits than may have been previously supposed.

3   Biodiversity of Toxigenic Fungi

Mycotoxins show a very high chemical diversity that reflects also the great genetic diversity of fungal species producing them and occurring worldwide. However, many other minor fungal species, genetically related to the main responsible species, can also be involved in the production of each mycotoxin mentioned above, showing that the risk related to the contamination of food commodities is not only often determined by a single producing species, but is also the result of a multispecies contamination that reflects the great biodiversity existing within the toxigenic fungi. The knowledge of toxigenic fungal biodiversity has arisen to great importance, not only for food safety, but also for the preservation of biodiversity itself. A polyphasic approach by morphological, molecular, and biochemical studies has been developed for many toxigenic fungi and has become clearly fundamental for developing a deep knowledge on the biodiversity of these very important fungi. The correct collection and evaluation of these different data have led to an integrated approach useful to not only identify interspecific differences among the strains belonging to the different toxigenic fungal species, but also deepen the knowledge of their eventual intraspecific genetic and biochemical differences. Moreover, phenotypic and metabolic plasticity of toxigenic fungi that threaten food safety allows these microorganisms to colonize a broad range of agriculturally important crops and to adapt to a range of environmental conditions which characterize various ecosystems. The knowledge of the main environmental parameters related to the growth and the mycotoxin production of toxigenic fungi is therefore of particular interest in biodiversity studies, since they can influence the evolution and the development of populations, the interaction with host plants, and the biosynthesis of mycotoxins in vivo. Nevertheless, the emerging problems related to the global climate change contribute to increasing the risks caused by toxigenic fungi due to the significant influence played by the environment on their distribution and production of related mycotoxins. New mycotoxin/commodity combinations are of further concern and provide evidence of a great capability of these fungi to continuously select new genotypes provided of higher aggressiveness and mycotoxin production. The increase in studies on molecular biodiversity of toxigenic fungi at global level, particularly those that address rapid detection systems, has shown a high intra- and interspecific genetic variability also revealing the existence of intra- and interspecific differences in mycotoxin biosynthetic gene clusters. For some of the most worrisome species belonging to the genera *Aspergillus* and *Fusarium*, differences in the biosynthetic gene clusters for individual families of mycotoxins have been detected, indicating that the differences could be related to specific evolutionary adaptation of each species/population [11]. Examples of these differences have been reported for (a) *A. flavus* and *A. parasiticus* with respect to the presence/absence of the aflatoxin biosynthetic pathway in two subpopulations of these species [12]; (b) *Fusarium* species with respect to the trichothecene pathway [13]; (c) the *A. niger* and *F. fujikuroi* species complexes with respect to the fumonisin biosynthetic gene clusters [14, 15]. Therefore, the increasing of the knowledge of toxigenic fungi molecular biodiversity is a key point to better understand host/pathogen and environment/fungus interactions and to prevent mycotoxin production at its biological origin along all critical points from preharvest to storage of crops. Since the environmental conditions are determinant in the expression of genes involved in biosynthetic pathways of mycotoxins, we could expect that in *Fusarium*, *Aspergillus*, and *Penicillium*, which include ubiquitous species and populations, a great number of unidentified taxa or biological entities should still exist and consequently a great genetic diversity of their mycotoxin profiles as a result of different distribution and location in the genome of their biosynthetic gene pathways.

4   The Economic Impact

The international trade in agricultural commodities amounts to hundreds of millions of tonnes each year. Many of these commodities run a high risk of mycotoxin contamination. Regulations on mycotoxins have been set and are strictly enforced by most importing countries, thus affecting international trade. For some developing countries, where usually agricultural commodities account for a high amount of the total national exports, the economic importance of mycotoxins is considerable, since this contamination is the main cause, as an example, of food commodities rejection by the EU authorities. Moreover, in developing countries, the impact of export losses is worsened by the situation that these countries are forced to export their highest quality maize and retain the poorer grains for domestic use, often at high mycotoxin contamination exposure risk, with an increase of health negative impact on populations and consequent further economical costs. Indeed, the human health impacts of mycotoxins are the most difficult to quantify. These negative effects of mycotoxins are due to acute (single exposure) toxicoses by mycotoxins, as well as chronic (repeated low exposure) effects. In the past decade, several outbreaks of aflatoxicosis in Kenya have led to hundreds of fatalities, while over 98 % of individuals tested in several West African countries were positive for aflatoxin exposure [9]. However, unfortunately, reports on the economical costs due to impact on the human health in developing countries are poorly available, although, due to the elevated levels of mycotoxins, especially aflatoxins, regularly found in the commodities, it is likely that losses consistently exceed those occurring in the Western countries. As an example, losses due to aflatoxins in three Asian countries (Indonesia, the Philippines, and Thailand) were estimated at 900 million US Dollars annually [16]. Of the reported 900 million US Dollar impact of aflatoxins in Southeast Asia, 500 million of the costs were related to human health effects. Thus, according to the National Academy of Sciences, mycotoxins probably contribute to human cancer rates, even in the USA. Therefore, on a global scale, human health is the most significant impact of mycotoxins, with significant losses in monetary terms (through health care costs and productivity loss) and in human lives lost. Furthermore, the evaluation of the economic losses due to mycotoxins is due to several factors such as yield loss due to diseases induced by toxigenic fungi, reduced crop value resulting from mycotoxin contamination, losses in animal productivity from mycotoxin-related health problems, and cost of management along the whole food chain. Reports on the costs of mycotoxins at worldwide level are mostly inconsistent, often limited and in general spotty. Estimates in the USA and Canada vary in a range from 0.5 to 5 billion US Dollars per year. In particular, aflatoxins in the USA have been estimated as 225 million US Dollars per year impact, in maize, while for peanuts the costs were calculated as over 26 million in losses per year during 1993–1996 in the USA and, internationally, the standard limit of 4 ppb (adopting the EU limit) for aflatoxins in peanuts has been estimated to cost about 450 million US Dollars, annually, in lost exports [16]. In another study, Mitchell et al. [17] estimated that aflatoxin contamination could cause losses to the maize industry ranging from 52 million to 1.7 billion US Dollars, annually, in the USA. Also for *Fusarium* mycotoxins, reports on the economical costs due to their contamination on cereals are available. However, these reports are mainly available from the USA where more accurate estimates have been calculated. In particular, in the Tri-State area of Minnesota, North Dakota, and South Dakota, the barley producers have calculated a total loss of 406 million US Dollars for the 6 years from 1993 through 1998 because of deoxynivalenol contamination of kernels. On the other hand, losses associated with deoxynivalenol in wheat kernels in the same States were estimated around 200 million US Dollars per year, in the period from 1993 to 2000, without including the costs of the secondary economic activity, meaning households, retail trade, finance, insurance and real estate, and personal business and professional services, which amount has been evaluated as an additional 2.10 US Dollars for each dollar of lost net revenues for the producer [18]. Finally, according to Windels [19], in the USA, losses of barley and wheat caused by Fusarium head blight epidemics, a common cereal disease related to deoxynivalenol accumulation in the kernels, were estimated in 3 billion US Dollars during the 1990s. Also maize growers can undergo dramatic costs for *Fusarium* mycotoxin contamination of the kernels. In particular, fumonisin contamination accounts for around 18 million dollars per year only for the swine industry in the USA, while the economic loss in Italy has been calculated in 800 million Euro only for the Italian maize business. However, an exact figure for world economic losses resulting from mycotoxin contamination is very difficult to be achieved since it is also very difficult to separate the costs due to the loss of products because of the reduced harvest and the loss of products because of the high level of mycotoxin contamination. Moreover, apart from the obvious losses of food and feed, there are losses caused by lower productivity; losses of valuable foreign exchange earnings; costs incurred by inspection, sampling, and analysis before and after shipments; losses attributable to compensation paid in case of claims; farmer subsidies to cover production losses; research and training; and costs of detoxification. The final combination of these costs may be extremely high.

5   Occurrence and International Control

Toxigenic fungi are extremely common, and they can grow on a wide range of substrates under a wide range of environmental conditions. However, the severity of crop contamination tends to vary from year to year based on weather and other environmental factors. More generally, mycotoxin problems increase whenever shipping, handling, and storage practices are conducive to growth of toxigenic fungi and production of related mycotoxins in final products. The levels of contamination that are recorded at global level can dramatically differ also according to the different geographical areas and they are also strongly related to their social and economical development. To this respect, in some African countries, such as Nigeria and Kenya, or Asian countries, such as India, several cases of acute toxicoses with death or hospitalization of several people periodically still occur [9]. This is due to several aspects: the extreme environmental conditions that often induce proliferation of toxigenic fungi and are conducive for the related mycotoxin production in the field; the uncorrected conditions of storage; and the poor availability of food that makes the waste of contaminated food not possible, since often no other food alternatives are available. On the other hand, in the so-called developed countries, where the availability of food is high, food heavily contaminated by toxigenic fungi is normally avoided; therefore dietary exposure to acute levels of mycotoxins rarely happens, if ever. However, since mycotoxins can resist to processing and can be accumulated into flours and meals at low levels, they can pose a significant chronic hazard to human health. Therefore, to date, based on the toxigenicity of several mycotoxins, regulatory levels have been set by many national governments and adopted for use in national and international food trade. Internationally, the Codex Alimentarius Commission (CAC), the EU, and other regional organizations have issued maximum levels in foods and feeds of some selected mycotoxins according to the provisional maximum TDI, used as a guideline for controlling contamination by mycotoxins, and preventing and reducing toxin contamination for the safety of consumers. CAC was founded in 1963 by the FAO and the World Health Organization (WHO) to develop CODEX standards, guidelines, and other documents pertaining to foods such as the *Code of Practice* for protecting the health of consumers and ensuring fair practices in food trade. The CAC comprises more than 180 member countries, representing 99% of the world's population. The Codex Committee on Food Additives and Contaminants has issued codes of practice for the prevention and reduction of mycotoxin contamination in several foods and feeds (see CAC/RCP issues). As consequences, currently, over 100 countries have regulations regarding mycotoxins or groups of mycotoxins which are of concern in the food and feed [20]. In particular, over 100 nations have aflatoxin regulations, which are intended to protect human and animal health, but also incur economic losses to nations that attempt to export maize and other aflatoxin-contaminated commodities [16]. In Europe, and in particular in the EU, regulatory and scientific interest in mycotoxins has undergone a development in the last 15 years from autonomous national activity toward more EU-driven activity with a structural and network character. Harmonized EU limits now exist for several mycotoxin–food combinations. However, although several national and international organizations and agencies have special committees and commissions that set recommended guidelines, develop standardized assay protocols, and maintain up-to-date information on regulatory statutes (among these, the Council for Agricultural Science and Technology, the FAO of the United Nations, the Institute of Public Health in Japan, and the US Food and Drug Administration Committee on Additives and Contaminants), mycotoxins are still a "largely ignored global health issue" [21]. Furthermore, several scientific associations on mycotoxins keep high the level of awareness on mycotoxin risks in food safety such as the International Society of Mycotoxicology, the Society for Mycotoxin Research of Germany, and the Japanese Association of Mycotoxicology. All these institutions aim to keep constant the evaluation of the occurrence of mycotoxins in foods and feeds. The guidelines used for establishing the tolerance limits are based on epidemiological data and extrapolations from animal models, taking into account the inherent uncertainties associated with both types of analysis. However, a complete elimination of any natural toxicant from foods is an unattainable objective. Therefore, despite the established guidelines around the world for safe doses of mycotoxins in food and feed, there is still a need for worldwide harmonization of mycotoxin regulations, since different sets of guidelines are used. The main efforts of both international scientific community and main international institutions are now addressed to obtain such harmonization.

References

1. Richard JL (2007) Some major mycotoxins and their mycotoxicoses—an overview. Int J Food Microbiol 119:3–10

2. Logrieco A, Bailey JA, Corazza L et al (2002) Mycotoxins in plant disease. Eur J Plant Pathol 108:594–734

3. FAO (Food and Agriculture Organization) (2004) Worldwide regulations for mycotoxins in foods and feeds in 2003. FAO Food and Nutrition Paper 81. Rome, Italy

4. WHO (2002) Evaluation of certain mycotoxins in food. Fifty-sixth report of the JointFAO/WHO expert committee on food additives. WHO Technical Report Series 906. World Health Organization, Ginevra, 62 pp

5. Marasas WFO, Gelderblom WCA, Vismer HF (2008) Mycotoxins: a global problem. In: Leslie JF, Bandyopadhayay R, Visconti A (eds) Mycotoxins. CABI, Oxfordshire, pp 29–40

6. Logrieco A, Moretti A, Solfrizzo M (2009) *Alternaria* mycotoxins: *Alternaria* toxins and plant diseases: an overview of origin, occurrence and risks. World Mycotoxin J 2:129–140

7. Kuiper-Goodman T (1994) Prevention of human mycotoxicoses through risk assessment and risk management. In: Miller JD, Trenholm HL (eds) Mycotoxins in grain: compounds other than aflatoxin. Eagan Press, St. Paul, pp 439–469

8. IARC (International Agency for Cancer Research) (1993) Some naturally occurring substances: food items and constituents, heterocyclic aromatic amines and mycotoxins. IARC Monogr Eval Carcinog Risks Hum 56:1–599

9. Wu F (2013) Aflatoxin exposure and chronic human diseases: estimates of burden of disease. In: Unnevehr L, Grace D (eds) Aflatoxins: finding solutions for improved food safety. International Food Policy Research Institute, Washington, DC, Focus 20, Brief 3

10. Kabak B, Dobson ADW, Var I (2006) Strategies to prevent mycotoxin contamination of food and animal feed: a review. Crit Rev Food Sci Nutr 46:593–619

11. Moretti A, Susca A, Mulé G et al (2013) Molecular biodiversity of mycotoxigenic fungi that threaten food safety. Int J Food Microbiol 167:57–66

12. Gallo A, Stea G, Battilani P et al (2012) Molecular characterization of an *Aspergillus flavus* population isolated from maize during the first outbreak of aflatoxin contamination in Italy. Phytopathol Mediterr 51:198–206

13. Proctor RH, McCormick SP, Alexander NJ et al (2009) Evidence that a secondary metabolic biosynthetic gene cluster has grown by gene relocation during evolution of the filamentous fungus *Fusarium*. Mol Microbiol 74:1128–1142

14. Susca A, Proctor RH, Butchko RAE et al (2014) Variation in the fumonisin biosynthetic gene cluster in fumonisin-producing and non-producing black aspergilli. Fungal Genet Biol 73:39–52

15. Proctor RH, Van Hove F, Susca A et al (2013) Birth, death, and horizontal transfer of the fumonisin biosynthetic gene cluster during the evolutionary diversification of *Fusarium*. Mol Microbiol 90:290–306

16. Wu F (2015) Global impacts of aflatoxin in maize: trade and human health. World Mycotoxin J 8:137–142

17. Mitchell NJ, Bowers E, Hurburgh C et al (2016) Potential economic losses to the US corn industry from aflatoxin contamination. Food Addit Contam Part A Chem Anal Control Expo Risk Assess 33:540–550

18. Robens J, Cardwell K (2003) The costs of mycotoxin management to the USA: management of aflatoxins in the United States. J Toxicol Toxin Rev 22:139–152

19. Windels CE (2000) Economic and social impacts of Fusarium head blight: changing farms and rural communities in the Northern great plains. Phytopathology 90:17–21

20. van Egmond HP, Schothorst RC, Jonker MA (2007) Regulations relating to mycotoxins in food: perspectives in a global and European context. Anal Bioanal Chem 389:147–157

21. Wild CP, Gong YY (2010) Mycotoxins and human disease: a largely ignored global health issue. Carcinogenesis 31:71–82

Chapter 2

Alternaria Species and Their Associated Mycotoxins

Virginia Elena Fernández Pinto and Andrea Patriarca

Abstract

The genus *Alternaria* includes more than 250 species. The traditional methods for identification of *Alternaria* species are based on morphological characteristics of the reproductive structures and sporulation patterns under controlled culture conditions. Cladistics analyses of "housekeeping genes" commonly used for other genera, failed to discriminate among the small-spored *Alternaria* species. The development of molecular methods achieving a better agreement with morphological differences is still needed. The production of secondary metabolites has also been used as a means of classification and identification. *Alternaria* spp. can produce a wide variety of toxic metabolites. These metabolites belong principally to three different structural groups: (1) the dibenzopyrone derivatives, alternariol (AOH), alternariol monomethyl ether (AME), and altenuene (ALT); (2) the perylene derivative altertoxins (ATX-I, ATX-II, and ATX II); and (3) the tetramic acid derivative, tenuazonic acid (TeA). TeA, AOH, AME, ALT, and ATX-I are the main. Certain species in the genus *Alternaria* produce host-specific toxins (HSTs) that contribute to their pathogenicity and virulence. *Alternaria* species are plant pathogens that cause spoilage of agricultural commodities with consequent mycotoxin accumulation and economic losses. Vegetable foods infected by *Alternaria* rot could introduce high amounts of these toxins to the human diet. More investigations on the toxic potential of these toxins and their hazard for human consumption are needed to make a reliable risk assessment of dietary exposure.

Key words *Alternaria* species, Taxonomy, Mycotoxins, Grains, Fruits, Vegetables

1 Introduction

The genus *Alternaria* includes more than 250 species of ubiquitous dematiaceous hyphomycetes [1–4]. It is widely distributed in the environment and its spores can be isolated from several different habitats. Some saprotrophic species are commonly found in soil, air, or indoor environments [5]. However, most are plant pathogens that cause pre- and postharvest damage to agricultural products including cereal grains, fruits, and vegetables [6]. The genus can infect more than 4000 host plants. Its spores are among the most common and potent airborne allergens and sensitization to *Alternaria* allergens has been determined as an important onset of childhood asthma in arid regions [7].

Antonio Moretti and Antonia Susca (eds.), *Mycotoxigenic Fungi: Methods and Protocols*, Methods in Molecular Biology, vol. 1542, DOI 10.1007/978-1-4939-6707-0_2, © Springer Science+Business Media LLC 2017

2 Taxonomy

2.1 Morpho-Taxonomy

The genus was first described by Nees [8] with *A. tenuis* as the type. It is characterized by the production of large brown or dark conidia with both longitudinal and transverse septa (phaeodictyospores), borne from inconspicuous conidiophores, and with a distinct conical narrowing or "beak" at the apical end. These structures can be solitary or produced in various patterns of chains. Several subsequent descriptions of additional *Alternaria* species have been made by Elliot [9], Wiltshire [10], Neergaard [11], Joly [12], Simmons [13], and Ellis [1, 2]. The traditional methods for identification of *Alternaria* species are primarily based on morphological characteristics of the reproductive structures, including shape, color, size, septation, and ornamentation. However, due to the wide diversity of species and the complexity of these structures, identification solely based on these characteristics can be extremely laborious and time consuming, becoming restricted to experts in this field.

Several attempts to organize the genus in subgeneric groups to simplify its classification have been proposed, either formally or informally [14]. A common segregation consists in the distinction of two groups according to conidia size, the "large-spored" (conidia size 60–100 μ) and "small-spored" (conidia <60 μ) *Alternaria*. The small-spored species are cosmopolitan saprotrophs, plant pathogens, allergens, and mycotoxin producers, being the most commonly reported group in foods. Its taxonomy is still under revision, and there is a need for their accurate identification in a broad range of disciplines.

More recently, Simmons [3] developed a classification based on the species group concept, organizing the genus into a number of species groups distinguished by sporulation patterns and conidia morphology, each of which is typified by a representative species, for instance the *A. alternata*, *A. tenuissima*, *A. infectoria*, *A. porri*, or *A. brassicicola* species group. This subgeneric level classification arranges the morphologically diverse assemblage of *Alternaria* spp. and allows a generalized discussion of morphologically similar species.

A further attempt to simplify the identification of *Alternaria* species was introduced by Simmons and Roberts [15]. Their study involved a large number of small-spored *Alternaria* with the utilization of the three-dimensional sporulation pattern as a tool for categorizing species group. They described six major sporulation groups (1–6), each one associated with a representative species. The definition of stable sporulation patterns under controlled culture conditions and the grouping of similar species have been particularly valuable among the small-spored catenulate *Alternaria*, which represent the most challenging in terms of accurate diagnostics due to their complex three-dimensional sporulation patterns [6].

Simmons has intended to cover the entire genus in his series of taxonomic essays in *Alternaria Themes and Variations* [16–20], describing at least 296 taxa sufficiently distinctive to be maintained in an initial assembly of named species. His identification manual [4] summarizes descriptions and illustrations of the maintained species based on the examination of stable isolates in axenic culture.

There are still discrepancies among the use of morphological characters as criteria of identification for small-spored *Alternaria* species. Those classifications based on conidial size as the primary taxonomic criterion concluded that all isolates whose spore dimensions fall within the range described for *A. alternata* should be considered to belong to this species. Nishimura et al. [21] proposed naming all pathogen species indistinguishable from *A. alternata* by conidial size, which were host-specific toxin producers, as pathotypes of *A. alternata*. Thus, several species were included in this collective group, such as *A. gaisen* (Japanese pear pathotype), *A. citri* (citrus pathotype), and *A. mali* (apple pathotype), as shown in Table 1. Rotem [22] named these pathotypes as special forms of *A. alternata* (e.g., *A. alternata* f. sp. *lycopersici* for the tomato pathotype). Several adverse consequences of these approaches have been pointed out in many subsequent scientific works. They criticized the inclusion of large amounts of discriminating data in the literature under a single nondiscriminating name [23]. Moreover, it has been demonstrated that some pathotypes can spontaneously lose the capacity of producing the host-specific toxin, with a consequent loss of pathogenicity. It has also been suggested that lateral

Table 1
Host-specific toxins of plant pathogen *Alternaria* species

Disease	Pathotype	Species (synonym)	Toxins
Black spot of Japanese pear	*A. alternata* Japanese pear pathotype	*Alternaria gaisen* Nagano (*A. kikuchiana* Tanaka)	AK
Black spot of strawberries	*A. alternata* strawberry pathotype	*A. alternata* f. sp. *fragariae* Dingley	AF
Brown spot of tangerine	*A. alternata* tangerine pathotype	*A. tangelonis* Simmons (*A. citri* tangerine pathotype)	ACT
Leaf spot of rough lemon	*A. alternata* rough lemon pathotype	*A. limoniasperae* Simmons (*A. citri* rough lemon pathotype)	ACR
Brown spot of tobacco	*A. alternata* tobacco pathotype	*A. longipes* Mason	AT
Alternaria blotch of apple	*A. alternata* apple pathotype	*A. mali* Roberts	AM
Stem canker of tomato	*A. alternata* tomato pathotype	*A. arborescens* Simmons (*A. alternata* f. sp. *lycopersici* Keissl)	AAL

gene transfer of toxin genes might occur, indicating that toxin production is not a stable character. Thus a system for classifying the small-spored *Alternaria* species based on pathotype is not a practical or desirable system for *Alternaria* taxonomy [24]. The extended use of this system has led to the general belief that *A. alternata* is the most abundant small-spore taxon in nature.

2.2 Molecular Taxonomy

With the advancement of molecular techniques, several studies have examined taxonomic relationships among *Alternaria* spp. using a variety of methods in an attempt to establish consensus with contemporary morphological based species. Most of them have been focused on small-spored catenulate *Alternaria*, which show little resolution in their molecular phylogeny. However, cladistics analyses of "housekeeping genes" commonly used for other genera, such as the mitochondrial large subunit (mtLSU) ribosomal DNA, internal transcribed spacer (ITS), β-tubulin, translation elongation factor α, calmodulin, actin, and chitin synthetase, failed to discriminate among the small-spored species, except for the *A. infectoria* species group. Analyses of RAPD and PCR-RFLP data were effective to distinguish small-spored from large-spored species, such as *A. porri* or *A. solani* [25, 26], and provided resolution among some of the most common small-spored species groups. Pryor and Michailides [27] obtained separate clusters for *A. infectoria*, *A. arborescens*, and a combined *A. alternata/A. tenuissima* cluster. These last two species groups have proved to be the most difficult to discriminate by molecular techniques, although they can be distinguished by culture in standardized conditions. Roberts et al. [28] reported that RAPD analyses resolved the small-spored morphological groups or species *A. gaisen*, *A. longipes*, *A. tenuissima* sp.-grp., *A. arborescens* sp.-grp., and *A. infectoria* sp.-grp. Peever et al. [29] found that an endopolygalacturonase (endoPG) gen and two anonymous loci were sufficiently variable to differentiate members of the *A. alternata* sp.-grp., with general agreement, but not strict congruence between morphological classification and the phylogeny. This research was expanded by Andrew et al. [24], using OPA1-3, OPA10-2, and endoPG, founding strict agreement between morphology and phylogenetic lineage for isolates classified in the *A. arborescens* group, but not for the *A. alternata* and *A. tenuissima* groups.

More recently, Lawrence et al. [7] attempted to assess the phylogenetic relationships among *Alternaria* and closely related genera, using a larger sample of taxa. Based on the analysis of five loci (gpd, *Alt a1*, actin, plasma membrane ATPase, calmodulin) they introduced two new species groups, *A. panax* and *A. gypsophila*, and proposed to elevate eight asexual species groups to the taxonomic status of sections within *Alternaria*, since morphological features of the species groups were not congruent with molecular data. According to their results, the sexual phylogenetic

Alternaria lineage, the *A. infectoria* sp.-grp., did not get the status of section. In another recent work, Woundenberg et al. [30] intended to delineate phylogenetic lineages within *Alternaria* and allied genera based on nucleotide sequence data of parts of the 18S nrDNA (SSU), 28S nrDNA (LSU), the internal transcribed spacer regions 1 and 2 and intervening 5.8S nrDNA (ITS), glyceraldehyde-3-phosphate dehydrogenase (GAPDH), RNA polymerase second largest subunit (RPB2), and translation elongation factor 1-alpha (TEF1) gene regions. Species of *Alternaria* were assigned to 24 *Alternaria* sections, of which 16 were newly described, and 6 monotypic lineages. As a result of this proposed classification many of the most common small-spored species groups described by Simmons [4], such as *A. gaisen*, *A. tenuissima*, *A. arborescens*, *A. longipes*, among others, were enclosed together into the *Alternata* section of which *Alternaria alternata* (Fr. Keissl) was described as the type species. *A. infectoria* remained differentiated from the asexual small-spored species groups as the type species of the *Infectoriae* section, in which other members of the *A. infectoria* sp.-grp. described by Simmons were included.

In a study on the *Alternaria* species causing brown spot of citrus, Stewart et al. [31] found that morphospecies described as citrus pathogens were poorly supported by molecular analyses, sequencing endoPG gen, two anonymous, noncoding SCAR markers OPA1-3 and OPA2-1, and one noncoding microsatellite flanking region Flank-F3. According to their results, citrus brown spot is caused by a maximum of two species of *Alternaria*, and they suggested that taxonomic revision of *Alternaria*-infecting citrus, based on congruent morphological and genetic analyses, is needed. Characterization of *Alternaria* species by morphological and molecular analyses is important in making a correct identification, but might not be sufficient to differentiate between closely related species groups. Lineage sorting, recombination, and horizontal transfer make phylogenetic analyses and species delimitation among small-spored *Alternaria* challenging. Sequencing of "housekeeping genes" or some functional genes has not provided segregation among the small-spored *Alternaria* species. However, this lack of resolution does not necessarily imply that they all belong to the same species; it might indicate that there is little diversity among the isolates on the particular sequences under study. Techniques such as RAPD, which characterizes random priming sites across the entire genome, provided better resolution for the small-spored species. There is still the need for molecular methods that could achieve a better agreement between morphological differences.

2.3 Chemo-Taxonomy

In addition to morphology and molecular analysis, the production of secondary metabolites has been used as a means of classification and identification, taking advantage of the enormous potential of this genus to biosynthesize secondary metabolites.

Chromatographic methods such as thin-layer chromatography, initially, and high-performance liquid chromatography with diode array detection (HPLC-DAD) or combined with mass spectrometry (HPLC-MS), later, have been used in several scientific works to determine the profiles of metabolites produced on standardized laboratory media [32]. Gas chromatography combined with mass spectrometry (GC-MS) has been used for volatile secondary metabolites in *Penicillium* taxonomy [33]. Nowadays, the method of preference for fungal chemotaxonomical studies is HPLC-DAD-MS [34, 35].

Extraction methods are easy to use, less time consuming than morphological characterization, and relatively economic, and they have been successful in differentiating between species in other genera such as *Aspergillus*, *Fusarium*, and *Penicillium* [32]. Secondary metabolite data can be statistically analyzed to determine a characteristic profile for a species or species group, or they can be used to determine species-specific metabolites that could be adopted as chemotaxonomic markers in taxon identification.

Andersen and Thrane [36] reported that the chemical profiles for the *A. infectoria* sp.-grp. contained unique metabolites not identical to any of the known *Alternaria* metabolites, and it could be useful to distinguish between this and the *A. alternata* sp.-grp. Andersen et al. [37] extended their studies demonstrating the chemical and morphological segregation of *Alternaria alternata*, *A. gaisen* (Japanese pear pathotype), and *A. longipes* (tobacco pathotype), when the cultures were grown under standardized conditions. Based on these results they recommended the use of the species names instead of their corresponding *A. alternata* pathotype. In a further work, Andersen et al. [38] found that the secondary metabolite profile from the *A. infectoria* sp.-grp. is chemically very different from both the *A. arborescens* and the *A. tenuissima* sp.-grp. with only a few metabolites in common. *A. arborescens* and *A. tenuissima* sp.-grp. had most of the known *Alternaria* metabolites in common, but they also produced a number of metabolites by which the two species groups can be distinguished. Furthermore, by combining morphological and chemical data obtained by two different methods (HPLC-UV and MS) more host-specific toxin-producing *Alternaria* isolates could be segregated from *A. alternata*. Andersen et al. [23] showed that the analyses of cultural and chemical data allowed to segregate *A. alternata*, *A. longipes* (tobacco pathotype), *A. gaisen* (Japanese pear pathotype, syn. *A. kikuchiana*), *A. tangelonis*, *A. turkisafria*, and *A. limoniasperae*, which have been segregated as new species from the citrus pathogen complex, but regarded as pathotypes of *A alternata*. The metabolite profile of *A. alternata* was different from those of the five species that are commonly described as *A. alternata* pathotypes. In another work from Andersen et al. [39], chemotaxonomy proved useful to discriminate between *A. dauci*, *A. solani*, and *A. tomatophila* and sets of species-specific metabolites could be selected for each of these species as chemotaxonomic markers.

2.4 Polyphasic Taxonomy

Most recently, a polyphasic approach, which integrates three sets of data, such as morphological characteristics, molecular analyses, and secondary metabolite profiling, has been applied to segregate *Alternaria* species, especially in an attempt of differentiate the complicated small-spored species. The combination of all the information provided by different perspectives represents a powerful tool for classification of this complex genus.

The polyphasic approach was applied by Andersen et al. [40] to characterize strains from the *A. infectoria* sp.-grp. and closely related species. The *A. infectoria* sp.-grp. could be separated from *Embellisia abundans*, *Chalastospora cetera*, and *Alternaria malorum* based on morphology, secondary metabolite profiles, and molecular classification. From the chemical analysis, the main factor segregating *A. infectoria* was the capability of producing altertoxins and novae-zelandins. Sequence analyses of ITS, *gpd*, and translocation elongation factor 1α showed two clades, one with all the *A. infectoria* sp.-grp. strains and one with the rest of the species tested. This polyphasic approach revealed that *A. malorum* var. *polymorpha* and *A. malorum* strains do not belong in *Alternaria*, but in the *Chalastospora* genus, as several distinct species.

Another polyphasic study was carried on to characterize endophytic *Alternaria* strains isolated from grapevine [41]. A pooled cluster analysis was obtained by combining morphological, molecular, and chemical data. The species were morphologically identified as members of the *A. arborescens* and *A. tenuissima* species group, and the RAPD analysis confirmed these results and showed that they were molecularly distinct from strains belonging to the *A. alternata* sp.-grp. The strains were also grouped in the same way by chemotaxonomy, with strains producing metabolites typical of these species groups.

3   *Alternaria* Toxins

Alternaria spp. can produce a wide variety of toxic metabolites which play an important role in plant pathogenesis. About 70 toxic metabolites of *Alternaria spp.* have been characterized to date. These bioactive compounds with different chemical structure also exhibit different biological activities and functions and under certain conditions of temperature and humidity could accumulate in vegetable foods and be harmful to humans and animals [42, 43]. These metabolites belong principally to three different structural groups: (1) the dibenzopyrone derivatives, alternariol (AOH), alternariol monomethyl ether (AME), and altenuene (ALT); (2) the perylene derivative altertoxins (ATX-I, ATX-II, and ATX II); and (3) the tetramic acid derivative, tenuazonic acid (TeA). TeA, AOH, AME, ALT, and ATX-I are the main *Alternaria* mycotoxins that can be found as contaminants of food commodities. Of particular health concern is the association found between *Alternaria* contamination in cereal grains and the high levels of human esophageal cancer in China [44, 45].

3.1 Alternariol, Alternariol Monomethyl Ether, and Altenuene

The mutagenicity and carcinogenicity of AME and AOH, and their relevance to the etiology of human esophageal cancer, were studied. These mycotoxins were the main toxic compounds found in grains in an area with high incidence of esophageal cancer. AME and AOH might cause cell mutagenicity and transformation, and could combine with the DNA isolated from human fetal esophageal epithelium and promote proliferation of human fetal esophageal epithelium in vitro. Also, squamous cell carcinoma of the fetal oesophagus could be induced by AOH [42, 45]. The mutagenicity of AOH in Chinese hamster V79 cells and in mouse lymphoma L5178Y tk+/− cells (MLC) was investigated. The mutagenic potency of AOH was about 50-fold lower than that of the established mutagen 4-nitroquinoline-*N*-oxide in both cell lines. The mutagenicity of AOH may have an incidence on the carcinogenicity of this mycotoxin [46].

AOH inhibited metabolic activity and cellular proliferation of porcine granulosa cells. In the regulation of female fertility the hormone progesterone (P4) plays an important role. AOH and AME inhibited P4 secretion in cultured porcine granulosa cells, so their reproductive cycles in pig and other mammalian species may be affected [47].

Cell proliferation studies on human endometrial adenocarcinoma cell line (Ishikawa) and Chinese hamster V79 cells indicated that AOH inhibited cell proliferation by interfering with the cell cycle [48]. AOH induces oxidative DNA damage and DNA strand breaks [49]. AOH and AME act as topoisomerase poisons, which contribute to their genotoxic properties and might cause DNA damage in human colon carcinoma cells. DNA topoisomerases are enzymes regulating DNA topology during transcription, replication, chromosome condensation, and maintenance of genome stability. When interference with the activity of topoisomerases occurs the DNA integrity could be affected [50].

There are very few toxicological data on altenuene, indicating that it has a low acute toxicity and a low-to-moderate antimicrobial activity [51, 52].

3.2 Altertoxins

ATXs are mutagenic in the Ames test when Salmonella strains TA98 and TA100 were used.

ATX-I, ATX-II, and ATX-III are more potent mutagens and acute toxins to mice than AOH and AME [42, 53]. ATX-I was studied by Schrader et al. [54] with and without nitrosylation, using Ames Salmonella strains TA97, TA102, and TA104. ATX-I was mutagenic in strain TA102 and weakly mutagenic in strain TA104. Nitrosylation of ATX-I enhanced mutagenicity. ATX-I was also assessed for mammalian mutagenicity in Chinese hamster V79 lung fibroblasts and rat hepatoma H4IIE cells. ATX-I was not mutagenic in either V79 cells or H4IIE cells, but nitrosylated ATX-I was also directly mutagenic in mammalian test systems.

ATX-II is highly mutagenic in the Ames test and is a potent mutagen in cultured Chinese hamster V79 cells. ATX-II is at least 50 times more potent as a mutagen than AOH and AME. ATX-II does not affect the cell cycle but causes DNA strand breaks of V79 cells [55].

ATX-I and -II have been studied in the Caco-2 cell system, which is a widely accepted in vitro model for human intestinal absorption and metabolism. Caco-2 cells are derived from a human colonic tumor and form a monolayer with tight junctions similar to the human intestinal epithelium. ATX-I was well absorbed from the intestinal lumen and ATX-II intestinal absorption was very low. It must be expected that ATX-II will act primarily in the digestive tract and that ATX-I will reach blood circulation and act systemically [56].

3.3 Tenuazonic Acid

TeA is toxic to several animal species, e.g., mice, chicken, and dogs. In dogs, it caused hemorrhages in several organs. Increasing TeA doses in chicken feed suppressed weight gain and increased internal hemorrhages. TeA is more toxic than AOH, AME, and ALT. TeA is not mutagenic in bacterial systems [42, 53, 57]. Precancerous changes were observed in esophageal mucosa of mice [58]. Sorghum grain colonized by *Phoma sorghina* that contained TeA was associated with the human hematological disorder known as Onyalai in Southern Africa [42].

Using *Chlamydomonas reinhardtii*, *Vicia faba* root tip, and three mammalian normal cell lines, toxicity of TeA was examined. The growth and chlorophyll concentration of *C. reinhardtii* were inhibited. TeA also inhibited the proliferation of 3 T3 mouse fibroblasts (3 T3 cells), Chinese hamster lung cells (CHL cells), and human hepatocytes (L-O2 cells). These results suggested that TA inhibited protein biosynthesis in the cells [59].

3.4 Host-Specific Toxins

Certain species in the genus *Alternaria* produce low-molecular-weight compounds known as host-specific toxins (HSTs) that contribute to their pathogenicity and virulence. Plants that are susceptible to the pathogen are sensitive to the toxin and all isolates that fail to produce HSTs lose pathogenicity to the plants. These host-specific forms have been earlier designed as pathotypes of *A. alternata*, as it is mentioned above, but this classification has not been accepted widely because of difficulties in the discrimination of small-spored *Alternaria* species with few morphological characteristics [60, 61]. In more recent works they were assigned to other species, as it is shown in Table 1.

Simmons and Roberts [15], based in three-dimensional conidiation patterns for differentiating similar species in the *Alternaria* small-spored groups, sorted the isolates from black spot lesions of Japanese pear into six conidiation groups or species groups. Molecular phylogenetic studies have failed in resolving species groups and host association within the small-spored *Alternaria* species [24].

Chemical structures of HSTs have been determined. Toxins of the Japanese pear, strawberry, and tangerine pathotypes were found to be similar metabolites that are esters of the epoxydecatrienoic acid (EDA). The Japanese pear pathotype produces AK toxins I and II. Both toxins exhibit toxicity only on susceptible pear cultivars. The strawberry pathotype affects strawberry-susceptible cultivars. This pathotype was also pathogenic to susceptible Japanese pear in laboratory and produces AF toxins I, II, and III. AF toxin I is toxic to both strawberry and pear, AF toxin II is toxic only to pear, and toxin III is highly toxic to strawberry and slightly to pear. The tangerine pathotype affects tangerines and mandarins and was also found pathogenic to Japanese pear cultivars. The tangerine pathotype produces ACT toxins I and II. ACT toxin I is toxic to both citrus and pear.

The chemical structure of AM toxin I from the apple pathotype was elucidated as a cyclic tetrapeptide and the rough lemon pathotype produces ACR toxins. The major toxin, ACR toxin I, is a C19 polyalcohol with a dihydropyrone ring.

The tomato pathotype produces AAL toxins which are similar to fumonisins. It is known that fumonisins, very toxic mycotoxins produced by *Fusarium* species, can cause leukoencephalomalacia and pulmonary edema syndrome in animals and are associated to human esophageal cancer and neural tube defects. Fumonisins and AAL toxins together are called sphinganine analog mycotoxins (SAMT) due to their structural similarity to sphinganine, which is the backbone precursor of sphingolipids. AAL toxins and fumonisins show similar toxicity to plants and mammalian cells and also exhibited inhibitory activity to ceramide synthase, which is involved in sphingolipid biosynthesis. AAL toxins are produced by the tomato pathogen.

The mechanism for SAMT to execute their toxicity is through the competitive inhibition of sphinganine N-acetyltransferase (ceramide synthase). This leads to the obstruction of complex sphingolipid biosynthesis, such as the important second messenger ceramide in animal systems, and the accumulation of sphinganine. The inhibition of this enzyme leads to various diseases in animals and humans as ceramides and sphingolipids are ubiquitous constituents of eukaryotic cells and involved in crucial signal transduction of numerous cellular processes. SAMT are also found to induce apoptosis. In addition to their animal toxicity, AAL toxins are known as the causal agent of stem canker in tomato.

The gene clusters involved in HST production have been identified from the Japanese pear pathotype (*AKT* genes), strawberry pathotype (*AFT* genes), tangerine pathotype *(ACT* genes), apple pathotype (*AMT* genes), rough lemon pathotype (*ACRT* genes), and tomato pathotype (*ALT* genes). There is evidence that these biosynthetic genes were clustered in small chromosomes of <2.0 Mb. These chromosomes appear to be conditionally dispensable (CD) chromosomes, which are not required for growth but that are essential to produce toxin and to cause disease. CD chromosomes, which nonpathogenic strains do not have, suggest that the ability to produce HSTs in the pathotypes could be acquired by intraspecies transfer of CD chromosomes. Protoplast fusion experiments provided evidence for intraspecies transfer of CD chromosomes in *A. alternata*. Hybrid strains between the tomato and apple pathotypes and between the tomato and strawberry pathotypes were made by protoplast fusion [62, 63]. The fusants synthesized two toxins produced by the parental strains and showed pathogenicity to both plants affected by the toxins. The fusants carried two CD chromosomes, one derived from each of the parental strains. It seems that *A. alternata* is able to accept and maintain a small, exogenous chromosome in its genome. This fact could indicate that pathogenicity could be acquired by strains by horizontal transfer of an entire pathogenicity chromosome and this could provide a possible mechanism by which new pathogens arise in nature [60–63].

3.5 Tentoxin

Tentoxin is a cyclic tetrapeptide from plant pathogen *Alternaria spp.* that inhibits chloroplast with the development of chlorotic symptoms on infected tissues. There is no direct effect of tentoxin on chlorophyll synthesis. Two fundamental processes are linked with this fact. The first one is inhibition of energy transfer of the chloroplast-localized CF1 ATPase. This process alone could not be responsible for the chlorosis because tentoxin also completely inhibits the transport of nuclear enzyme polyphenol oxidase (PPO) into the plastid even in etioplasts which should have no CF1 ATPase activity. Without this action PPO has no enzyme activity. Inhibition of these two steps seems to be linked, and both are inhibited in vivo in tentoxin-sensitive plant species and not affected in insensitive species. Tentoxin was also responsible for chlorophyll accumulation through overenergization of thylakoids, but this fact does not explain its effects on PPO processing in etioplasts without thylakoid membranes. The linkage of the β-subunit of proton ATPase to PPO processing remains unexplained [43, 64].

4 Natural Occurrence of *Alternaria* Toxins in Food and Feed

Alternaria species are plant pathogens that cause spoilage of agricultural commodities with consequent mycotoxin accumulation and economic losses. Mycotoxin accumulation in fruits and vegetables may occur in the field and during harvest, postharvest, and storage (Table 2).

Table 2
Natural occurrence of *Alternaria* toxins in food and feed

Food/feed	Mycotoxin range (µg/kg) (No. positive samples/no. total samples)					Country	References
	TeA	AOH	AME	ALT	ATX-I		
Wheat	Max 4224[a] (322/1064)	Max 832[a] (86/1064)	Max 905[a] (33/1064)	Max 197[a] (7/1064)	–	Germany	Muller et al. [65]
Wheat	1001-8814 (12/64)	645-1348 (4/64)	546-7451 (15/64)	–	–	Argentina	Azcarate et al. [66]
Feeding wheat	–	0.3-29 (21)[b]	0.3-133 (21)[b]	ND	–	Czech. Rep.	Zachariasova et al. [67]
Feeding maize	–	0.3-37 (8)[b]	0.3-34 (8)[b]	ND	–	Czech. Rep.	Zachariasova et al. [67]
Feeding oat	–	295-523 (3)[b]	223-444 (3)[b]	ND	–	Czech. Rep.	Zachariasova et al. [67]
Soya beans	–	25-211 (23/50)	62-1153 (22/50)	–	–	Argentina	Oviedo et al. [68]
Tomato sauces	ND	4-33 (11/17)	1-9 (12/17)	ND	ND	Switzerland	Noser et al. [69]
Tomato sauces	ND	4.0-6.8 (5/10)	ND	3.8-4.8 (8/10)	–	Italy	Prelle et al. [70]
Ketchup	10.2-1787 (31/31)	2.5-300 (14/31)	0.32-38 (28/31)	–	–	China	Zhao et al. [71]
Ketchup	ND	4-5 (3/19)	1 (3/19)	ND	ND	Switzerland	Noser et al. [69]
Tomato pure	29-4012 (29/80)	187-8.8 (6/80)	84-1.7 (26/80)	–	–	Argentina	Terminiello et al. [72]
Tomato pure	ND	4-10 (8/24)	1-4 (7/24)	ND	ND	Switzerland	Noser et al. [69]
Red wine	–	0.36-7.5 (5/5)	0.04-0.15 (5/5)	–	-	Germany	Asam et al. [75]
Red wine	–	0.03-7.41 (20/25)	0.01-0.23 (20/25)	–	–	Canada	Scott et al. [75]
White wine	–	0.10-7.59 (6/6)	ND	–	–	Germany	Asam et al. [73]
White wine	–	0.67-1.48 (2/23)	0.02-0.06 (2/23)	–	–	Canada	Scott et al. [75]
Grape juice	–	0.10-1.05 (5/5)	ND	–	–	Germany	Asam et al. [73]

						Country	Reference
Grape juice	–	0.03-0.46 (5/10)	0.01-39.5 (5/10)	–	–	Canada	Scott et al. [75]
Apple juice	–	0.16-0.22 (3/4)	ND	–	–	Germany	Asam et al. [73]
Apple juice	24.3-45.3 (2/10)	ND	ND	45.6 (1/10)	–	Italy	Prelle et al. [70]
Orange juice	–	0.16-0.24 (2/2)	ND	–	–	Germany	Asam et al. [73]
Citrus juice	1.21-4.3 (9/36)	ND	0.11-0.20 (4/36)	–	–	China	Zhao et al. [71]
Dried wine berries	4-18973 (10/13)	52-1308 (11/13)	776-26 (10/13)	4120-48 (7/13)	7.7-159 (11/13)	Slovakia	Mikusova et al. [76]

ND not detected, below the detection limit

– not determined

[a]range not reported

[b]total number of samples

number of positive samples not available in the reference

Vegetable foods infected by *Alternaria* rot could introduce high amounts of these toxins to the human diet if moldy fruit is not removed before processing.

4.1 Tomatoes

Tomatoes are susceptible to fungal decay because of their soft skin. *Alternaria* is responsible of the disease known as "black mold of tomato." Typical lesions are dark brown to black areas, with firm texture that can become several centimeters in diameter. Fruits become more susceptible to fungal invasion during ripening. The disease is favored by warm and rainy weather. Temperature is one of the major factors that affect the shelf life of tomato fruits, and, to control mold growth and toxin accumulation in tomatoes, the temperature should be maintained below 6 °C to avoid infection.

Alternaria mycotoxin occurrence has been reported in tomatoes. TeA was the major toxin produced in naturally infected fruits. Lower levels of AOH and AME were also recorded.

Moldy tomatoes could be used for processed tomato products with the consequent accumulation of toxins in these products. TeA, AOH, AME, ALT, and ALTX were detected in tomato paste, tomato pulp, and tomato puree samples, occasionally in very high amounts [57, 72, 77].

4.2 Apples

Moldy core rot is a factor that reduces apple fruit quality and it is a worldwide problem occurring in most countries where apples are grown. The disease is produced by *Alternaria* spp. Infection occurs via the open calyces, into the core or carpel regions, during fruit ripening and storage or by fungal spores on the fruit surface that enter through wounds formed during harvesting and handling. *Alternaria* strains isolated from rotten apples produced AOH and AME in the whole fruits after inoculation. High levels of mycotoxins were found in processed apple products made with apples affected by moldy core. The natural occurrence of AOH, AME, TeA, ALT, and ALTX in samples of apple juice and apple juice concentrate was reported in several countries [70, 78, 79].

4.3 Citrus Fruits

"Black heart rot" of oranges and lemons caused by *Alternaria* species is described as internal blackening of the fruit. Fruit with these defects should not be used to produce juice because the accumulation of toxins could occur.

Alternaria brown spot is a disease of mandarins, tangerines, and various tangerine hybrids. The pathogen causes necrotic lesions in mature fruit that are unacceptable to consumers. TeA, AME, and AOH were found in rotten samples [71, 79].

4.4 Cereal Grains

Alternaria is the most common genus found in cereal grains in several regions of the world. References from many countries about prevalence of this fungus in cereals indicate a very high incidence with more than 90 % of the grains affected. Infected grains develop a disease called "black point" consisting of a discoloration of the germ and the seed due to mycelial and conidial masses. Small grain cereals such as wheat, triticale, barley, and oats are frequently infected, whereas rice and maize are less susceptible. Black point is known to affect grain quality, giving a grayish color to the flour and by-products with great economic losses. Several *Alternaria* species have been involved. *A. triticina* is the major cause of wheat leaf blight. The *A. infectoria* species group is the casual agent of black point in certain wheat cultivars in Argentina, Australia, North America, and several European countries. Small grain cereals are frequently contaminated with *Alternaria* mycotoxins. Natural occurrence of AOH, AME, and TeA has been reported worldwide in wheat, barley, and oats [57, 65, 66, 80].

4.5   Other Foods

Olives are often affected by *Alternaria*, particularly if the fruits remain in the soil for a long time after ripening. Several *Alternaria* toxins were also found in olive oil as well as in other edible oils (rapeseed, sesame, and sunflower).

Alternaria mycotoxins have been reported in many other vegetable foods that are frequently infected by the fungus, such as peppers, melons, mangoes, sunflower, soya beans, raspberries, pecans, and Japanese pears. AOH and AME were detected in several fruit beverages such as grape juices, cranberry nectar, raspberry juice, red wine, and prune nectar [42, 57, 68, 79].

5   *Alternaria* Secondary Metabolite Profiles

The *Alternaria* genus is characterized by its enormous capacity of biosynthesizing secondary metabolites; many of them are known mycotoxins, others are phytotoxins, but the toxicity of most of them is still to be investigated.

It is known that the *A. infectoria* species group has a secondary metabolite profile completely different from the other small-spored species groups. Several works have showed that none of the isolates belonging to the *A. infectoria* sp.-grp. was able to produce any of the known *Alternaria* metabolites, such as alternariols, altenuene, tentoxin, tenuazonic acid, altersolanols, and AAL toxins [38, 40]. These isolates were instead producers of infectopyrone, 4Z-infectopyrone, novae-zelandin A, and novae-zelandin B, metabolites that could be used as chemotaxonomic markers for the *A. infectoria* sp.-grp. [81].

The metabolites confirmed to be synthesized by *A. alternata* include altenuene, alternariol, alternariol monomethyl ether, and altertoxins, but not tenuazonic acid [37, 38]. Although several works in the literature reported the production of tenuazonic acid by *A. alternata* the discrepancies in this genus taxonomy could have led to most of the small-spored *Alternaria* species identified

Table 3
Secondary metabolites most frequently produced by small-spored *Alternaria* species

Metabolite	A. alternata	A. tenuissima	A. arborescens	A. longipes	A. gaisen	A. tangelonis	A. turkisafria	A. limonia sperae	A. mali
Altenuene	+	+[b]	+[b]	−	−	−	−	−	+
Alternariol	+	+	+	-	+	+	+	+	+
Alternariol monomethyl ether	+	+	+	-	+	+	+	+	+
Altersetin	−	+	+	−	+	+	+	+[c]	+[a]
Altertoxin I	+[a]	+	+[b]	+	+	+	+	+	+
Tentoxin	+[b]	+[b]	+[c]	−	+	+	+	−	+
Tenuazonic acid	−	+[a]	+[a]	+	+	+	+	+	+

(+) >90 % isolates
[a]70–90 % isolates
[b]<30 % isolates
[c]<10 % isolates

Table 4
Secondary metabolites most frequently produced by large-spored *Alternaria* species

Metabolite	A. dauci	A. porri	A. solani	A. tomatophila
Altenuene	−	−	−	−
Alternariol	+	−	+	+[c]
Alterporriol	−	+	+	−
Altersolanol A	−	+	+	+[b]
Altertoxin	−	−	+[a]	+
Macrosporin	−	+	+	+[b]
Tentoxin	−	+	+/−*	−
Tenuazonic acid	−	−	−	−
Zinniol	+	+	+	−

*discrepant data in literature
[a]70–90 % isolates
[b]50–70 % isolates
[c]<10 % isolates

as *A. alternata*; thus, other small-spored species, whose morphology is closely related to this species, could have been responsible for tenuazonic acid production. Table 3 shows the secondary metabolites most frequently produced by small-spored plant pathogenic and food-contaminant *Alternaria* species.

Large-spored *Alternaria* species can be easily distinguished from the small-spored ones by chemotaxonomy since they have few metabolites in common with them. Alterporriol, altersolanol, and macrosporin are the most frequent compounds biosynthesized by these species. Table 4 shows the most common compounds produced by some plant pathogenic large-spored *Alternaria* species.

6 Conclusions

Species delimitation is important within the *Alternaria* genus, which includes a large number of human and plant pathogenic species, most of them producing a wide range of active metabolites. The correct segregation of species plays a critical role due to the economic importance of *Alternaria* species, especially the small-spored ones, which can contaminate crops of agricultural relevance. Furthermore, for the unambiguous identification of species it is necessary to track the movement of plant pathogens in global trade of foods. The threat of introducing a new pathogen to a different habitat around the world has resulted in rejection of exported crops [31]. The presence of a certain pathogen in food crops is associated with the possible occurrence of secondary metabolites representing a health risk to humans and animals. Thus, incorrect naming of new species or the misidentification of a species could mean significant economic losses.

At present, there are no specific regulations for any of the *Alternaria* toxins in foods. However, these mycotoxins should not be underestimated since they are produced by several *Alternaria* species frequently associated with a wide range of agricultural products and processed plant foods of relevant value in the human diet. More investigations on the toxic potential of these toxins and their hazard for human consumption are needed to make a reliable risk assessment of dietary exposure and better define eventual guidelines on *Alternaria* mycotoxin limits in foods [74].

References

1. Ellis MB (1971) Dematiaceous hyphomycetes. Commonwealth Mycological Institute, Kew
2. Ellis MB (1976) More dematiaceous hyphomycetes. Commonwealth Mycological Institute, Kew
3. Simmons EG (1992) *Alternaria* taxonomy: current status, viewpoint, challenge. In: Chelkowski J, Visconti A (eds) *Alternaria* biology, plant diseases and metabolites. Elsevier Science Publishers, Amsterdam, pp 1–35
4. Simmons EG (2007) *Alternaria*. An identification manual. CBS Fungal Biodiversity Centre, Utrecht

5. Samson RA, Houbraken J, Thrane U et al (2010) Food and indoor fungi. CBS-KNAW Fungal Biodiversity Centre, Utrecht

6. Patriarca A, Vaamonde G, Pinto VF (2014) *Alternaria*. In: Batt CA, Tortorello ML (eds) Encyclopedia of food microbiology, vol 1. Academic Press, Elsevier, London, pp 54–60

7. Lawrence DP, Gannibal PB, Peever TL et al (2013) The sections of *Alternaria*: formalizing species-group concepts. Mycologia 105(3):530–546

8. Nees von Esenbeck CG (1816–17) Das System der Pilze und Schwämme. Stahelschen Buchhandlung, Würzburg

9. Elliott JA (1917) Taxonomic characters of the genera *Alternaria* and *Macrosporium*. Am J Bot 4:439–476

10. Wiltshire SP (1933) The foundation species of *Alternaria* and *Macrosporium*. Trans Br Mycol Soc 18:135–160

11. Neergaard P (1945) Danish species of *Alternaria* and *Stemphylium*. Oxford University Press, London

12. Joly P (1964) Le genre *Alternaria*. Encyclopedie mycologique. P. Lechevalier, Paris

13. Simmons EG (1967) Typification of *Alternaria*, *Stemphylium*, and *Ulocladium*. Mycologia 59:67–92

14. Pryor BM, Gilbertson RL (2000) Molecular phylogenetic relationships amongst *Alternaria* species and related fungi based upon analysis of nuclear ITS and mt SSU rDNA sequences. Mycol Res 104(11):1312–1321

15. Simmons EG, Roberts RG (1993) *Alternaria* themes and variations (73). Mycotaxon 48:109–140

16. Simmons EG (1981) *Alternaria* themes and variations (1–6). Mycotaxon 13:16–34

17. Simmons EG (1993) *Alternaria* themes and variations (63–72). Mycotaxon 48:91–107

18. Simmons EG (1999) *Alternaria* themes and variations (226–235). Classification of citrus pathogens. Mycotaxon 70:263–323

19. Simmons EG (1999) *Alternaria* themes and variations (236–243). Host-specific toxin producers. Mycotaxon 70:325–369

20. Simmons EG (2003) *Alternaria* themes and variations (310–335). Species on Malvaceae. Mycotaxon 88:163–217

21. Nishimura S, Sugihara M, Kohmoto K, Otani H (1978) Two different phases in pathogenicity of the Alternaria pathogen causing black spot disease of Japanese pear. J Fac Agric Tottori Univ 13:1–10

22. Rotem J (1994) The genus *Alternaria*: biology, epidemiology, and pathogenicity. APS Press, St Paul

23. Andersen B, Hansen ME, Smedsgaard J (2005) Automated and unbiased image analyses as tools in phenotypic classification of small-spored Alternaria spp. Phytopathology 95:1021–1029

24. Andrew M, Peever TL, Pryor BM (2009) An expanded multilocus phylogeny does not resolve morphological species within the small-spored *Alternaria* species complex. Mycologia 101(1):95–109

25. Kusaba M, Tsuge T (1995) Phylogeny of *Alternaria* fungi known to produce host-specific toxins on the basis of variation in internal transcribed spacers of ribosomal DNA. Curr Genet 28:491–498

26. Weir TL, Huff DR, Christ BJ, Peter Romaine C (1998) RAPD-PCR analysis of genetic variation among isolates of *Alternaria solani* and *Alternaria alternata* from potato and tomato. Mycologia 90:813–821

27. Pryor BM, Michailides TJ (2002) Morphological, pathogenic, and molecular characterization of *Alternaria* isolates associated with alternaria late blight of pistachio. Phytopathology 92(4):406–416

28. Roberts RG, Reymond ST, Andersen B (2000) RAPD fragment pattern analysis and morphological segregation of small-spored *Alternaria* species and species groups. Mycol Res 104(2):151–160

29. Peever TL, Su G, Carpenter-Boggs L, Timmer LW (2004) Molecular systematics of citrus-associated *Alternaria* species. Mycologia 96:119–134

30. Woudenberg JHC, Groenewald JZ, Binder M, Crous PW (2013) *Alternaria* redefined. Stud Mycol 75:171–212

31. Stewart JE, Timmer LW, Lawrence CB, Pryor BM, Peever TL (2014) Discord between morphological and phylogenetic species boundaries: incomplete lineage sorting and recombination results in fuzzy species boundaries in an asexual fungal pathogen. BMC Evol Biol 14(1):38

32. Frisvad JC, Andersen B, Thrane U (2008) The use of secondary metabolite profiling in chemotaxonomy of filamentous fungi. Mycol Res 112:231–240

33. Larsen TO, Frisvad JC (1994) A simple method for collection of volatile metabolites from fungi based on diffusive sampling from Petri dishes. J Microbiol Methods 19:297–305

34. Nielsen KF, Smedsgaard J (2003) Fungal metabolite screening: database of 474 mycotoxins and fungal metabolites for dereplication by standardised liquid chromatography–UV-mass spectrometry methodology. J Chromatogr A 1002:111–136

35. Smedsgaard J, Nielsen J (2005) Metabolite profiling of fungi and yeast: from phenotype to metabolome by MS and informatics. J Exp Bot 56:273–286

36. Andersen B, Thrane U (1996) Differentiation of *Alternaria infectoria* and *Alternaria alternata* based on morphology, metabolite profiles, and cultural characteristics. Can J Microbiol 42:685–689

37. Andersen B, Krøger E, Roberts RG (2001) Chemical and morphological segregation of *Alternaria alternata*, *A. gaisen* and *A. longipes*. Mycol Res 105(3):291–299

38. Andersen B, Krøger E, Roberts RG (2002) Chemical and morphological segregation of *Alternaria arborescens*, *A. infectoria* and *A. tenuissima* species-groups. Mycol Res 106(2):170–182

39. Andersen B, Dongo A, Pryor BM (2008) Secondary metabolite profiling of *Alternaria dauci*, *A. porri*, *A. solani*, and *A. tomatophila*. Mycol Res 112:241–250

40. Andersen B, Sørensen JL, Nielsen KF et al (2009) A polyphasic approach to the taxonomy of the *Alternaria infectoria* species-group. Fungal Genet Biol 46:642–656

41. Polizzotto R, Andersen B, Martini M et al (2012) A polyphasic approach for the characterization of endophytic *Alternaria* strains isolated from grapevines. J Microbiol Methods 88:162–171

42. Ostry V (2008) *Alternaria* mycotoxins: an overview of chemical characterization, producers, toxicity, analysis and occurrence in foods. World Mycotoxin J 1(2):175–188

43. Lou J, Fu L, Peng Y, Zhou L (2013) Metabolites from *Alternaria* fungi and their bioactivities. Molecules 18:5891–5935

44. Liu GT, Qian YZ, Zhang P et al (1991) Relationships between *Alternaria alternata* and oesophageal cancer. IARC Sci Publ 105:258–262

45. Liu GT, Qian YZ, Zhang P, Dong WH et al (1992) Etiologic role of Alternaria alternata in human esophageal cancer. Chin Med J 105:394–400

46. Brugger EM, Wagner J, Schumacher DM et al (2006) Mutagenicity of the mycotoxin alternariol in cultured mammalian cells. Toxicol Lett 164:221–230

47. Tiemann U, Tomek W, Schneider F et al (2009) The mycotoxins alternariol and alternariol methyl ether negatively affect progesterone synthesis in porcine granulosa cells in vitro. Toxicol Lett 86:139–145

48. Lehmann L, Wagner J, Metzler M (2006) Estrogenic and clastogenic potential of the mycotoxin alternariol in cultured mammalian cells. Food Chem Toxicol 44:398–408

49. Soulhaug A, Vines LL, Ivanova L et al (2011) Mechanisms involved in alternariol-induced cell cycle arrest. Mutat Res 738–739:1–11

50. Fehr M, Pahlke G, Fritz J et al (2009) Alternariol acts as a topoisomerase poison, preferentially affecting the II isoform. Mol Nutr Food Res 53:441–451

51. Siegel D, Feist M, Proske M et al (2010) Degradation of the *Alternaria* mycotoxins alternariol, alternariol monomethyl ether, and altenuene upon bread baking. J Agric Food Chem 58:9622–9630

52. Wang Y, Yang M, Wang X et al (2014) Bioactive metabolites from the endophytic fungus *Alternaria alternata*. Fitoterapia 99:153–158

53. Schrader TJ, Cherry W, Soper K et al (2001) Examination of *Alternaria alternata* mutagenicity and effects of nitrosylation using the Ames Salmonella test. Teratog Carcinog Mutagen 21:261–274

54. Schrader TJ, Cherry W, Soper K, Langlois I (2006) Further examination of the effects of nitrosylation on *Alternaria alternata* mycotoxin mutagenicity in vitro. Mutat Res 606:61–71

55. Fleck SC, Burkhardt B, Pfeiffer E, Metzler M (2012) *Alternaria* toxins: altertoxin II is a much stronger mutagen and DNA strand breaking mycotoxin than alternariol and its methyl ether in cultured mammalian cells. Toxicol Lett 214:27–32

56. Fleck S, Pfeiffer E, Podlech J, Metzler M (2014) Epoxide reduction to an alcohol: a novel metabolic pathway for perylene quinone-type *Alternaria* mycotoxins in mammalian cells. Chem Res Toxicol 27:247–253

57. Logrieco A, Moretti A, Solfrizzo M (2009) *Alternaria* toxins and plant diseases: an overview of origin, occurrence and risks. World Mycotoxin J 2:129–140

58. Yekeler HB, Bitmiş K, Özçelık N et al (2001) Analysis of toxic effects of *Alternaria* toxins on esophagus of mice by light and electron microscopy. Toxicol Pathol 29(4):492–497

59. Zhou B, Qiang S (2008) Environmental, genetic and cellular toxicity of tenuazonic acid isolated from *Alternaria alternata*. Afr J Biotechnol 7(8):1151–1156

60. Akimitsu K, Tsuge T, Kodama M et al (2014) *Alternaria* host-selective toxins: determinant factors of plant disease. J Gen Plant Pathol 80:109–122

61. Tsuge T, Harimoto Y, Akimitsu K et al (2013) Host-selective toxins produced by the plant pathogenic fungus *Alternaria alternata*. FEMS Microbiol Rev 37:44–66

62. Akagi Y, Akamatsu H, Otani H, Kodama M (2009) Horizontal chromosome transfer, a mechanism for the evolution and differentia-

tion of a plant-pathogenic fungus. Eukaryot Cell 8:1732–1738

63. Akagi Y, Taga M, Yamamoto M et al (2009) Chromosome constitution of hybrid strains constructed by protoplast fusion between the tomato and strawberry pathotypes of *Alternaria alternata*. J Gen Plant Pathol 75:101–109

64. Duke S, Dayan F (2011) Modes of action of microbially-produced phytotoxins. Toxins 3:1038–1064

65. Müller M, Korn U (2013) *Alternaria* mycotoxins in wheat—A 10 years survey in the Northeast of Germany. Food Control 34:191–197

66. Azcarate MP, Patriarca A, Terminiello L, Fernández Pinto V (2008) *Alternaria* toxins in wheat during the 2004 to 2005 Argentinean harvest. J Food Prot 71:1262–1265

67. Zachariasova M, Dzuman Z, Veprikova Z et al (2014) Occurrence of multiple mycotoxins in European feeding stuffs, assessment of dietary intake by farm animals. Anim Feed Sci Technol 193:124–140

68. Oviedo MS, Barros G, Chulze S, Ramirez ML (2012) Natural occurrence of alternariol and alternariol monomethyl ether in soya beans. Mycotoxin Res 28:169–174

69. Noser J, Schneider P, Rother M, Schmutz H (2011) Determination of six *Alternaria* toxins with UPLC-MS/MS and their occurrence in tomatoes and tomato products from the Swiss market. Mycotoxin Res 27:265–271

70. Prelle A, Spadaro D, Garibaldi A, Gullino ML (2013) A new method for detection of five *Alternaria* toxins in food matrices based on LC–APCI-MS. Food Chem 140:161–167

71. Zhao K, Shao B, Yang D, Li F (2015) Natural occurrence of four *Alternaria* mycotoxins in tomato and citrus-based foods in China. J Agric Food Chem 63:343–348

72. Terminiello L, Patriarca A, Pose G, Fernandez Pinto V (2006) Occurrence of alternariol, alternariol monomethyl ether and tenuazonic acid in Argentinean tomato puree. Mycotoxin Res 22(4):236–240

73. Asam S, Konitzer K, Schieberle P, Rychlik M (2009) Stable isotope dilution assays of alternariol and alternariol monomethyl ether in beverages. J Agric Food Chem 57:5152–5160

74. European Food Safety Authority (EFSA), Panel on Contaminants in the Food Chain (CONTAM) (2011) Scientific opinion on the risks for animal and public health related to the presence of *Alternaria* toxins in feed and food. EFSA J 9(10):2407 (97 pp)

75. Scott PM, Lawrence GA, Lau BP (2006) Analysis of wines, grape juices and cranberry juices for *Alternaria* toxins. Mycotoxin Res 22(2):142–147

76. Mikusova P, Santin A, Ritieni A et al (2012) Berries contamination by microfungi in Slovakia vineyard regions: impact of climate conditions on microfungi biodiversity. Rev Iberoam Micol 29(3):126–131

77. Pose G, Patriarca A, Kyanko V et al (2010) Water activity and temperature effects on mycotoxin production by *Alternaria alternata* on a synthetic tomato medium. Int J Food Microbiol 142:348–353

78. Robiglio AL, Lopez SE (1995) Mycotoxin production by *Alternaria alternata* strains isolated from red delicious apples in Argentina. Int J Food Microbiol 24:413–417

79. Barkai-Golan R, Paster N (2008) Mouldy fruits and vegetables as a source of mycotoxins: part 1. World Mycotoxin J 1:147–159

80. Patriarca A, Azcarate MP, Terminiello L, Fernández Pinto V (2007) Mycotoxin production by *Alternaria* strains isolated from Argentinean wheat. Int J Food Microbiol 119:684–695

81. Christensen KB, Van Klink JW, Weavers RT et al (2005) Novel chemotaxonomic markers of the *Alternaria infectoria* species-group. J Agric Food Chem 53:9431–9435

Chapter 3

Aspergillus Species and Their Associated Mycotoxins

Giancarlo Perrone and Antonia Gallo

Abstract

The genus *Aspergillus* is among the most abundant and widely distributed organism on earth, and at the moment comprises 339 known species. It is one of the most important economically fungal genus and the biotechnological use of *Aspergillus* species is related to production of soy sauce, of different hydrolytic enzymes (amylases, lipases) and organic acid (citric acid, gluconic acid), as well as biologically active metabolites such as lovastatin. Although they are not considered to be major cause of plant diseases, *Aspergillus* species are responsible for several disorders in various plants and plant products, especially as opportunistic storage moulds. The notable consequence of their presence is contamination of foods and feeds by mycotoxins, among which the most important are aflatoxins, ochratoxin A, and, at a less extent, fumonisins. Aflatoxins B_1, B_2, G_1, G_2 are the most toxic and carcinogenic mycotoxins, due to their extreme hepatocarcinogenicity; ochratoxin A is a potent nephrotoxin, it is also carcinogenic, teratogenic, and immunotoxic in rats and possibly in humans; fumonisins are hepatotoxic and nephrotoxic with potential carcinogenic effects on rat and mice. In this chapter we summarize the main aspects of morphology, ecology, epidemiology, and toxigenicity of *Aspergillus* foodborne pathogens which belong to sections *Flavi*, *Circumdati*, and *Nigri*, occurring in several agricultural products and responsible of aflatoxin, ochratoxin A, and fumonisins contamination of food and feed.

Key words *Aspergillus* Sect. *Cirmundati*, Sect. *Flavi*, Sect. *Nigri*, Aflatoxins, Ochratoxins, Fumonisins

1 Introduction

The first description of *Aspergillus* dates from 1729, when P. A. Micheli describing the genus named it *Aspergillus*, seeing that characteristic spore-bearing structure of the genus resembled an aspergillum, a device used by the Catholic church to sprinkle holy water. The genus *Aspergillus* is among the most abundant and widely distributed organism on earth, with currently 4 subgenera and 19 sections accepted for a total of 339 known species. It represents one of the most economically important fungal genus [1]. *Aspergillus* spp. are widespread geographically and can be either beneficial or harmful microorganisms, however they have mainly a saprophytic lifestyle and predominantly grow on plant decaying materials. To adapt to the variety of niches they inhabit, they have

Antonio Moretti and Antonia Susca (eds.), *Mycotoxigenic Fungi: Methods and Protocols*, Methods in Molecular Biology, vol. 1542, DOI 10.1007/978-1-4939-6707-0_3, © Springer Science+Business Media LLC 2017

evolved a myriad of metabolites. Some of these have been exploited by humankind [2]. A number of *Aspergillus*-related patents have been issued for medical compounds, such as lovastatin, produced by *A. terreus*, which was one of the first commercially successful cholesterol-lowering drugs [3]. A number of antibiotic, antitumoral, and antifungal agents have been derived from *Aspergillus* metabolites. Strains of *Aspergillus* are diffusely used in industrial production like soy sauce, miso, sake (*A. oryzae* and *A. sojae*), several organic acids and enzymes (*A. niger, A. aculeatus, A. carbonarius*). Two of the most important industrial products produced by *Aspergilli* are amylase and citric acids [4]. Unfortunately, *Aspergilli* are one of the major causes of degradation of agricultural products, as they can contaminate foods and feeds at different stages including pre- and postharvest, processing, and handling [5]. In addition, mainly *A. niger, A. flavus*, and *A. fumigatus* species are also causes of animal and human diseases, like mycotoxicosis, noninvasive, and invasive infections in immune-compromised patients, and hypersensitive reactions (e.g., asthma, allergic alveolitis) due to exposure to fungal fragments. Differentially from the common specialized plant pathogens like rust, powdery mildew and some *Fusarium* species, *Aspergillus* species are opportunistic pathogens without host specialization, and frequently isolated as food contaminants. Only a limited number of *Aspergillus* species are able to invade living plant tissues, while most of the species are storage mold on plant products [6]. Agricultural products can be contaminated by *Aspergillus* species (Fig. 1), with changes of sensorial, nutritional, and qualitative nature like pigmentation, discoloration, rotting, and development of off-odors and off-flavors. Moreover, a number of pathogenic and saprophytic species produce toxigenic secondary metabolites on host tissue and plant products, so the most notable consequence of their presence is mycotoxin contamination of foods and feeds.

2   Main Aspergillus Mycotoxins

The main mycotoxins produced by species belonging to *Aspergillus* genus are aflatoxins (B_1, B_2, G_1, G_2); ochratoxin A; fumonisins (B_2 and B_4), patulin; sterigmatocystin; cyclopiazonic acid; penicillic acid; citrinin; cytochalasin E; verruculogen; and fumitremorgin A and B [6, 7]. Among these, the most important are **aflatoxins**, **ochratoxin A**, and **fumonisins** (Fig. 2).

2.1   Aflatoxins (AFB$_s$)

Aflatoxins are decaketide-derived secondary metabolites produced by a complex biosynthetic pathway which could lead to four different metabolites: aflatoxin B_1, B_2, G_1, and G_2 (AFs). Aflatoxins, mainly AFB_1, are the most toxic and carcinogenic naturally occurring mycotoxins. Aflatoxin B_1 exhibits hepatocarcinogenic and

Fig. 1 *Aspergillus* species on plant products: (**a**) *A. westerdijkiae* on dried fruit; (**b**) *A. flavus* on almond; (**c**) *A. flavus* on maize kernel; (**d**) *A. flavus* on ear of corn; (**e**) *A. niger* on ear of corn; (**f**) *A. carbonarius* on grape berries

hepatotoxic properties, and epidemiological data implicate AFB1 as a component of liver cancer in humans in certain parts of the world. In addition, its toxicity can lead to chronic aflatoxicosis in animals by consumption of aflatoxin-tainted foods [8]. Symptoms of aflatoxicosis include reduced weight gain, hemorrhage, and suppression of the immune system. In this respect, extensive research has been carried out on the natural occurrence, identification, characterization, biosynthesis, and genetic regulation of aflatoxins [7–9]. Aflatoxins pose a risk to human health because of their extensive pre-harvest contamination of corn, cotton, soybean, peanuts, and tree nuts, and because residues from contaminated feed may appear in milk. Several aflatoxin outbreaks in humans after consumption of contaminated grains have been documented; they occurred in several parts of Asia and Africa resulting in the death of hundred people [10, 11]. Recently, various papers have emphasized the effects of climate change on food safety in relation to aflatoxins producing fungi, whose habitat is expanding from tropical and subtropical countries to the Mediterranean and central Europe area [12]. The most important aflatoxin producing species belong mainly to

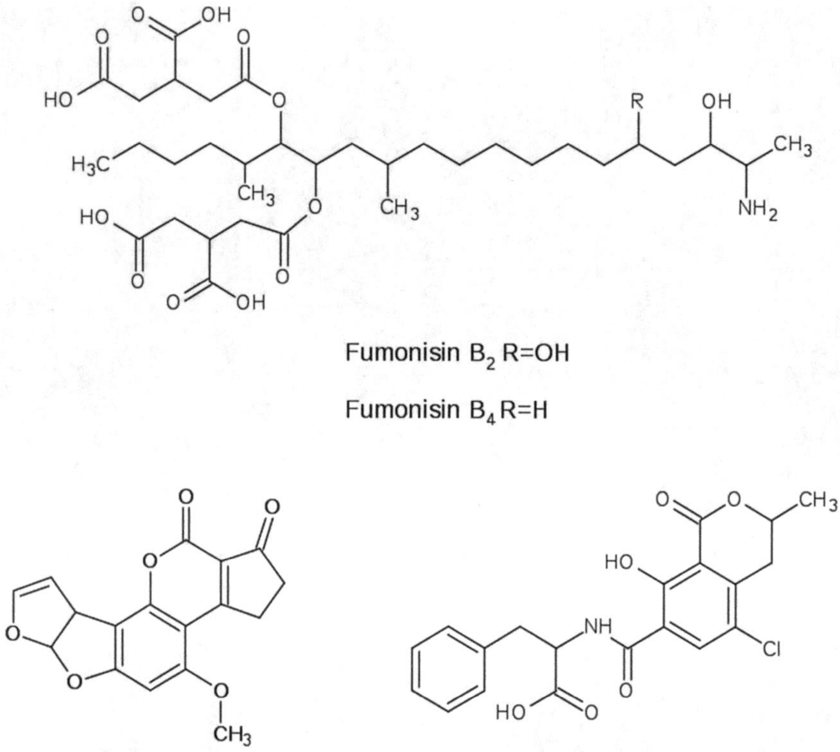

Fumonisin B$_2$ R=OH

Fumonisin B$_4$ R=H

Aflatoxin B$_1$ Ochratoxin A

Fig. 2 Chemical structures of the main mycotoxins produced by *Aspergillus* species

Aspergillus section *Flavi*, including *A. flavus*, *A. parasiticus*, and several other species whose importance and ecology are treated below; less importance has some aflatoxin-producing species belonging to sections *Ochraceorosei* and *Nidulantes* (Table 1).

2.2 Ochratoxin A (OTA)

Ochratoxins A is a potent pentaketide nephrotoxin diffusely distributed in food and feed products (grains, legumes, coffee, dried fruits, beer and wine, and meats); it is also carcinogenic; neurotoxic in vitro and in vivo in rats; teratogenic in mice, rats, and rabbits; and immunotoxic in rats and possibly in humans [13]. Several nephropathies affecting animals as well as humans have been attributed to OTA, it is the etiological agent of Danish porcine nephropathy, and it is cited as possible causative agent of Balkan endemic nephropathy. Numerous animal studies have shown that OTA is a potent nephrotoxin with the degree of renal injury depending on both toxin dose and exposure time; decreasing nephrotoxic sensitivity was observed from pig to rat, to mice [14]. Contamination of food commodities, including cereals and cereal products, pulses, coffee, beer, grape juice, dry vine fruits, and wine as well as cacao products, nuts, and spices, is diffusely reported from all over the world. In addition, contamination of

Table 1
Aspergillus mycotoxigenic species

Aspergillus species producing ochratoxin A (OTA)	
Sect. *CIRCUMDATI*	**Sect. *NIGRI***
A. affinis	*A. carbonarius*
A. cretensis	*A. lacticoffeatus*
A. flocculosus	*A. niger*
A. fresenii	*A. sclerotioniger*
A. muricatus	*A. welwitschiae*
A. occultus	
A. ochraceus	**Sect. Circumdati (*Weak OTA producers*)**
A. pseudoelegans	*A. melleus*
A. pulvericola	*A. ostianus*
A. roseoglobulosus	*A. persii*
A. steynii	*A. salwaensis*
A. westerdijkiae	*A. sclerotiorum*
	A. sesamicola
Sect. *FLAVI*	*A. subramanianii*
A. albertensis	*A. westlandensis*
A. alliaceus	
Aspergillus species producing aflatoxins (B and G type)	
Sect. *FLAVI*	*A. sergii* (AFB and AFG)
A. arachidicola (AFB and AFG)	*A. togoensis* (AFB)
A. bombycis (AFB and AFG)	*A. transmontanensis* (AFB and AFG)
A. flavus (syn. *A. toxicarius*) (AFB and AFG)	**Sect. *OCHRACEOROSEI***
A. minisclerotigenes (AFB and AFG)	*A. ochraceoroseus* (AFB)
A. mottae (AFB and AFG)	*A. rambelli* (AFB)
A. nomius (AFB and AFG)	
A. parasiticus (AFB and AFG)	**Sect. *NIDULANTES***
A. parvisclerotigenus (AFB and AFG)	*A. astellatus* (AFB)
A. pseudocelatus (AFB and AFG)	*A. venezuelensis* (AFB)
A. pseudonomius (AFB)	
A. pseudotamarii (AFB)	
Aspergillus species producing Fumonisins	
Sect *NIGRI*	
A. niger	
A. welwitschiae	

animal feeds with OTA may result in the presence of residues in edible offal and blood serum, whereas OTA contamination in meat, milk, and eggs is negligible. Despite efforts to reduce the amount of this mycotoxin in foods as consumed, a certain degree of contamination seems unavoidable at present. Then, OTA is receiving increasing attention worldwide due to data that show human exposure most likely coming from low level of OTA contamination of a wide range of different foods [15]. This concern was evidenced in a recent case study in South of Italy through urinary biomarkers showing a higher exposure to OTA (>6–100 times) respect to TDI (tolerable daily intake) [16]. Ochratoxins are produced mostly by *Penicillium* species in colder temperate climates, whereas a number of *Aspergillus* species are responsible of OTA production in warmer and tropical parts of the world. *Aspergillus* isolates usually produce both ochratoxin A and B (dechlorinated analogue of OTA), while Penicillia produce only OTA. The economically most important OTA producers belong to *Aspergillus* sections *Circumdati* and *Nigri*, with only two minor OTA-producing species in section *Flavi* (Table 1).

2.3 Fumonisins

Fumonisins are mycotoxins produced mainly by *Fusarium verticillioides* and *F. proliferatum*, which frequently contaminate maize and maize products worldwide. Fumonisin B (FB) analogs are the most common fumonisins, among which FB_1 predominates on FB_2 and FB_3, while FB4 is usually detected in insignificant amounts [17]. The IARC evaluated FB_1 as a Group 2B carcinogen [18]. However FB_2 was reported as more cytotoxic than FB_1 [19]. Limits for total fumonisins B_1 and B_2 have been set for cereals and cereal-based products [20]. Fumonisins are carcinogenic mycotoxins associated with high prevalence of human esophageal cancer in several parts of the world, including Transkei region in South Africa, Linxian province in China, Northern Italy, southeastern USA, India, Kenya, etc., and they were also involved in leukoencephalomalacia in horses, pulmonary edema in pigs, and liver cancer and neural tube defects in experimental rodents [21]. Since the genome sequencing of *A. niger* lead to the identification of fumonisin putative biosynthetic cluster in this species, various studies in the last years demonstrated the ability of *A. niger* and *A. welwitschiae* (formerly *A. awamori*) strains to produce FB_2 and FB_4 [22–24]. In this respect the natural occurrence of fumonisins in musts and dried vine fruits was widely demonstrated in various surveys [23, 25, 26]. In addition, *Aspergillus* species could also contribute to FBs contamination of maize [27]. In recent years, FB_2, produced by Aspergilli, was detected in coffee beans, beer, other grain-based products, barley, and wheat [28].

3 Main Aspergillus Mycotoxigenic Species

Below are summarized the main aspect of morphology, ecology and toxigenicity of *Aspergillus* mycotoxigenic fungi. They are currently grouped in the Subgenus *Circumdati* which comprises six sections of which Sect. *Circumdati, Flavi,* and *Nigri* are relevant for mycotoxin producing ability of some of their species (Table 1). In fact, species economically important for agro-food productions belong mainly to the above-mentioned sections with the exception of some minor important species belonging to the subgenus *Nidulantes* and *Ochraceorosei.* The most common species are *A. flavus* and *A. niger,* with its cryptic sister species *A. welwitschiae,* followed by *A. parasiticus, A. ochraceus, A. carbonarius, A. tubingensis, A. nomius, A. alliaceus (Petromyces alliaceus),* and recently also *A. westerdijkiae* and *A. stenyii.* Mycotoxins associated with plant products and main producing species are summarized in Table 2.

3.1 Section Circumdati

This section, named also *Aspergillus ochraceus* group, includes species with biseriate conidial heads in shades of yellow to ochre, responsible of production of several mycotoxins harmful for animals and humans including ochratoxin A, penicillic acid, xanthomegnin, and viomellein. The most important mycotoxin is ochratoxin A, named after the producer *A. ochraceus.* Some species of the section are utilized for the biochemical transformation of steroids and alkaloids, or as sources of proteolytic enzymes; while other produce several promising anticancer compounds [29].

Table 2
Aspergillus mycotoxins occurring on plant products and associated producing species

Mycotoxins	Agricultural products	Species[a]
Aflatoxins	Peanut, maize, cotton, spices, walnut,	*A. flavus, A. parasiticus, A. nomius*[b],
	Brazil nuts, almond, figs, pistachio nuts	*A. minisclerotigens, A. mottae, A. arachidicola,*
		A. transmontanensis, A. sergii
Ochratoxins	Cereals, grain	*A. westerdijkiae, A. steynii*
	Grape, wine	*A. carbonarius, A. welwitschiae, A. niger*
	Coffee, spices	*A. steynii, A. westerdijkiae, A. ochraceus,*
		A. niger, A. carbonarius
	Figs	*A. alliaceus, A. niger*
Fumonisins	Grape, raisins, figs, onion, maize	*A. welwitschiae, A. niger*

[a]Species in bold represent the main occurring on the relevant product
[b]Main occurring on brazil nut together with *A. flavus*

Recently, Visagie et al. [29] revised section *Circumdati* with 27 species accepted, and introduced seven new species: *A. occultus, A. pallidofulvus, A. pulvericola, A. salwaensis, A. sesamicola, A. subramanianii,* and *A. westlandensis.* This section is generally characterized by the production of some extrolites like orthosporins, aspyrones, and melleins. Eleven species produce large amounts of OTA: *A. cretensis, A. flocculosus* (syn: *A. ochraceopetaliformis*), *A. fresenii, A. muricatus, A. ochraceus, A. pseudoelegans, A. pulvericola, A. roseoglobulosus, A. sclerotiorum, A. steynii,* and *A. westerdijkiae,* while seven further species produce OTA in trace amounts: *A. ostianus, A. melleus, A. persii, A. salwaensis, A. sclerotiorum, A. subramanianii,* and *A. westlandensis.* The most important species regarding potential OTA production in coffee, rice, beverages, and other foodstuffs are *A. ochraceus, A. westerdijkiae,* and *A. steynii* [29, 30].

Until 2004, *A. ochraceus* was considered the main important species of this section relevant for the contamination of food by OTA, then the two new closely related species, *A westerdijkiae* and *A. steynii,* were characterized from strains previously identified as *A. ochraceus* and OTA producers [30]. The main morphological features of these three important species are: colonies ochre or pale yellow; large radiate and biseriate heads; closely packed metulae and small phialides, smooth or finely roughed conidia. *Aspergillus ochraceus* has a variable growth at 37 °C, and many isolates form pinkish-brown sclerotia; differently, *A. westerdijkiae,* which is possibly the main source of OTA from section *Circumdati,* doesn't grow at 37 °C, and the sclerotia are white/cream; *A. steynii* doesn't grow at 37 °C and has broadly ellipsoidal conidia "en masse" with pale yellow color on MEA (malt extract agar). The three species are very difficult to differentiate without the support of molecular and biochemical data. The other species closely related to *A. ochraceus* and good OTA producers are apparently rare and may be not important about potential mycotoxin contamination in foods and beverages [6].

Various ecophysiological studies have been made to identify conditions (incubation temperature, water activity, pH, different substrates) favoring growth, sporulation, and toxin production by potential ochratoxigenic species. However, some data resulted confusing or controversial depending on the criteria used for species identification and laboratory tests used.

In general, *A. ochraceus, A. westerdijkiae,* and *A. steynii* are reported as saprophytic storage fungi, growing between 8 and 37 °C, with the optimum at 24–31 °C, and optimal 0.95–0.99 a_w. *A. ochraceus* species group grows well between pH 3 and 10. The optimum of OTA production by these species is very variable on the basis of substrate (corn, rice, coffee, grapes, etc.) from 15 to 35 °C and 0.90–0.99 a_w, while the minimal a_w for OTA production is 0.80–0.85 depending on substrate; so a_w represents the most important critical control point (CCP) in storage of food and feed

[31]. In the last decades, it has been demonstrated that *A. wester-dijkiae* and *A. steynii* species are far more important OTA produc-ers than *A. ochraceus*. *Aspergillus steynii* was found to be able to grow and produce OTA in a wider set of conditions than *A. west-erdijkiae* and *A. ochraceus*, then posing a higher risk of OTA con-tamination in coffee and other food. Neither *A. steynii* nor *A. westerdijkiae* were able to grow at the lowest value of aw (0.89) evaluated and OTA production was extremely low at 0.91 aw [32].

As described, these three species are morphological and phe-notypical indistinguishable among them and also from some other species of section *Circumdati*. For these reasons, in the recent years a big amount of work has been done in developing molecular tools and strategies for correct identification and discrimination of potential ochratoxigenic *Circumdati* species in food commodities. However, several new species have been described and this fact requires new PCR-based diagnostic assays, against pure culture procedure, for the correct species assignment and a more effective detection in food and commodities [6].

3.2 *Section* Flavi   Members of this section, such as *A. flavus* and *A. parasiticus*, are the most widely investigated because are by far the most important producers of aflatoxins in food commodities; while their domesti-cated counterparts, *A. oryzae* and *A. sojae*, are used in oriental food fermentations and as hosts for heterologous gene expression. Although evidence suggests that *A. sojae* and *A. oryzae* are mor-phological variants of *A. parasiticus* and *A. flavus*, respectively, these species are separated because of the regulatory confusion that conspecificity might generate [33]. In general, section *Flavi* includes species with conidial heads in shades of yellow-green to brown, and dark sclerotia, and currently comprises 27 species and taxa according to new recently described species [34–37]. Only the aflatoxigenic species listed in Table 1 will be treated in this chapter.

Aspergillus flavus is the main important species of the section for its distribution and aflatoxigenicity. It is characterized by bright yellow-green colonies (sometimes yellow), most heads have metu-lae and phialides, heads are radiate, conidia are smooth to finely roughened and of variable size (globose to ellipsoidal). It grows rapidly at 37 °C and not all *A. flavus* isolates produce AFs and those that do, usually produce only B aflatoxins. *Aspergillus nomius* has conidia similar to *A. flavus*, small and elongated (bullet-shaped) sclerotia, and may be distinguished by production of both B and G aflatoxins; whereas its sister species, namely *A. pseudonomius*, pro-duces only aflatoxin B1 (but not G-type aflatoxins), chrysogine and kojic acid. Also the close rare species *A. bombycis* may produce both the aflatoxins and could be distinguished by its slow growth at 37 °C and smooth stipe walls.

Aspergillus parasiticus, the other relevant species of this section, exhibits dark green (never yellow) colonies, heads usually have only phialides (metulae occasionally), conidia are rough walled and usually more uniform in size than *A. flavus*. *Aspergillus parasiticus* comprises a higher percentage of toxigenic isolates, producing both aflatoxins B and G. *Aspergillus toxicarius* has been synonymized to *A. flavus* and closely related to *A. parasiticus*; it still results difficult to distinguish from both species and its species definition is not completely resolved [38].

Aspergillus tamarii and *A. pseudotamarii*, similar and closely related to *A. flavus*, also show rapid growth at 37 °C, they may be distinguished by colonies more brown than *A. flavus* and by conidia very rough to tuberculate, much rougher than *A. flavus* or *A. parasiticus*. *Aspergillus pseudotamarii* differs from *A. tamarii* for production of aflatoxins (B type).

The recently new described aflatoxin producing species *A. parvisclerotigenus* and *A. minisclerotigenes* are difficult to differentiate morphologically from *A. flavus*, and *A. arachidicola* from *A. parasiticus*. However, *A. parvisclerotigenus* and *A. minisclerotigenes* have both tinier sclerotia than *A. flavus* strains and produce both aflatoxins (B and G type), while they differs from each other only by molecular data and sclerotia extrolites production. *Aspergillus arachidicola* instead, is more similar to *A. parasiticus* from which differs for less dark green color of the culture, more biseriate conidiophores and the production of chrysogine [39]. The new aflatoxigenic species identified from almonds in Portugal—*A. mottae*, *A. sergii*, and *A. transmontanensis*—are very difficult to distinguish among the *A. flavus* group. *Aspergillus mottae* resembles *A. flavus*, in having yellow-green biseriate conidial heads; in addition produced numerous small dark sclerotia, such as *A. minisclerotigenes*. *Aspergillus sergii* most closely resembles *A. parasiticus* because of the rough conidia and the production of predominantly uniseriate conidial heads; it differs from *A. parasiticus* for the production of cyclopiazonic acid. *Aspergillus transmontanensis* is also very similar to *A. parasiticus* but it has primarily biseriate conidial heads, while *A. parasiticus* usually has primarily uniseriate conidial heads, and *A. transmontanensis* produces larger abundant brown sclerotia than *A. parasiticus* [35]. Finally, the rarely occurring aflatoxin species *A. pseudocaelatus* is represented by a single isolate collected from an Arachis burkartii leaf in Argentina. It is closely related to the non-aflatoxin producing *A. caelatus*, and produces aflatoxins B and G, cyclopiazonic acid and kojic acid [37].

The diversity of ecological niches occupied by members of *Aspergillus* Sect. *Flavi* and the ability of some species to produce aflatoxin make this group of fungi one of the most highly studied to date. Species in this section occur in nature as saprophytes in the soil and on decaying plant material or as parasites on plants, insects and animals. *Aspergillus* Sect. *Flavi* species in general appear to be most abundant in subtropical and warm temperate regions, particularly in agricultural and desert soils, and decrease in density and species diversity with increasing of latitude; although the climate changes expanded the latitude of occurrence of this species in the last decade [12]. Several species of this section are able to produce aflatoxins, but in crop and food they are mainly produced by *A. flavus* and *A. parasiticus* which coexist and grow on almost any crop or food. In nature, *A. flavus* is one of the most abundant and widely distributed soil-borne molds, its optimal growth is at 28–37 °C and 0.90–0.99 a_w, but it can grows also at temperature from 12 to 48 °C and a_w 0.77. Optimum for production of AFs is 28–30 °C and 0.99 a_w, limit conditions are 15 °C and 0.83 a_w [40]. *Aspergillus flavus* is a saprophytic fungus capable of surviving on many organic nutrient sources, however it is also a weak opportunistic pathogen of many agricultural crops such as corn, cotton, peanuts, and tree nuts. Its great importance as plant pathogen is due to aflatoxin production in the seeds of several crops both before and after harvest, causing health hazard for animals and humans. The percentage of toxigenic *A. flavus* varies with strain, substrate, and geographic origin. Whereas *A. flavus* is broadly spread on soil and various crops, *A. parasiticus* is generally less abundant than *A. flavus*, it infects primarily peanutsand is uncommon in aerial crops. Almost all *A. parasiticus* isolates produce aflatoxins B and G. *Aspergillus parasiticus* has a lower temperature for seed invasion than *A. flavus* and seems more adapted to soil survival, explaining its preference for peanuts compared to *A. flavus*. *Aspergillus parasiticus* has similar temperature and a_w limit to *A. flavus* for fungal growth and AF production, with aflatoxin production optimum between 24 and 30 °C and high water activities (0.95–0.96) [41]. In general, ecology and epidemiology have been widely investigated for *A. flavus* and *A. parasiticus* with respect to the other Section *Flavi* species, which are of minor importance for agro-food system.

However, population analyses in section *Flavi* have evidenced great variability in morphological characters (phenotype) which renders difficult the identification and discrimination of the species; then the presence of other aflatoxigenic species could be currently underestimated. In particular, the morphologically indistinguishable new species *A. minisclerotigenes*, *A. arachidicola*, and *A. parvisclerotigenus*. In this regard, since in the last years the surveys of aflatoxigenic species were more often supported by molecular and biochemical tools, new aflatoxin producing species have been reported in crop and food: *A. nomius* in various surveys from brazil nut [42]; *A. minisclerotigenes* from maize in Portugal and spices in Morocco [35, 43]; *A. mottae*, *A. sergii*, and *A. transmontanensis* from almonds and maize [35].

In this respect, the molecular and phylogenetic analysis of Sect. *Flavi* have evidenced three main clades: "*A. flavus*," "*A. tamarii*," and "*A. alliaceus*," and minor clades in which are scattered other aflatoxin-producing species like *A. nomius*, *A. bombycis*, and *A. pseudotamarii*. This indicates that the aflatoxin-producing ability was probably lost (or gained) several times during evolution. Another aflatoxin-producing species, *A. ochraceoroseus*, was found to be not related to any of the species belonging to section *Flavi* [44]. This species, together with the recently described species *A. rambelli* [38], are the only known to accumulate aflatoxin B_1 and sterigmatocystin simultaneously and belong to the section *Ochraceorosei*, a sister group of the section *Flavi* (Table 1). Additionally, AF production has also been recently observed in two species outside of section *Flavi*, *A. venezuelensis* (syn: *Emericella venezuelensis*) and *A. astellatus* (syn: *E. astellata*) belonging to *Aspergillus* section *Nidulantes* [38].

Finally, with regard to the Sect. *Flavi* species OTA-producing *A. alliaceus* and *A. albertensis* listed in Table 1, they are considered widely distributed, but not common and never identified as contributor of OTA contamination of vegetable products or food; only *A. alliaceus* has been rarely isolated in figs and tree nuts in California [45]. About their ecophysiology no data are available, except information about an higher toxigenicity compared to *A. ochraceus* and *A. melleus* species, which are the dominant species on fig orchards.

3.3 Section Nigri

Named also "black aspergilli," they have a significant impact on modern society as they cause food spoilage, and are used in biotechnology for the production of (extracellular) enzymes, organic acids, vitamins, and antibiotics applied in food fermentations such as awamori liquors, koj fermentation, and Puerth tea. This section includes species with biseriate or uniseriate conidial heads in shades of brown-violet to black, and sclerotia with different color and size in 15 out of 27 species considered. The taxonomy of the Section is still not completely resolved, especially within the *A. niger* species aggregate (a group of morphologically indistinguishable species), leading often to misidentification of the species distribution in food. The taxonomy of section *Nigri* and the classification of strains belonging to this section have been studied various times since the introduction of molecular techniques and currently 27 species are accepted [1]. Five of these produce ochratoxin A (Table 1). *Aspergillus niger* is the representative species of the section and it is the most frequently reported species in food together with *A. carbonarius*, *A. japonicus* and *A. aculeatus*. Recently, *A. tubingensis*, *A. uvarum*, and *A. welwitschiae* have also been found as food contaminating species [46–48]. In general, Aspergilli known as black- and white-koji molds, that are used for food and beverage fermentations (e.g., awamori, shochu, makgeolli), are reported in literature as *A. luchuensis*, *A. awamori*, *A. kawachii*, and *A. acidus*. The taxonomic position of these species was investigated and *A. acidus* and *A. kawachii* were placed in synonymy with *A. luchuensis* based on priority [24]. In the same study a reassessment of the taxonomy of the *A. awamori* species was made because the type strains of this species was erroneously associated with "black koji" fermentations and "awamori" production. Accordingly, it was evidenced that all the strains identified as *A. awamori* belong to a neotype strain isolated from *Welwitschia mirabilis* and are also potential mycotoxin producers. Finally, *A. awamori* strains associated with plant product and mycotoxin production were renamed in *A. welwitschiae*, while the name *A. awamori* species remains a synonym of *A. niger* or *A. luchuensis*, two species commonly found in awamori liquors [24]. The various review of taxonomic names and accepted species in this *Aspergillus* section often make confusion, in fact several not accepted names are still in use like *A. citricus*, *A. foetidus*, and *A. usamii* as synonymous of *A. niger*, *A. saitoi*, and *A. pulverulentus* synonymous of *A. tubingensis*; an overview of this invalid name was made by Houbraken et al. (2014) [49]. As previously mentioned, *A. niger* is one of the most important industrial filamentous fungal species used in biotechnology. This species has been considered to be nontoxic for years, and its safety under industrial conditions was also demonstrated, but natural strains of *A. niger*, together with its sister species *A. welwitschiae*, can produce OTA and fumonisins [50].

Aspergillus niger is characterized by very dark brown to black colonies, radiate and biseriate conidial heads with wide and spherical vesicles, and globose conidia irregularly roughened with ridges and bars. Optimal growth conditions are 35–37 °C. However, *A. niger* group comprises currently an aggregate of eleven species of which most are morphological similar or indistinguishable. Among these, the most frequently isolated are *A. niger*, *A. tubingensis*, and *A. welwitschiae* that could be toxigenic on plant products. Other minor species as food borne could be *A. brasiliensis* detected at lower frequencies, while new described species like *A. piperis*, *A. lacticoffeatus*, *A. costaricaensis*, and *A. vadensis* are rarely found in food commodities [51, 52]. The main important species in this section, either for its high capacity of producing OTA or for high percentage of toxigenic strains, is *Aspergillus carbonarius*. It has optimal growth at 32–35 °C and it can easily be distinguished from other biseriate species due to its big and spiny conidia and the stipes up to several millimeters long. A high percentage of strains of this species (98–100%) have been shown to produce OTA. Other biseriate species similar to *A. carbonarius*, but difficult to distinguish, are *A. ibericus*, with light smaller conidia than *A. carbonarius* and no OTA production, and the producing species *A. sclerotioniger* with yellow mycelium, orange to brown sclerotia, and smooth to verruculose conidia. However, this latter species has been only found as a single strain on Arabica coffee in India [51, 53].

Section *Nigri* has also represented by a consistent group of "uniseriate" species which comprises eleven different taxa of which *A. japonicus* and *A. aculeatus*, together with the recently described from grapes *A. uvarum*, are the most common species isolated from food. Although, they are not distinguishable by morphology, and none of this group resulted to be mycotoxigenic [54].

They have uniseriate heads, conidia usually rough, from sub-globose to ellipsoidal and echinulate with evenly spaced spines. *Aspergillus aculeatus* has larger conidial heads and conidia more ellipsoidal in shape. *Aspergillus uvarum* is more similar to *A. japonicus*, it grows more slowly at 37 °C than *A. japonicus* or *A. aculeatus*; they are distinguished only by molecular and biochemical data [46].

The ecophysiology of this group of fungi has been widely investigated due to the risks that they posed in coffee and cocoa products and in grape and by-product contamination in the last 15 years;. More recently, their occurrence has been also associated to risk for fumonisins contamination of maize and other cereal grains. Among the *A. niger* species aggregate, *A. niger*, *A. welwitschiae*, and *A. tubingensis* could widely occur on plant products with a low percentage of OTA-producing strains (5–10%). Optimal growth conditions are 35–37 °C and 0.93–0.98 a_w (min 6–8 °C and max 47 °C) for these isolates. OTA production by *A. niger* species aggregate normally occurs at 20–25 °C and 0.95/0.98 a_w [52]. Instead, the main responsible of OTA accumulation in grapes, and at less extent in cocoa and coffee, is *A. carbonarius* [52, 55, 56]; it has optimal growth conditions at temperatures between 25 and 35 °C (min 10 °C and max 42 °C) and 0.95–0.98 a_w. Optimal conditions for OTA production by *A. carbonarius* are at 20 °C and 0.95/0.98 a_w [57].

With regard to fumonisin production by *A. niger* and *A. welwitschiae*, Frisvad et al. (2001) found about 80% of the producing FB_2 strains when 180 strains of *A. niger* from various sources were studied [58]; in other reports the percentage of FB_2 producing strains varying from 40 to 65%. In particular, Frisvad et al. (2011) showed that some of the industrially used *A. niger* strains can produce OTA and fumonisin at conditions mimicking industrial citric acid production conditions. Then, other black *Aspergilli* used in food fermentation, citric acid, and enzyme production such as *A. aculeatus*, *A. brasiliensis*, *A. japonicus*, *A. luchuensis*, and *A. tubingensis* do not produce either OTA or fumonisins, and they might be better candidates for biotechnological use than *A. niger*. In general, the data on production and contamination of fumonisins in grape products suggest that it is a minimal risk respect to OTA occurrence in these products, unless additional studies are needed for a better evaluation of the risk. Recently, Logrieco et al. (2014) showed that black aspergilli, and in particular strains belonging to *A. niger* and *A. welwitschiae*, could contribute at same extent to the contamination of fumonisins in maize in association to *F. verticillioides* [27].

4   Conclusion

In this chapter we wished to give an overview on the biodiversity, toxigenic potential and ecology of main *Aspergillus* mycotoxigenic species. In addition, the main mycotoxins produced and their toxigenicity and harmfulness towards man and animal was also summarized. The new phylogenetic and biochemical methods, applied as a whole polyphasic approach, together with morphological and ecological features of the strains influenced and enormously changed the taxonomical definition and the number of the species in genus *Aspergillus*. In this regard, this chapter evidenced the importance of a correct identification and classification of the fungal species for reducing misidentifications and subsequently misinterpretation of results. In recent years the availability of multilocus sequences and fully sequenced genomes resulted in large amounts of sequence data, and will inevitably also have an impact on taxonomy. Genomics can help the taxonomy by serving as a source of novel and unprecedented quantitative comparative data and could improve the actual molecular tools for a more accurate delineation of species boundaries [49]. In the last decades, the potential of PCR systems to differentiate between morphologically similar species having different toxicological profile has proven to be useful and applicable in various situation of the food chain. Thus, these methods represent useful tools for the objective assessment of food safety by identifying mycotoxins producing *Aspergillus* species that are difficult to characterize. PCR methods may be applied to the screening of agricultural commodities for the presence of mycotoxins producers prior or even after processing, and negative results may indicate that the sample is virtually free of mycotoxins. Complexity and diversity of *Aspergillus* mycotoxigenic species, as described in this chapter, explain the vast number of PCR methods which have been developed in the recent years for a rapid and robust identification of these potential harmful species.

References

1. Samson RA, Visagie CM, Houbraken J et al (2014) Phylogeny, identification and nomenclature of the genus *Aspergillus*. Stud Mycol 78:141–173
2. Powell KA, Renwick A, Peberdy JF (1994) The genus aspergillus: from taxonomy and genetics to industrial applications. Plenum Press, New York
3. Lam TY (1983) US Patent 4,376,863
4. Bennett JW, Klich MA (1992) Aspergillus: biology and industrial applications. Butterworth-Heinemann, Boston
5. Varga J, Juhász Á, Kevei F et al (2004) Molecular diversity of agriculturally important *Aspergillus* species. Eur J Plant Pathol 110:627–640
6. Perrone G, Gallo A, Susca A (2009) Molecular detection of foodborne Aspergillus in agricultural products. In: Liu D (ed) Molecular detection of foodborne pathogens. CRC press, Boca Raton, pp 529–548
7. Bennett JW, Klich M (2003) Mycotoxins. Clin Microbiol Rev 16:497–516
8. Payne GA, Brown MP (1998) Genetics and physiology of aflatoxin biosynthesis. Annu Rev Phytopathol 36:329–362
9. Yu J, Chang PK, Ehrlich KC et al (2004) Clustered pathway genes in aflatoxin biosynthesis. Appl Environ Microbiol 70:1253–1262
10. Krishnamachari KA, Bhat RV, Nagarajan V et al (1975) Hepatitis due to aflatoxicosis.

An outbreak in western India. Lancet 1:1061–1063

11. Azziz-Baumgartner E, Lindblade K, Gieseker K et al (2005) Case-control study of an acute aflatoxicosis outbreak, Kenya 2004. Environ Health Perspect 113:1779–1783

12. Perrone G, Gallo A, Logrieco AF (2014) Biodiversity of Aspergillus section Flavi in Europe in relation to the management of aflatoxin risk. Front Microbiol 5:377. doi:10.3389/fmicb.2014.00377

13. IARC Monographs (1993) Eval. Carcinog. Risks. Hum. Some naturally occurring substances: food items and constituents, heterocyclic aromatic amines and mycotoxins. International Agency for Research on Cancer, Lyon, France, 56, 489, 1993

14. Petzinger E, Weidenbach A (2002) Mycotoxins in the food chain: the role of ochratoxins. Livest Prod Sci 75:245–250

15. EFSA European Food Safety Authority (2006) Opinion of the scientific panel on contaminants in the food chain of the EFSA on a request from the commission related to ochratoxin A in food. EFSA J. 365, 1. http://www.efsa.europa.eu/en/scdocs/doc/contam_op_ej365_ochratoxin_a_food_en.pdf. Accessed 4 Aug 2015

16. Solfrizzo M, Gambacorta L, Visconti A (2014) Assessment of multi-mycotoxin exposure in Southern Italy by urinary multi-biomarker determination. Toxins 6:523–538

17. Gutleb AC, Morrison E, Murk AJ (2002) Cytotoxicity assays for mycotoxins produced by Fusarium strains: a review. Environ Toxicol Pharmacol 11:309–320

18. IARC Monographs (2002) Fumonisin B1 on the evaluation of carcinogenic risks to humans: some traditional medicines, some mycotoxins, naphthalene and styrene. IARC 82:301–366

19. Rheeder JP, Marasas WFO, Vismer HF (2002) Production of fumonisin analogs by Fusarium species. Appl Environ Microbiol 68:2101–2105

20. European Commission (2007) Commission Regulation (EC) No 1126/2007 of 28 September 2007, amending Regulation (EC) No 1881/2006 setting maximum levels for certain contaminants in foodstuffs as regards Fusarium toxins in maize and maize products. Off J Eur Un L 255:14–17

21. Stockmann-Juvala H, Savolainen K (2008) A review of the toxic effects and mechanisms of action of fumonisin B1. Hum Exp Toxicol 27:799–809

22. Frisvad JC, Smedsgaard J, Samson RA et al (2007) Fumonisin B2 production by Aspergillus niger. J Agric Food Chem 55:9727–9732

23. Mogensen JM, Frisvad JC, Thrane U et al (2010) Production of fumonisin B2 and B4 by Aspergillus niger on grapes and raisins. J Agric Food Chem 58:954–958

24. Hong SB, Lee M, Kim DH et al (2013) Aspergillus luchuensis, an industrially important black Aspergillus in East Asia. PLoS One 8:e63769. doi:10.1371/journal.pone.0063769

25. Logrieco A, Ferracane R, Haidukowsky M et al (2009) Fumonisin B2 production by Aspergillus niger from grapes and natural occurrence in must. Food Addit Contam Part A Chem Anal Control Expo Risk Assess 26:1495–1500

26. Varga J, Kocsubé S, Suri K et al (2010) Fumonisin contamination and fumonisin producing black Aspergilli in dried vine fruits of different origin. Int J Food Microbiol 143:143–149

27. Logrieco AF, Haidukowski M, Susca A et al (2014) Aspergillus section Nigri as contributor of fumonisin B2 contamination in maize. Food Addit Contam Part A Chem Anal Control Expo Risk Assess 31:149–155

28. Scott PM (2012) Recent research on fumonisins: a review. Food Addit Contam Part A Chem Anal Control Expo Risk Assess 29:242–248

29. Visagie CM, Varga J, Houbraken J et al (2014) Ochratoxin production and taxonomy of the yellow aspergilli (Aspergillus sectionCircumdati). Stud Mycol 78:1–61

30. Frisvad JC, Frank JM, Houbraken J et al (2004) New ochratoxin A producing species of Aspergillus section Circumdati. Stud Mycol 50:23–43

31. Pardo E, Marin S, Ramos AJ et al (2006) Ecophysiology of ochratoxigenic Aspergillus ochraceus and Penicillium verrucosum isolates. Predictive models for fungal spoilage prevention – a review. Food Addit Contam 23:398–410

32. Gil-Serna J, Vazquez C, Sandino FG et al (2014) Evaluation of growth and ochratoxin A production by Aspergillus steynii and Aspergillus westerdijkiae in green-coffee based medium under different environmental conditions. Food Res Int 61:127–131

33. Geiser DM, Pitt JI, Taylor JW (1998) Cryptic speciation and recombination in the aflatoxin-producing fungus Aspergillus flavus. Proc Natl Acad Sci U S A 95:388–393

34. Gonçalves SS, Stchigel AM, Cano JF et al (2012) Aspergillus novoparasiticus: a new clinical species of the section Flavi. Med Mycol 50:152–160

35. Soare C, Rodrigues P, Peterson SW et al (2012) Three new species of Aspergillus section Flavi isolated from almonds and maize in Portugal. Mycologia 104:682–697

36. Taniwaki MH, Pitt JI, Iamanaka BT et al (2012) Aspergillus bertholletius sp. nov. from brazil nuts. PLoS One 7:e42480. doi:10.1371/journal.pone.0042480

37. Varga J, Frisvad JC, Samson RA (2011) Two new aflatoxin producing species, and an overview of *Aspergillus* section *Flavi*. Stud Mycol 69:57–80

38. Frisvad JC, Skouboe P, Samson RA (2005) Taxonomic comparison of three different groups of aflatoxin producers and a new efficient producer of aflatoxin B$_1$, sterigmatocystin and 3-*O*-methylsterigmatocystin, *Aspergillus rambellii* sp. nov. Syst Appl Microbiol 28:442–453

39. Pildain MB, Frisvad JC, Vaamonde G et al (2008) Two novel aflatoxin-producing *Aspergillus* species from Argentinean peanuts. Int J Syst Evol Microbiol 58:725–735

40. Sanchis V, Magan N (2004) Environmental conditions affecting mycotoxins. In: Magan N, Olsen M (eds) Mycotoxins in Food. Woodhead Publishing Limited, Cambridge, pp 174–189

41. Klich M (2007) Environmental and developmental factors influencing aflatoxin production by *Aspergillus flavus* and *Aspergillus parasiticus*. Mycoscience 48:71–80

42. Calderari TO, Iamanaka BT, Frisvad JC et al (2013) The biodiversity of *Aspergillus* section *Flavi* in brazil nuts: from rainforest to consumer. Int J Food Microbiol 160:267–272

43. El Mahgubi A, Puel O, Bailly S et al (2013) Distribution and toxigenicity of *Aspergillus* section *Flavi* in spices marketed in Morocco. Food Control 32:143–148

44. Klich MA, Cary JW, Beltz SB et al (2003) Phylogenetic and morphological analysis of *Aspergillus ochraceoroseus*. Mycologia 95:1252–1260

45. Bayman P, Baker JL, Doster MA et al (2002) Ochratoxin production by the *Aspergillus ochraceus* group and *Aspergillus alliaceus*. Appl Environ Microbiol 68:2326–2329

46. Perrone G, Varga J, Susca A et al (2011) *Aspergillus uvarum* sp. nov., a uniseriate black *Aspergillus* species isolated from grapes in Europe. Int J Syst Evol Microbiol 58:1032–1039

47. Perrone G, Mulè G, Susca A et al (2006) Ochratoxin A production and AFLP analysis of *Aspergillus carbonarius, Aspergillus tubingensis,* and *Aspergillus niger* strains isolated from grapes in Italy. Appl Environ Microbiol 72:680–685

48. Perrone G, Stea G, Epifani F et al (2011) *Aspergillus niger* contains the cryptic phylogenetic species *A. awamori*. Fungal Biol 115:1138–1150

49. Houbraken J, de Vries RP, Samson RA (2014) Modern taxonomy of biotechnologically important *Aspergillus* and *Penicillium* species. Adv Appl Microbiol 86:199–249

50. Susca A, Moretti A, Stea G et al (2014) Comparison of species composition and fumonisin production in Aspergillus section Nigri populations in maize kernels from USA and Italy. Int J Food Microbiol 188:75–82

51. Samson RA, Houbraken J, Kuijpers AF et al (2004) New ochratoxin A or sclerotium producing species in *Aspergillus* section *Nigri*. Stud Mycol 50:45–61

52. Perrone G, Susca A, Cozzi G et al (2007) Biodiversity of *Aspergillus* species in some important agricultural products. Stud Mycol 59:53–66

53. Serra R, Cabañes FJ, Perrone G et al (2006) *Aspergillus ibericus:* a new species of the Section *Nigri* isolated from grapes. Mycologia 98(295):2006

54. Somma S, Perrone G, Logrieco AF (2012) Diversity of black Aspergilli and mycotoxin risks in grape, wine and dried vine fruits. Phytopathol Mediterr 51:131–147

55. Joosten HMLJ, Goetz J, Pittet A et al (2001) Production of ochratoxin A by *Aspergillus carbonarius* on coffe cherries. Int J Food Microbiol 65:39–44

56. Taniwaki MH, Pitt JI, Teixeira AA et al (2003) The source of ochratoxin A in Brazilian coffee and its formation in relation to processing methods. Int J Food Microbiol 82:173–179

57. Bellí N, Ramos AJ, Coronas I et al (2005) *Aspergillus carbonarius* growth and ochratoxin A production on a synthetic grape medium in relation to environmental factors. J Appl Microbiol 98:839–844

58. Frisvad JC, Larsen TO, Thrane U et al (2011) Fumonisin and ochratoxin production in industrial Aspergillus niger strains. PLoS One 6:e23496. doi:10.1371/journal.pone.0023496

Chapter 4

Fusarium Species and Their Associated Mycotoxins

Gary P. Munkvold

Abstract

The genus *Fusarium* includes numerous toxigenic species that are pathogenic to plants or humans, and are able to colonize a wide range of environments on earth. The genus comprises around 70 well-known species, identified by using a polyphasic approach, and as many as 300 putative species, according to phylogenetic species concepts; many putative species do not yet have formal names.

Fusarium is one of the most economically important fungal genera because of yield loss due to plant pathogenic activity; mycotoxin contamination of food and feed products which often render them unaccep for marketing; and health impacts to humans and livestock, due to consumption of mycotoxins. Among the most important mycotoxins produced by species of *Fusarium* are the trichothecenes and the fumonisins. Fumonisins cause fatal livestock diseases and are considered potentially carcinogenic mycotoxins for humans, while trichothecenes are potent inhibitors of protein synthesis. This chapter summarizes the main aspects of morphology, pathology, and toxigenicity of the main *Fusarium* species that colonize different agricultural crops and environments worldwide, and cause mycotoxin contamination of food and feed.

Key words *Fusarium sections*, Fumonisins, Trichothecenes, Taxonomy, Fusaric acid

1 Introduction

Fusarium is one of the most widely recognized genera of plant pathogenic fungi that produce economically important mycotoxins. Taken together, the impacts of *Fusarium* mycotoxins exceed that of any other toxin or group of toxins. *Fusarium* spp. produce a wide diversity of toxic compounds, but the greatest economic impacts can be associated with deoxynivalenol (DON) and its derivatives. Only the aflatoxins, produced by *Aspergillus* spp., have a greater impact on trade and animal and human health than DON. *Fusarium* mycotoxins have been the subject of numerous books and review papers [1–3]; more in-depth information can be found in those sources. In this chapter, I present information on toxin production among *Fusarium* species, summarized and updated from Desjardins [1], and discuss the occurrence and importance of the major *Fusarium* mycotoxins.

Antonio Moretti and Antonia Susca (eds.), *Mycotoxigenic Fungi: Methods and Protocols*, Methods in Molecular Biology, vol. 1542, DOI 10.1007/978-1-4939-6707-0_4, © Springer Science+Business Media LLC 2017

Wheat and maize are the crops in which *Fusarium* mycotoxins have the most frequent occurrence and greatest impact. However, toxigenic *Fusarium* species can occur in all small grain crops, as well as many other crops such as asparagus, figs, forage grasses, soybean and other legumes, spice plants and medicinal plants, and some nut crops such as pistachio [4, 5]. *Fusarium* toxins also can occur in spoiled food products made from plants that are not necessarily hosts for pathogenic infections. As a consequence of crop contamination, *Fusarium* toxins occur in prepared animal feeds and human food products, including fermented products such as beer. With a major portion of the maize crop in North America now being processed into fuel ethanol, a growing concern is the occurrence of *Fusarium* toxins in ethanol co-products, such as dried distillers' grains and solids (DDGS) [6]. Fumonisins may occur in crops and food products in the absence of *Fusarium* spp., due to their production by some *Aspergillus* species [7].

The importance of *Fusarium* as a threat to agriculture and human health is reflected in the number of species which have been subject to whole-genome sequencing (Table 1), which continues to include a growing list of *Fusarium* species.

Table 1

Publicly available *Fusarium* genome sequences

Species complex[a]	Species[b]	Strain(s)[c]	Source[d]	References
sambucinum	*F. culmorum*	CS7071	NCBI	Gardiner et al., unpublished
	F. graminearum	PH-1,	Broad, MIPS	[8]
		CS3005	NCBI	Gardiner et al., unpublished
	F. langsethiae	Fl201059	NCBI	Lysoe et al., unpublished
	F. pseudograminearum	CS3096	NCBI	[9]
Incarnatum-equiseti	FIESC 5	CS3069	NCBI	Gardiner et al., unpublished
tricinctum	*F. acuminatum*	CS5907	NCBI	Gardiner et al., unpublished
	F. avenaceum	Fa05001	NCBI	[10]
		FaLH03	NCBI	[10]
		FaLH27	NCBI	[10]

(continued)

Table 1
(continued)

Species complex[a]	Species[b]	Strain(s)[c]	Source[d]	References
fujikuroi	*F. circinatum*	FSP 34	NCBI	[11]
	F. fujikuroi	IMI 58289	MIPS	[11]
		B14	NCBI	[12]
		KSU 3368	NCBI	Fanelli et al., unpublished
		FGSC 8932	NCBI	Fanelli et al., unpublished
		KSU X-10626	NCBI	Fanelli et al., unpublished
	F. verticillioides	FRC M-3125 (=7600)	Broad	[13]
oxysporum	*F. oxysporum* f. sp. radices-lycopersici	CL57	Broad	[13]
	F. oxysporum (biocontrol isolate)	Fo47	Broad	[13]
	F. oxysporum f. sp. *cubense* race 1	Foc1	NCBI	[14]
	F. oxysporum f. sp. *cubense* race 4	Foc4	NCBI	[14]
	F. oxysporum f. sp. *lycopersici*	Fol4287	Broad	[13]
	F. oxysporum f. sp. *lycopersici*	Fo5176	Broad	[13]
	F. oxysporum f. sp. *pisi*	HDV247	Broad	[13]
	F. oxysporum f. sp. *cubense* race 4	II5	Broad	[13]
	F. oxysporum f. sp. *lycopersici* race 3	MN25	Broad	[13]
	F. oxysporum f. sp. *vasinfectum*	NRRL 25433	Broad	[13]
	F. oxysporum f. sp. *melonis*	NRRL 26406	Broad	[13]
	F. oxysporum (human isolate)	NRRL 32931	Broad	[13]
	F. oxysporum f. sp. *conglutinans*	PHW808	Broad	[13]
	F. oxysporum f. sp. *raphini*	PHW815	Broad	[13]
solani	*F. solani* f. sp. *pisi*	77-13-4	JGI	[15]
	F. virguliforme	Mont-1	NCBI	[16]

[a]Species complexes are as described previously [17, 18]. The *sambucinum* complex includes the previously described *F. graminearum* species complex

[b]FIESC is the abbreviation for *F. incarnatum-equiseti* species complex. Numbers after FIESC (e.g. FIESC 5 and FIESC 15) indicate phylogenetically distinct species based on multi locus sequence typing as previously described [19]

[c]Strain designations: ITEM indicates strains from the National Research Council, Institute of Sciences of Food Production in Bari, Italy; CS indicates strains from the Commonwealth Scientific and Industrial Research Organisation (CSIRO) in Brisbane, Australia; FRC indicates strains from the *Fusarium* Research Center at Pennsylvania State University in University Park Pennsylvania

[d]Genome sequence data can be downloaded from the following sources: Broad, Broad Institute at the Massachusetts Institute of Technology; JGI, Joint Genome Institute; MIPS, Munich Information Center for Protein Sequences—Helmholtz Zentrum München, German Research Center for Environmental Health; NCBI, National Center for Biotechnology Information

1.1 Taxonomy of Fusarium

Species concepts in the genus *Fusarium* have been revised numerous times. The genus was first described in 1809 by Link [20] and during the next 100 years, more than 1000 species were described. The first major work to bring criteria to a *Fusarium* species concept was undertaken by Wollenweber and Reinking in the 1930s and their 1935 publication [21] was the starting point for modern *Fusarium* taxonomic systems. In this publication they organized 65 species with 77 subspecific varieties and forms into 16 sections. In several publications during the 1940s and 1950s, Snyder and Hansen [22–24] reduced the number of species to only nine, based on the idea that morphological characters from a single-spored isolate could vary enough to fit more than one species under Wollenweber and Reinking's system. The scheme of Snyder and Hansen simplified species identification, but lumped together many genotypes of widely varying phenotypes and phylogenetic origins. During the following decades, variations on the Snyder and Hansen system, or hybrids between the two systems, were published by several other researchers; notably Gordon, who published a series of papers on *Fusarium* spp. in Canada [25] using a system that recognized 26 species, using concepts of both Wollenweber and Reinking and Snyder and Hansen. This was followed by the publication in 1971 of "The Genus *Fusarium*," by Booth [26], which recognized 44 species, primarily following Wollenweber and Reinking's concepts. The trend toward recognizing more species continued with the publication of "The Genus *Fusarium*: a Pictorial Atlas" by Gerlach and Nirenberg, which recognized more than 90 species and varieties. A contrast to their system was the widely accepted taxonomic scheme presented by Nelson et al. [27] that recognized 30 "well-documented" species and 16 additional "insufficiently documented" *Fusarium* species, retaining the Wollenweber and Reinking sections, reduced from 16 to 12. Although this popular system did not recognize many of the Gerlach and Nirenberg species, recent phylogenetic analyses have often supported the more complicated species concepts of Gerlach and Nirenberg. Both of these publications affirmed that the simplified system introduced by Snyder and Hansen was inadequate, and in combination, these two works resolved much of the confusion introduced by the nine-species concept (with the exceptions of the *F. oxysporum* and *F. solani* species complexes).

All of the aforementioned systems relied entirely on morphological characters to distinguish species. Subsequent research during the late twentieth century introduced biological (based on sexual compatibility) and phylogenetic (based on evolutionary relatedness) species concepts into *Fusarium* taxonomic systems. The only comprehensive publication that encompasses morphological, biological, and (to some extent) phylogenetic species concepts in *Fusarium* is "The *Fusarium* Laboratory Manual" by Leslie and Summerell [28]. This publication recognizes 70 species, and includes detailed information regarding the various taxonomic systems and species concepts in the genus, as well as valuable information about pathogenicity and toxigenicity for each of the described species. In this volume and most subsequent publications on *Fusarium* taxonomy, the section concept has been de-emphasized or abandoned. Since the publication of this book in 2006, many new species have been described, often following phylogenetic species concepts based on DNA sequence analysis. This approach has led to further de-emphasis of the section concept in favor of phylogenetic clade or species complex designations for grouping species [17, 18]. It also has led to a proliferation of species as well as controversies over the appropriate emphasis on phylogenetic concepts in species distinctions in *Fusarium*. Recently, prominent *Fusarium* researchers have suggested that the true number of *Fusarium* species is more than 300, mostly based on molecular phylogenetic analyses [17, 18, 29]. Several known species have been separated into numerous new species, and many of the putative 300 species do not yet have formal names.

As a result of the numerous taxonomic revisions and varying species concepts in *Fusarium*, the total number of toxigenic species is uncertain. The volume by Marasas et al. [2], "Toxigenic *Fusarium* species: Identity and Mycotoxicology," was a seminal work on the association of *Fusarium* species with their toxins and toxicoses; the authors discussed 24 species but only 13 were believed by the authors to be verified as toxigenic species. This book, along with its companion volume by Nelson et al. [27], brought much clarity to the situation regarding erroneous nomenclature and unsubstantiated reports of toxin production. Of course, many new species have been described since 1983. In 2006, two books were published that represented updates to Nelson et al. [27] and Marasas et al. [2], respectively. Leslie and Summerell [28] clarified contemporary species concepts and "*Fusarium* Mycotoxins: Chemistry, Genetics, and Biology" [1], critically reviewed verifiable mycotoxin production among the species. The Desjardins book lists 45 toxigenic species. Marasas et al. [2] and Desjardins [1] are the most comprehensive texts available that discuss important toxigenic species and document the species-toxin associations that are supported by strong evidence. However, this is a rapidly changing field and no text can be universal. As a result of the recent proliferation of species descriptions, the toxigenic capabilities of many *Fusarium* species are unknown and our knowledge in this field has tremendous room for growth.

1.2 Teleomorphs of Fusarium spp.

Many *Fusarium* species do not have known teleomorphs, but many economically important species have teleomorphs that have been described from nature or from laboratory crosses. Nearly all of the described teleomorph genera are in the order Hypocreales of the Ascomycetes [28]. Most of these have been described in the genus

Gibberella [30], while others have had teleomorph descriptions in *Nectria* or closely related genera. The generic concept for *Gibberella* and its connection with *Fusarium* anamorphs has been stable, but many of the "*Nectria*-like" generic teleomorph connections have changed over time, leading to some confusion. Even with stable teleomorph-anamorph nomenclature, it is widely recognized that the use of two names for the same fungus can lead to misunderstanding. The International Code of Botanical Nomenclature (ICBN Article 59.1) previously directed that teleomorph names take precedence in cases where two names existed. Nevertheless, *Fusarium* remained in use, even for species with known teleomorphs. This has occurred for several reasons, including the fact that fewer than 20% of *Fusarium* spp. have a known teleomorph, and the teleomorph is never seen in nature for most important plant-pathogenic *Fusarium* spp. [29]. More recently, changes in the International Code of Nomenclature direct that Article 59 no longer applies after January 1, 2013 [31], so that creation of both anamorph and teleomorph names is not allowed, existing teleomorph-anamorph names are to be unified, and teleomorph names no longer take priority. This has led to a consensus among most mycologists and *Fusarium* researchers to conserve the genus name *Fusarium* for this group of fungi [17], with a genus concept that includes the "*Gibberella* clade" and the *F. solani* species complex, but excludes some other clades that have *Fusarium*-like anamorphs, but are not plant or human pathogens. Consistent with this approach, in this chapter the *Fusarium* names are used; however, teleomorph names also are mentioned, in order to connect *Fusarium* species to previous literature in which teleomorph names were used.

1.3 Fusarium spp. as Plant Pathogens

Fusarium spp. are important to humankind primarily in their role as plant pathogens, which in turn leads to their role as toxicological risks to humans and our domestic animal species. A wide range of plant diseases is associated with *Fusarium*, and a thorough review is beyond the scope of this chapter; books by Nelson et al. [32], Leslie and Summerell [28], and Desjardins [1] are good resources. A recent review paper by Aoki et al. [29] presents a concise assessment of the major groups of plant-pathogenic *Fusarium* spp., and their approach is summarized here. Phylogenetic analysis based on gene sequencing data allows for the identification of numerous species complexes in the genus, but according to Aoki et al. [29], most plant-pathogenic *Fusarium* spp. are grouped in four species complexes as defined by RNA polymerase II subunit gene sequence phylogeny:

- *Fusarium fujikuroi* (formerly *Gibberella fujikuroi*) species complex (FFSC)—members of this species complex cause important diseases in maize, sorghum, rice, sugarcane, mangoes, and other crops. They are also responsible for fumonisin contamination of grain and other crops. The FFSC corresponds roughly to the species concept of *F. moniliforme* as described by Snyder and Hansen, or Section *Liseola* as defined by Wollenweber and Reinking. The biological species concept has been applied most extensively in this group, leading to identification of at least 13 biological species, but more than 50 phylogenetic species have been suggested [29]. Important plant diseases caused by members of the FFSC include ear rot, stalk rot, and seedling disease of maize, caused by *F. proliferatum, F. subglutinans, F. temperatum,* and *F. verticillioides*; pitch canker of pine trees, caused by *F. circinatum*; Bakanae disease of rice, caused by *F. fujikuroi*; pokkah boeng disease of sugarcane, caused by *F. sacchari*; and mango malformation caused by *F. mangiferae* and other species.

- *Fusarium graminearum* species complex (FGSC)—members of this complex cause important diseases in small grain cereals and maize, and are responsible for contamination of grain with Type B trichothecene mycotoxins and zearalenone. Most of these species would have been considered *F. roseum* according to Snyder and Hansen or section *Discolor* of Wollenweber and Reinking. Fusarium head blight of small grain cereals is the most important disease caused by the FGSC; yield losses and mycotoxin contamination due to Fusarium head blight have had very significant economic, societal, and human health impacts. *Fusarium graminearum sensu stricto* is the most important cause of Fusarium head blight worldwide, but several other species such as *F. culmorum* and *F. crookwellense* are important, and head blight can be caused by a few species which are outside the FGSC. Ear rot and stalk rot of maize also are caused by *F. graminearum* and related species. Seedling blights and root rots of many crops, including small grains, maize, soybean, and some vegetable crops, also are attributable to this species complex. Most phylogenetic species in the FGSC have been formally named. See the section on FGSC in this chapter for more details on individual species.

- *Fusarium oxysporum* species complex (FOSC)—members of this group are most well-known for causing vascular wilt disease, but also are important root rot pathogens; however, they are not considered major contributors to mycotoxin exposure in humans or livestock. The FOSC corresponds to section *Elegans* of Wollenweber and Reinking and the Snyder and Hansen species concept of *F. oxysporum*. There are more than 100 *formae speciales* and races described for *F. oxysporum;* they cause significant diseases on a very wide range of crops. Vegetative compatibility groups (VCG) have been useful in the FOSC to characterize strains with similar pathogenic properties [29], and the identification of effectors (small proteins that are related to pathogenicity) and their genetic basis is an active area of research on the

FOSC [33]. Important *formae speciales* include *F. oxysporum* f.sp. *cubense,* which causes Panama disease of bananas, *F. oxysporum* f.sp. *vasinfectum* on cotton, *F. oxysporum* f.sp. *phaseoli* on common bean, *F. oxysporum* f.sp. *lycopersici* on tomato, and *F. oxysporum* f.sp. *lactucae* on lettuce. *Fusarium oxysporum sensu lato* is a common root rot and seedling pathogen of many grain crops and legumes, but many morphologically indistinguishable FOSC strains are non-pathogenic saprophytes.

- *Fusarium solani* species complex (FSSC)—members of this group cause foot and root rot of numerous crops, but are not major contributors to mycotoxin exposure in humans or livestock. The FSSC, however, contains the majority of *Fusarium* species that are pathogenic to humans and other animals. The species in the FSSC would be considered in section *Martiella* of Wollenweber and Reinking, but the Snyder and Hansen concept of *F. solani* was broader, also including section *Ventricosum.* The species complex includes numerous *F. solani formae speciales* for pathogens causing root rot, foot rot, or various blights and fruit rots of bean, cucurbits, peppers, potato, sweet potato, mulberry, pea, and others. *F. solani sensu lato* is a common root rot and seedling pathogen of many grain crops and legumes. Most members of this complex are still known as *F. solani* because most of the numerous phylogenetic species have not been formally named. Another important disease, sudden death syndrome of soybean, is caused by *F. virguliforme* and a few other species in the FSSC.

Outside of these four clades, there are several other important plant pathogenic *Fusarium* species, including those that produce Type A trichothecenes, such as *F. langsethiae, F. poae,* and *F. sporotrichioides.* These three, along with *F. avenaceum,* are all associated with Fusarium head blight. *Fusarium lateritium* is an important pathogen of several woody plant species, and dry rot of potato, caused by *F. sambucinum* and other species is an important storage problem and toxicological concern.

2   *Fusarium* Mycotoxins

Several hundred compounds have been described as toxic or potentially toxic secondary metabolites of *Fusarium* spp. Toxicity of many of these compounds has been demonstrated in bioassays or feeding studies. A relatively small number of these compounds has been definitively linked with toxicoses occurring in humans or livestock animals. Comprehensive reviews of *Fusarium* mycotoxins have been published that include details of chemical and toxicological properties [1, 2, 34–36]. During the past two decades, significant strides have been made in the understanding of mycotoxin biosynthesis and its genetic basis, particularly for trichothecenes and fumonisins. Those details are beyond the scope of this chapter, but several reviews have been published [37–42]. Many reports have been published regarding the incidence of contamination and range of contamination levels for *Fusarium* mycotoxins in various geographies. Reviews including data on the most important compounds have been published periodically [1, 3, 43–47]. In this chapter the major groups of mycotoxins are discussed, with emphasis on those that are believed to be important with respect to human health and/or livestock agriculture.

2.1 Trichothecenes

Trichothecenes are the most economically important mycotoxins produced by *Fusarium* species. Among the *Fusarium* mycotoxins, trichothecenes have most often been associated with human toxicoses and widespread livestock health issues. They are produced by several genera of fungi in the order Hypocreales, and were named after the fungus *Trichothecium roseum*, from which the first member of this class of toxins was isolated. They are sesquiterpenoid compounds with a tricyclic 12,13-epoxytrichothec-9-ene ring that can be chemically substituted at several positions, resulting in multiple derivatives. There are more than 200 trichothecenes, which have been classified into four groups (types A, B, C, and D) based on substitutions at C-8 and other positions around the core structure. Trichothecenes produced by *Fusarium* spp. are Type A, which have a hydroxyl or ester or no oxygen substitution at C-8, or type B, which have a keto (carbonyl) group at C-8. Type B trichothecenes are generally produced in higher amounts, but are less toxic than type A compounds. Type C trichothecenes are a minor group of toxins produced by several other genera of fungi, and type D includes compounds produced by species of *Stachybotrys* (and other genera) that are important as indoor mold hazards [39].

Important type A trichothecenes produced by *Fusarium* spp. include diacetoxyscirpenol (DAS), HT-2, T-2, and neosolaniol. The most acutely toxic trichothecene in animals is T-2, although sensitivity to the various compounds varies among animal species [48, 49]. The most important producers of Type A trichothecenes are *F. armeniacum* (often reported as *F. acuminatum*), *F. langsethiae*, *F. poae*, *F. sambucinum*, *F. sporotrichioides*, and *F. venenatum*. They are also produced by other species (Table 2). The common type B trichothecenes are deoxynivalenol (DON), nivalenol, (NIV) and their acetylated derivatives. The most important type B trichothecene-producing species are in the *F. graminearum* species complex, including *F. graminearum sensu lato*, *F. culmorum*, *F. crookwellense*, and *F. pseudograminearum*. All the recently described species under the *Fusarium graminearum* species complex can produce some trichothecenes, but toxin profiles differ among and within phylogenetic species. Type B trichothecenes are also produced by other species (Table 2), but are generally absent from the

Table 2
Reported mycotoxin production for selected *Fusarium* spp.[a]

													Trichothecenes						
	Aurofusarin	Beauvericin	Butenolide	Chlamydosporol	Culmorin	Cyclonerodiol	Enniatins	Fumonisins	Fusaproliferin	Fusaric acid	Fusarins	Moniliformin	Trichothecenes[b]	Diacetoxy scirpenol	Deoxynivalenol	Nivalenol	HT-2	T-2	Zearalenone
Fusarium acuminatum	X	X		X			X				X	X	X	X			X	X	?
Fusarium acaciae-mearnsiic													X		X	X			
Fusarium acutatum		X					X	?				X							
Fusarium andiyazic								X											
Fusarium anthophilum	X	X					?	X	X			X							
Fusarium armeniacum	X	X									X		X				X	X	
Fusarium asiaticumc													X		X	X			
Fusarium austroamericanumc													X		X	X			
Fusarium avenaceum	X	X		X			X				X	X	X						
Fusarium begoniae								?	X			X							
Fusarium beomiforme		X										X	X						
Fusarium boothiic													X		X				
Fusarium brasilicumc													X						
Fusarium chlamydosporum				X			X					X							
Fusarium circinatum		X							X										
Fusarium compactum							X						X						
Fusarium concentricum		X					X		X			X							
Fusarium cortaderiaec	X												X			X			
Fusarium crookwellense	X		X		X	X				X	X		?			X			X

Species													
Fusarium culmorum	X	X	X	X	X	X	X	X	X	X	X	X	X
Fusarium denticulatum	X				X	?		?					
Fusarium dlamini	X			X			X		X				
Fusarium equiseti	X	?				X	X	X	X	X	X	X	X
Fusarium fujikuroi	X				X	X			X	X	X		
Fusarium globosum	X				X	X	X						
Fusarium graminearum	X	X	X	X	X	X	X	X	?	X	X	X	?
Fusarium guttiforme	X						X						
Fusarium konzum	X			X		X	X	X					
Fusarium kyushuense				X				X	X	X	X	X	X
Fusarium lactis						X							
Fusarium langsethiaec	X			X			X	X	X	X	X	X	X
Fusarium lateritium				X			X						
Fusarium meridionalec							X	X	X				
Fusarium mesoamericanumc						X	X	X	X				
Fusarium musae				X									
Fusarium napiforme				X		X	X	X					
Fusarium nygamai	X			X		X	X	X					
Fusarium oxysporum	X			X		X	?	X	X	X	X	X	X
Fusarium phyllophilum	X			X			X	X					
Fusarium poae	X	X	X	X		X	X	X	X	X	X	X	X
Fusarium proliferatum	X	X	X	X	X	X	X	X	X	X	X	X	X
Fusarium pseudograminearum							X	X	X	X	X	X	X
Fusarium sacchari	X	X		X		X	X	X					
Fusarium sambucinum	X	X	X	X		X	X	X	X	X	X	X	X

(continued)

Table 2
(continued)

	Auro fusarin	Beau vericin	Buten olide	Chlamy dosporol	Culmorin	Cyclone rodiol	Ennia tins	Fumon isins	Fusaproli ferin	Fusaric acid	Fusarins	Monili formin	Trichothecenes						
													Trichothe cenes[b]	Diacetoxy scirpenol	Deoxyni valenol	Nivalenol	HT-2	T-2	Zearale none
Fusarium semitectum		X										X	X	X				?	X
Fusarium solani										X			?						
Fusarium sporotrichioides	X	X	X		X		X				X	X	X	X			X	X	
Fusarium subglutinans		?						X	X	X		X							
Fusarium temperatumc		X						X	X			X							
Fusarium thapsinum								X		X		X							
Fusarium torulosum			X				X					X							
Fusarium tricinctum			X	X			X				X	X							
Fusarium venenatum	X		X		X		X				X		X	X					
Fusarium verticillioides		X						X		X	X								

[a]References: [1–3, 28, 50–56]

[b]Includes diacetoxyscirpenol, deoxynivalenol, nivalenol, HT-2, T-2, and all other known trichothecenes

[c]Mycotoxin production has not been thoroughly investigated

? = Reports are questionable

F. fujikuroi (*Gibberella fujikuroi*), *F. oxysporum,* and *F. solani* species complexes.

Trichothecenes have toxic effects on animals and plants; risks to humans and other animals have been reviewed in detail [4, 34, 35, 49, 57]. Toxicity occurs through several mechanisms, including inhibition of ribosomal protein synthesis, DNA and RNA biosynthesis, and mitochondrial function [58, 59]. At the cellular level, these effects lead to cellular oxidative stress, cell cycle arrest, apoptosis and cell membrane dysfunction [58, 60]. Trichothecenes have been shown to cause a variety of acute and chronic symptoms in experimental and livestock animals, including growth retardation, reproductive disorders, immune system suppression, feed refusal (anorexia), vomiting, hemorrhaging, diarrhea, and death [58, 61–63]. Monogastric mammals are the most susceptible, while ruminant mammals and poultry can tolerate higher levels. Human exposure to trichothecenes in the diet has been associated with nausea, diarrhea, abdominal pain, and fever [64]. Due to the toxic effects of trichothecenes on humans, tolerable daily intake (TDI), or provisional maximum tolerable daily intake (PMTDI) values have been set by the U.N. Food and Agriculture Organization/World Health Organization Joint Expert Committee on Food Additives (JECFA) and/or the European Food Safety Authority (EFSA) (Table 3). Trichothecenes are heat stable at temperatures experienced during food and feed processing, ethanol production, or autoclaving [58], and thus remain hazardous in processed foods and beverages and in co-products of fuel ethanol production.

Several outbreaks of mycotoxicoses in humans and livestock have been attributed to trichothecenes. The strongest evidence is related to the occurrence of feed refusal in swine, due to DON

Table 3
Recommended maximum tolerable daily intake for major *Fusarium* mycotoxins

Compound	TDI or PMTDI[a]	Type[a]	Basis	References
Deoxynivalenol and acetylated derivatives	1.0 μg/kg body weight/day	PMTDI	Rodent nephrotoxicity	[65]
Fumonisins $B_1 + B_2 + B_3$	2.0 μg/kg body weight/day	PMTDI	Body weight reduction—mice	[36]
Nivalenol	1.2 μg/kg body weight/day	TDI	White blood cell reduction—rats	[57]
T-2 + HT-2	0.06 μg/kg body weight/day	PMTDI	White and red blood cell reductions—swine	[35]
Zearalenone	0.5 μg/kg body weight/day	PMTDI	Hormonal effects—swine	[66]

[a]*TDI* tolerable daily intake; *PMTDI* Provisional maximum tolerable daily intake [67]

contamination. Symptoms of feed refusal, vomiting, poor weight gain or weight reduction, and even death, have repeatedly been associated with consumption of DON-contaminated grain [1]. These symptoms have been reproduced in studies with *F. graminearum* culture material and with pure DON, although these is some indication that DON alone does not account for all the observed symptoms [1]. Feed refusal and vomiting can occur in swine with levels of DON as low as 2 mg/kg in feed. In humans, trichothecenes have been associated with disease outbreaks known as alimentary toxic aleukia (ATA) in Russia and central Asia and akakabi-byo in Japan. ATA outbreaks occurred during the 1930s and 1940s, characterized by often fatal symptoms that included nausea, vomiting, diarrhea, skin rashes, reduced white blood cell counts, and necrotic lesions of internal organs. There was a close association between disease occurrence and consumption of over-wintered grain that was found to be infected with *F. sporotrichioides* and *F. poae* (and probably *F. langsethiae*). Symptoms could be reproduced using culture material of *F. sporotrichioides*, but no specific toxic compound was identified. In 1968, T-2 toxin was described and later shown to be capable of inducing symptoms typical of ATA [2]. In Japan, outbreaks of akakabi-byo or "red mold disease" were recorded numerous times during the late nineteenth to mid-twentieth centuries, associated with the consumption of *Fusarium*-contaminated foods. Red mold disease was characterized by nausea, vomiting, diarrhea, headaches, dizziness, and trembling; however, the condition was not generally fatal [68]. *Fusarium* strains isolated from implicated grains were identified as *F. graminearum* and *F. kyushuense* (initially identified as *F. sporotrichioides*). Deoxynivalenol was first described from Japanese barley in 1970, and symptoms typical of akakabi-byo were reproduced using culture material of DON-producing strains of *F. graminearum* or nivalenol-producing strains of *F. kyushuense*. Subsequent studies have shown that these symptoms can be induced by ingestion of DON or nivalenol [2]. For both ATA and akakabi-byo, there is convincing, but not definitive, evidence for the role of trichothecenes, mainly because the putative causal toxins were not known at the time of the major outbreaks.

Phytotoxicity has been demonstrated for most trichothecenes that have been tested. Effects include reduced seed germination, stunting of coleoptiles, roots, and shoots, chlorosis, wilting, and necrosis [58]. Toxicity to plants seems to be higher for DON than for other tested trichothecenes [69]. Deoxynivalenol has been shown to have a role in diseases caused by *F. graminearum*. In studies on small grain cereals and maize, mutant strains of *F. graminearum* that did not produce DON are consistently less aggressive as head blight and ear rot pathogens [70, 71]; similar studies have indicated that DON influences seedling disease in a crop- and cultivar-dependent manner [72, 73]. In wheat heads, initial infection was not influenced by the ability to produce DON, but only DON-producing strains were able to spread from spikelet to spikelet [74, 75]. Therefore there is significant evidence that DON contributes to pathogenicity of *F. graminearum*.

Occurrence of trichothecenes in crops in different geographical regions was reviewed by Desjardins [1]. The most important crops that are contaminated by trichothecene mycotoxins are the small grain cereals and maize. Trichothecenes can be hazardous contaminants of wheat, barley, oats, and triticale anywhere in the world where they are grown. Pathogens that cause Fusarium head blight (or infect cereals without symptoms) can produce either type A or type B trichothecenes, although the occurrence of type B is more widespread, due to infection by members of the *F. graminearum* species complex. Maize contamination by trichothecenes is more common in cooler temperate areas such as the Great Lakes region of the USA and Canada, northern Europe and northern China, but it can occur in other areas around the world. In maize, type B trichothecenes are more common, due to infection by members of the *F. graminearum* species complex, but type A trichothecenes can also occur in maize [1, 76]. Trichothecenes contamination also occurs in potatoes, rice, and other crops or food products molded by *Fusarium* spp. Because of the hazards posed to humans and livestock by trichothecenes in grains and foods, many countries have imposed guidelines or regulatory limits on allowable levels of DON and other trichothecenes (Table 4) [77].

2.2 Zearalenone and Related Compounds

Zearalenone and related compounds are estrogenic mycotoxins with low acute toxicity that do not cause fatal toxicoses. However, zearalenone is considered among the most economically important mycotoxins because of its association with reproductive abnormalities, primarily in swine. Zearalenone was first purified from a culture of *F. graminearum* and originally was designated as "fermentation estrogenic substance F-2." In 1966, it was structurally characterized and named zearalenone [78]. Zearalenone is produced by many of the same fungi that produce type B trichothecenes, and therefore it is frequently found together with DON or NIV. Fungi in the *F. graminearum* species complex are the most important ZEA producers, although there are other species that have been reported to produce ZEA (Table 2). Fungi in the *F. oxysporum*, *F. solani*, and *F. fujikuroi* species complexes are not considered to be ZEA producers. Zearalenone and its relatives are cyclic molecules with a basic resorcyclic acid lactone structure and variations in presence, reduction state and acetylation of hydroxyl groups around the core structure. The most well-known compounds are zearalenone and zearalenol, but several other compounds are known [1, 79]. Hormonal effects of some of these compounds make then useful for pharmaceutical purposes, and α-zearalenol has been marketed as a livestock growth stimulant [80].

Table 4
Recommendations and regulations for safe limits on mycotoxin concentrations in grain in the USA and European Union

Mycotoxin	Grain for human food		Grain for animal feed	
	USA[a]	EU[b]	USA[a]	EU[b]
Deoxynivalenol	1000 ppb	750 ppb	5000–10,000 ppb[d]	1750 ppb
Fumonisins	2000–4000 ppb[c]	1000 ppb	5000–100,000 ppb[d]	4000 ppb
T-2	No guidance levels, case by case basis	50–200[e]	No guidance levels, case by case basis	100–200 ppb[f]
Zearalenone	No guidance levels, case-by-case basis	75–100 ppb[c]	No guidance levels, case-by-case basis	100–350 ppb[d]

[a]Reference [3]
[b]Commission Regulation (EC) No 1126/2007 or 576/2006
[c]Varies among specific food items
[d]Varies among livestock species
[e]Varies among grain types
[f]Varies among grain types; up to 1000 ppb for oats with husks

The primary effects of zearalenone on animals are due to estrogenic activity. Estrogen-receptor relationships with zearalenone and related compounds have been reviewed [66, 81]. These estrogenic effects lead to reproductive abnormalities, including enlarged mammary glands and genitalia, atrophy of ovaries or testes, infertility, reduced litter size, and reduced weight of offspring. Swine are the most sensitive livestock species, and zearalenone was discovered in association with an outbreak of estrogenic syndrome in swine. Numerous such outbreaks have been associated with zearalenone [2]. Symptoms of typical of estrogenic syndrome have been reproduced in swine and mice by feeding pure culture material of *F. graminearum*, and subsequently the symptoms have been induced in many studies with pure zearalenone; this compound has clearly been shown to cause estrogenic syndrome in swine. Due to these hormonal effects, JECFA has established a PMTDI for zearalenone ingestion by humans (Table 3). Low levels of zearalenone (as low as 3 µg/g in the diet) can cause symptoms of delayed or prolonged estrus or pseudopregnancy in sows [82]. Loss of pregnancy and increased fetal mortality have been reported, but these symptoms have not been widely confirmed. Livestock other than swine are more resistant and can tolerate zearalenone at levels up to 30 µg/g. Zearalenone and related compounds do not exhibit acute toxicity toward animals; there is limited evidence for carcinogenic, hepatotoxic, and immunotoxic effects [1, 67, 83]. A 1982 report by the US National Toxicology Program (NTP) concluded that zearalenone was carcinogenic to mice but not rats [84].

Zearalenone does not appear to be phytotoxic or to play a role in pathogenicity of zearalenone-producing fungi.

Because it is produced by many of the same fungi that produce type B trichothecenes, the geographic and crop-species occurrence of zearalenone is similar to DON and NIV. Affected crops include maize (grain and silage), wheat, barley, oats, rye, sorghum, millet, and rice, and very occasionally soybean. Zearalenone can be secreted in milk if dairy cows ingest high concentrations, and it can be found in beer made with contaminated barley or other grains [66]. Problematic levels of zearalenone occur more often in maize than in small grains, partially due to the frequent use of maize grain in swine diets. Zearalenone can be found in these crops worldwide. Outbreaks of swine estrogenic syndrome have occurred most frequently in N. America, but also have been relatively widespread in Europe. High levels of zearalenone have been reported from China and other Asian countries [83], but information on livestock impacts in that region is not abundant. Occurrence and human exposure to zearalenone have been reviewed [83]. Zearalenone often co-occurs with trichothecenes, but the two groups of mycotoxins have different temperature optima for biosynthesis [85], so it is not uncommon for grain to have high levels of trichothecenes without appreciable zearalenone contamination, or vice versa. Because of the hazards posed to humans and livestock by zearalenone in grains and foods, many countries have imposed guidelines or regulatory limits on allowable levels (Table 4) [83].

2.3 Fumonisins

After the trichothecenes, fumonisins are the *Fusarium* mycotoxins most often associated with toxicoses in humans and livestock. They are the most common mycotoxins found in maize, and they are found in several other crops. Fumonisins are produced primarily by *Fusarium* spp. in the *F. fujikuroi (G. fujikuroi)* species complex, although minor amounts of fumonisin B_2 can be produced by some species in *Aspergillus* section *Nigri* [86]. The most important fumonisin-producing species are *F. verticillioides* and *F. proliferatum*; this class of mycotoxins was named for *F. moniliforme*, the name under which both of these species were formerly known. Several (but not all) species in the *F. fujikuroi* complex also produce fumonisins, and minor amounts have been reported from a few strains of *F. oxysporum* [41]. The fumonisins are a family at least 28 polyketide mycotoxins with a linear 18- to 20-carbon backbone. The fumonisin family consists of A, B, C, and P-series compounds. Fumonisins B_1, B_2, and B_3 are the most common and most intensively studied; B-series fumonisins have a 20-carbon backbone with a terminal amine function, several hydroxyl functions, and two propane-1,2,3-tricarboxylate esters at various positions. A-series and P-series fumonisins differ due to alteration or replacement of the terminal amine group, while C-series fumonisins have a 19-carbon backbone [87].

Fumonisins have toxic and carcinogenic effects on animals, and these have been reviewed in detail previously [36, 87]. The primary mechanism of toxicity for fumonisins is interference with sphingolipid biosynthesis, which leads to a wide variety of organ-specific outcomes in different animal species. By inhibiting ceramide synthase, fumonisins cause accumulation of sphinganine and other toxic intermediates of sphingolipid metabolism and depletion of complex sphingolipids, which interferes with cell membrane structure and function and disrupts numerous cell regulatory pathways. Free sphinganine leads to liver and kidney lesions, including disorganization of hepatic cords, hepatocellular vacuolation, megalocytosis, apoptosis, necrosis, and cell proliferation [36]. Disruption of sphingolipid metabolism also results in the gross symptoms associated with fumonisin toxicoses in horses and swine. Various cellular processes are disrupted, leading to decreased cardiovascular function in both horses and swine, and this in turn is believed to contribute to neurologic symptoms in horses and pulmonary symptoms in swine [88, 89]. Due to its nephrotoxic effects, JECFA has established a PMTDI for fumonisin ingestion by humans (Table 3); some researchers have suggested that a PMTDI based on other toxicological endpoints would suggest a lower value than the one adopted by JECFA. Several studies have demonstrated carcinogenic activity of fumonisins in livers and kidneys of mice and rats [36, 90], and IARC declared fumonisin B_1, the most naturally abundant of the fumonisins, to be a group 2B carcinogen [91].

Fumonisins have been shown to be phytotoxic to maize and other plant species. They inhibit the growth of shoots and roots, and cause wilting, chlorosis, and necrosis in a range of plant species. However, the role of fumonisins, if any, in plant disease is not clear. In an early study of naturally occurring strains that differed in fumonisin-producing capacity, virulence of *F. verticillioides* strains toward maize seedlings was correlated with fumonisin-producing capability [92]. Subsequent studies using isogenic fumonisin-producing and non-producing strains showed that lack of fumonisin production did not influence ability of strains cause maize ear rot [93]. Fumonisins have been reported as probable pathogenicity factors in maize seedling disease caused by *F. verticillioides* [94, 95], but other studies have shown that fumonisin non-producing mutants infect and colonize maize seedlings as effectively as their corresponding wild-type, fumonisin-producing progenitors [73]. It is likely that some of the symptoms in maize seedlings inoculated with *F. verticillioides* are caused by fumonisin phytotoxicity, but it is not clear how this relates to overall impacts of seedling disease.

Several toxicoses have been attributed to fumonisins in humans and livestock animals. Fumonisins were discovered in the late 1980s and first reported in 1988 [96] as a result of investigations into the cause of high rates of human esophageal cancer in the former Transkei region of South Africa. In samples collected from households in the affected district, high levels of fumonisin B_1 and B_2 were found, and these levels correlated with elevated incidence of esophageal cancer [97, 98]. Similar correlations have been reported in parts of China [99, 100], northeastern Italy [101], and Iran [102]. However, esophageal symptoms have not been experimentally reproduced [103, 104] and the role of fumonisins in this disease has not been definitively shown. Another possible human health impact is increased incidence of neural tube birth defects (NTDs) in births by women consuming fumonisin-contaminated maize. Associations between fumonisin ingestion and elevated incidence of NTDs have been observed along the US-Mexico border, in Guatemala, China, and South Africa [105, 106]. In studies with mice, fumonisins have been shown to cause neural tube defects and other similar birth defects by interfering with folate receptors in embryonic neuroepithelial cells [106, 107]. Therefore there is strong evidence that maternal fumonisin exposure contributes to the incidence of NTDs in some human populations.

In the late 1980s, just as South African researchers had uncovered the characteristics of fumonisins, outbreaks of equine leukoencephalomalacia (ELEM) were occurring in the USA and elsewhere. This disease is characterized by lethargy, tremors, paralysis, and convulsions, followed quickly by death. Brain tissue of affected animals becomes swollen, hemorrhages, and liquefies [108]. ELEM had been observed during the nineteenth and throughout the early twentieth century, and had long been associated with consumption of moldy grain. *Fusarium verticillioides* was first described (as *F. moniliforme*) in 1904 from maize grain associated with the disease [109], and symptoms were reproduced by feeding moldy grain to healthy horses. Subsequently, symptoms also were reproduced by feeding culture material of *F. verticillioides* [110]. Following the identification of fumonisins, researchers in South Africa then demonstrated that ELEM could be induced by injection or ingestion of pure fumonisins [108], and this has been confirmed multiple times [1]. During the same period, an outbreak of pulmonary edema in swine occurred in the USA; symptoms of this disease are weakness, breathing problems, cyanosis, and death. The lungs and thoracic cavity of affected animals fills with fluid. The disease was associated with the consumption of maize screenings (broken and damaged kernels discarded from high-quality grain lots) that later were found to have high levels of fumonisins [111]. Similar symptoms had been reproduced earlier in South Africa by feeding culture material of *F. verticillioides* [110]. Symptoms of pulmonary edema were reproduced by injection or oral dosing with pure fumonisins [112, 113], and this has been confirmed multiple times [114]. In addition to these acute effects, chronic exposure to fumonisins is associated with delayed development and poor weight gain in swine [115, 116].

Fumonisins occur in maize crops and maize-derived foods and feeds worldwide; In North America, higher levels of contamination typically are found at lower latitudes, but fumonisins are the most common mycotoxins found in the "Corn Belt" of the Central USA and they also occur in parts of Canada. Similarly, in Europe the highest levels have been found in Italy, southern France, Spain, and Portugal, while fumonisins are less common in northern Europe [117]. In South America, most data come from Brazil and Argentina, where significant levels of fumonisins can be found in the major maize-growing regions. Reports of fumonisins in maize also have been published from Colombia, Ecuador, Peru, Uruguay, and Venezuela [1]. Fumonisins also are important contaminants of maize in Central American countries such as Guatemala [118]. Fumonisins were first discovered in Africa and there continue to be reports of high incidences and concentrations of fumonisins in parts of South Africa. Fumonisins can be found where maize is grown throughout Africa. Data from African countries other than South Africa are less extensive; they generally indicate high incidences but lower concentrations than those reported from problem areas such as the Transkei region of South Africa [1]. Fumonisins are common, sometimes at high levels, in maize grown in China, where they are associated with elevated human esophageal cancer rates. There are multiple reports of fumonisin occurrence in other Asian countries, including Indonesia, Japan, Korea, the Philippines, Taiwan, Thailand, and Vietnam, but in those countries the incidence is often high but the mean levels of contamination are relatively low [1], with the exceptions of Iran and Nepal, where higher levels have been reported. Insect injury to maize increases contamination by fumonisins [119]. Because of the hazards posed to humans and livestock by fumonisins in grains and foods, many countries have imposed guidelines or regulatory limits on allowable levels of fumonisins (Table 4) [120].

2.4 Fusaric Acids

Fusaric acid (5-butyl picolinic acid) is a common mycotoxin produced by several *Fusarium* species. Several derivatives also have been reported. Fusaric acid is phytotoxic and was first discovered based on its phytotoxicity to rice seedlings. In some *Fusarium* species, particularly some *formae speciales* of *F. oxysporum*, fusaric acid plays a role in pathogenicity to plants, by inducing wilt symptoms [121]. The biosynthetic pathway is partially described [122, 123], and pathogenicity is attenuated in strains with disrupted fusaric acid biosynthetic genes. It was first isolated in 1934 in Japan from a strain of *F. heterosporum* (*F. fujikuroi*). Other species in the *F. fujikuroi* species complex also produce fusaric acid, and it is produced by strains in the *F. solani* species complex, *F. crookwellense*, and *F. sambucinum* (Table 2). In rats, fusaric acid caused changes in brain weight and body weight. It also is an inhibitor of DNA synthesis, but acute toxicity of fusaric acid seems to be very low. However, it appears to enhance the activity of several other *Fusarium* mycotoxins, including fumonisins, zearalenone, deoxynivalenol, diacetoxyscirpenol, and T-2 toxin [124]. Thus, the greatest significance of fusaric acid as a mycotoxin may be its interactions with other, more acutely toxic compounds. Because of its relatively low toxicity, little effort has been made to document the occurrence of fusaric acid in crops and foods. Some fusaric acid-producing species such as *F. proliferatum* and *F. oxysporum* have wide host ranges and it is likely that fusaric acid is common in various crops and food or feed products. Limited data from Canada and the USA indicated a high incidence (85–100%) of fusaric acid in a relatively small number of samples, with concentrations as high as 136 µg/g [1].

2.5 Moniliformin

Moniliformin is the potassium or sodium salt of 1-hydroxycyclobut-1-ene-3,4 dione, and it is a metabolite unique to *Fusarium* fungi. It has been implicated in various field mycotoxicoses, especially in poultry. It is acutely toxic and has an LD50 of 20.9 to 29.1 mg/kg body weight for intraperitoneal injection in mice. Moniliformin production is widespread among *Fusarium* species; it was named for *F. moniliforme*, which was the identification of the fungus from which it was first isolated. Strains originally associated with moniliformin production have subsequently been re-identified as *F. proliferatum* and *F. nygamai*. Many strains within the *F. fujikuroi* species complex produce moniliformin, but its production is very rare in *F. verticillioides*. Some isolates in the *F. oxysporum* species complex also produce moniliformin, but it seems to be absent from the *F. solani* species complex. Several trichothecene-producing species produce moniliformin (Table 2), including *F. acuminatum*, *F. culmorum*, *F. equiseti*, and *F. sporotrichioides*, but it has not been reported from *F. graminearum sensu stricto*.

Moniliformin can cause rapid death in poultry and swine. Toxicological effects include inhibition of protein synthesis, cytotoxicity, and chromosome damage. The primary mode of action appears to be inhibition of mitochondrial respiration [124], and damage to myocardial mitochondria, resulting, in some cases, in acute congestive heart failure. Immunosuppression also has been reported in poultry. Symptoms induced in laboratory studies with experimental mammals and poultry include reduced body weight, intestinal hemorrhaging, coma, and death [50]. Moniliformin is suspected of involvement in several distinct diseases in poultry, including ascites, round heart disease, and "spiking mortality syndrome." Mammals appear to be significantly less sensitive than poultry. However, Chinese researchers have reported a possible connection to Keshan disease, a serious endemic heart disease in some parts of China [124]. Insect injury to maize increases contamination by moniliformin [125]. Moniliformin has been reported to be phytotoxic, but little information is available and its toxicity to maize seedlings seems to be low [1].

Occurrence of moniliformin has not been widely studied, but reports are summarized by Jestoi [50]. It has been reported from maize in Canada, China, South Africa, the USA, and Europe, and from small grain cereals in Austria, Canada, Finland, Norway, and Poland, and rice in China and Iran [1, 50].

2.6   Fusarins

Fusarins are a group of mycotoxins that share a polyketide backbone with various substitutions at the 2-pyrrolidone moiety. Fusarin C was the first compound characterized in this group, followed by fusarins A, D, E, and F [1]. Fusarin C was identified from *F. verticillioides* (as *F. moniliforme*) based on bacterial mutagenic activity. Fusarins are produced by a few species scattered in several of the major species complexes (Table 2), but they have not been reported in the *F. solani* species complex. Fusarin C production by *F. oxysporum* has been reported, but rarely [28]. Fusarins can co-occur with both trichothecenes and fumonisins [1, 126]. Fusarins have not been associated with any known mycotoxicoses in humans or livestock, but fusarin C has been shown to be mutagenic and to cause disruption of chromosomes in mammalian cell cultures [1]. Fusarin C was acutely toxic to rats when administered at a high dose (100 mg/kg body weight) [124]. There is limited evidence for carcinogenicity and immunotoxicity [124]. Fusarin C has been reported to be unstable at ambient temperatures [124], and this may limit its role in toxicological outcomes. A cluster of nine genes has been identified for biosynthesis of fusarins, though only four genes, *fus1*, *fus2*, *fus8*, and *fus9*, are necessary for fusarin C production [127]. There is little information about the natural occurrence of fusarins in grains, feeds, or foods, but since they are produced by *Fusarium* species that are common in maize and small grain cereals, it is likely that they occur widely. They have been reported from maize in the United States, South Africa, and China [124]. In a study of maize and maize-based food samples in Germany, 80 % of samples had detectable fusarins, at levels up to 83 mg/kg [128].

2.7   Beauvericin and Enniatins

Beauvericin and enniatins are closely related cyclic hexadepsipeptides that can be produced by a wide range of *Fusarium* spp. There are at least 29 enniatin analogues, but not all of these are produced by *Fusarium* spp. At least seven analogues (enniatins A, A1, B, B1, B2, B3, and B4) have been found in cereals and are produced by *Fusarium;* the most common ones are enniatins A, A1, B, and B1. Beauvericin originally was identified from the entomopathogenic fungus *Beauveria bassiana*, but it has been identified as a common metabolite in *Fusarium*. Many members of the *F. fujikuroi* species complex [129, 130] can produce beauvericin, but *F. verticillioides* typically does not produce this compound. Other species also produce beauvericin, including *F. acuminatum, F. armeniacum, F. langsethiae, F. poae, F. sambucinum,* and *F. sporotrichioides* (Table 2).

Thus beauvericin is produced by fumonisin-producing and trichothecene-producing species. It does not appear to be produced by species in the *F. solani* complex, and it seems to be rare in the *F. oxysporum* complex, although it has been reported [51]. Beauvericin and enniatins have antimicrobial activity, are toxic to several insect species, and are cytotoxic in laboratory studies with cell lines of insects and humans [50]. Acute toxicity to rodents and other experimental mammals is low, and several studies on poultry showed low levels of sensitivity to beauvericin and enniatins [50, 131]. Effects on plants have not been demonstrated. Occurrence of beauvericin in grain crops was first reported in 1993 from maize in Poland infected by *F. subglutinans*. Subsequently, it has been reported from maize in other European countries, the USA (also from livestock feed), Mexico, South Africa, and Argentina, from small grain cereals in several European countries, and in rice in Iran and North Africa. Enniatins have primarily been reported from small grain cereals in Europe, sometimes at high incidences [50]. In a study by EFSA [131] including 12 European countries, beauvericin was detected in 54% of unprocessed grain samples, 21% of feed samples, and 20% of food samples. Enniatins were detected in 76% of unprocessed grain samples, 68% of feed samples, and 37% of food samples. Occurrence of beauvericin and the four enniatins was highly correlated. Indeed, in most reports, beauvericin has occurred together with enniatins and frequently with other mycotoxins, especially moniliformin and fumonisins, but also trichothecenes, fusaproliferin, and zearalenone [50]. In the EFSA study, exposure to beauvericin and enniatins was estimated for humans and several livestock species; the authors concluded that there was no indication for human health concern and that toxic effects on livestock were unlikely at the estimated exposure levels, but emphasized that the lack of data contributes to considerable uncertainty regarding the risks associated with these compounds [131].

2.8 Fusaproliferin

Fusaproliferin is a bicyclic sesterterpene, the first sesterterpene reported from *Fusarium* [132]. A deacetylated form also is a naturally occurring metabolite. The name is derived from *F. proliferatum*, the first species recognized to produce the compound. Other fusaproliferin-producing species are in the *F. fujikuroi* species complex, although *F. verticillioides* is not known to produce it (Table 2). Fusaproliferin production also has been reported outside the genus *Fusarium* [1]. Toxicity of fusaproliferin has been evaluated primarily in vitro. It is toxic to brine shrimp larvae [55], and cytotoxic to cell lines from an insect (*Spodoptera frugiperda*) and from humans (IARC/LCL 171). Fusaproliferin also is teratogenic to chicken embryos [124]. This result may explain certain pathological effects of *F. proliferatum* culture extracts on chicken embryos in previous studies. The role, if any, of fusaproliferin on naturally occurring mycotoxicoses is not clear, but it is a contaminant of feed samples associated with feed refusal by swine, where no deoxynivalenol or other trichothecenes could be detected [133]. Some phytotoxic effects of fusaproliferin and 24-diacetyl-fusaproliferin have been reported in maize [134] and ryegrass [1]. Fusaproliferin production is quite common in isolates of *F. proliferatum* and *F. subglutinans*, which suggests that maize could be frequently contaminated with these compounds anywhere in the world. Data on fusaproliferin occurrence is not extensive, but it has been reported from maize in the United States (also animal feeds), South Africa, and several European countries, as well as small grains in Europe and rice and breakfast cereals in Morocco [50, 135]. Fusaproliferin often occurs in maize together with beauvericin, enniatins, or fumonisins, but also occasionally with trichothecenes.

Fusarium species produce many additional less common or less well-known mycotoxins [2] that can be found in maize or cereal grains, and may have significant toxicological implications. Details of these mycotoxins are beyond the scope of this chapter, but several of these toxins have been reviewed [124].

3   Toxigenic Properties of Important *Fusarium* Species

This section describes key characteristics of selected toxigenic *Fusarium* species, including nomenclature, basic morphology, importance as plant pathogens, major mycotoxins produced, and association with human and animal toxicoses. Table 2 is a more inclusive list of species and their known toxic metabolites.

3.1   *F. acuminatum* and *F. armeniacum*

Fusarium acuminatum Ellis and Everhart was described in 1895 from potato stems in New York, USA. It has been reported from a wide range of plants, primarily in the northern hemisphere. In culture on PDA, *F. acuminatum* grows more slowly than most *Fusarium* spp. and produces white mycelium that often turns pink with age. The colony undersurface is pink to deep red. Macroconidia are long and thin, and usually 5-sepate, with a long, curved apical cell. They can appear similar to those of *F. equiseti*, but the two species can usually be easily distinguished by colony morphology on PDA. Microconidia are not typically produced and chlamydospores are rare. The teleomorph was described as *Gibberella acuminata* Wollenw., which has been observed only in the laboratory. *Fusarium acuminatum* is generally considered as a saprophyte, but it can cause severe root rot in some legumes, such as soybean [136]. *Fusarium acuminatum* also has been associated with head blight of small grains. When found on small grain cereals or maize, it is typically a minor component of the *Fusarium* population associated with head blight or ear rot symptoms [1, 28]. *Fusarium acuminatum* has not been associated with animal or human toxicoses, but it has been reported to produce a wide range of mycotoxins, including T-2, HT-2, DAS, neosolaniol, beauvericin, fusarins, enniatins, and moniliformin [1]. Recently, DON production has been reported by *F. acuminatum* [137]. Two forms of *F. acuminatum* were long recognized, based on differences in T-2 production and some morphological characters. Strains that produced high levels of T-2 were designated as *F. acuminatum* subsp. *armeniacum*, but subsequently elevated to species level as *F. armeniacum* (Forbes, Windels, and Burgess) Burgess and Summerell. Morphology of *F. armeniacum* is similar to *F. acuminatum* but *F. armeniacum* grows more rapidly, has longer macroconidia, and produces apricot-colored sporodochia. Subsequent research demonstrated that the trichodiene synthase gene, *Tri5*, which is an important trichothecene biosynthesis gene, is present in *F. armeniacum* but not in *F. acuminatum* [28]. In fact, *F. armeniacum* is more closely related to *F. sporotrichioides* than it is to *F. acuminatum*, according to phylogenetic analysis [138]. Some strains of *F. armeniacum* produce very high levels of T-2 toxin and other trichothecenes [1]. Zearalenone production by *F. acuminatum* is not confirmed, but it has been reported for *F. armeniacum*, along with several of the other toxins already attributed to *F. acuminatum* [139]. *F. armeniacum* also has been reported as a root rot pathogen of soybean [140]. A whole genome sequence is available for *F. acuminatum* (Table 1).

3.2 F. avenaceum

Fusarium avenaceum (Fr.) Sacc. was first described in 1822 from oats in Germany as *Fusisporium avenaceum,* and was transferred to *Fusarium* in 1886. It is known as a significant head blight pathogen in small grains. Cultural characteristics of *F. avenaceum* are highly variable. Mycelium on PDA varies from white to pink or brown, with pale orange to brown sporodochia. The colony undersurface may be peach to burgundy or brown. Macroconidia are long and slender, thin-walled, and usually 5-septate, varying from strongly curved to nearly straight. Microconidia can be produced on monophialides or polyphialides, but they are rare; chlamydospores are not produced. The teleomorph, *Gibberella avenacea* R.J. Cook, was reported from wheat stalks in 1967. *Fusarium avenaceum* has an extremely wide host range, causing seedling diseases in many crops, dry rot of potatoes, and root rot and head blight in small grains [1]. Its main importance is as a major cause of Fusarium head blight in cooler areas where small grains are grown on several continents. In some areas, such as Finland and Alberta, Canada, it can be the predominant head blight species [1]. It also has been associated with Fusarium ear of maize in Europe [117]. *Fusarium avenaceum* is a major producer of moniliformin, and also produces aurofusarin, beauvericin, chlamydosporol, and enniatins. Reports of trichothecene production by *F. avenaceum* are probably incorrect. This species has not been associated with any known toxicoses, although culture material was toxic to chickens, most likely due to moniliformin [1]. Whole genome sequences are available for several strains of *F. avenaceum* (Table 1).

3.3 *F. crookwellense*

Fusarium crookwellense L.W. Burgess, P.E. Nelson, and Toussoun was first isolated from diseased potato tubers in 1971 and then described in 1982, based on a large collection of strains. *Fusarium crookwellense* is closely related to *F. culmorum* and *F. graminearum* and it has similar cultural morphology, particularly to *F. culmorum*; mycelium is white to yellow or tan and may become red at the colony edge. The undersurface is usually red. Macroconidia are usually 5-septate, thick-walled and sickle-shaped. Microconidia are not produced, but chlamydospores may be abundant. No teleomorph has been described. *Fusarium cerealis* (Cooke) Sacc. is considered a synonym of *F. crookwellense*. *Fusarium crookwellense* is widely distributed but is more common in cooler temperate areas and often is isolated from root or crown tissue of various plant species. It can cause root and foot rot of wheat and is a minor component of the Fusarium head blight and maize ear rot complexes in Europe and Japan. This species has not been associated with any known toxicoses, but can produce a wide range of mycotoxins including fusaric acid, fusarin C, and NIV and its acetylated derivatives, but not DON, beauvericin, or enniatins [1]. Diacetoxyscirpenol has been associated with *F. crookwellense* [117], but most known DAS-producing strains have been reclassified as *F. venenatum* [28].

3.4 *F. culmorum*

Fusarium culmorum (W.G. Smith) Sacc. was described in 1884 as *Fusisporium culmorum* and then transferred to *Fusarium* in 1895. It is one of the most important species causing Fusarium head blight in cool climates. *Fusarium culmorum* is closely related to *F. crookwellense* and *F. graminearum* and it has similar cultural morphology, particularly to *F. crookwellense*. In culture on PDA, *F. culmorum* produces abundant aerial mycelium that is initially white but turns yellow or pink with age. The undersurface is usually deep red. Numerous orange sporodochia usually form in the center of the colony. Macroconidia are relatively short and blunt, slightly curved, usually 3- to 4-septate, with thick walls and rounded apical cells. There are no microconidia but chlamydospores are numerous [1]. No teleomorph has been described, but evidence suggests that genetic recombination occurs at a similar frequency as in *F. graminearum* [141]. In northern Europe, *F. culmorum* has been known as the predominant species causing Fusarium head blight; however, in recent years, frequency of *F. graminearum* has gradually surpassed that of *F. culmorum*, and *F. graminearum* s.s. is now the most common head blight species [142]. *Fusarium culmorum* also occurs as a head blight pathogen in China, Canada, the northern USA, and eastern Australia, but has not been the dominant species in those areas. It can cause foot rot in small grain cereals, and also can occur as an ear rot pathogen of maize in Europe [117] and other cool climates. *Fusarium culmorum* has also been reported as a pathogen on several other plant species, including hops, strawberries, leeks, and Norway spruce [28]. The mycotoxin profile of *F. culmorum* is similar to *F. graminearum*; both species produce high levels of type B trichothecenes and range of other toxic metabolites (Table 2) including fusarins, moniliformin, and zearalenone, but not beauvericin. As with *F. graminearum*, there are DON-producing and NIV-producing chemotypes of *F. culmorum*; NIV-producing chemotypes of *F. culmorum* are common in Europe but not in North America. Toxicoses in cattle and swine have been associated with *F. culmorum*, and culture materials are very toxic in laboratory studies. A whole genome sequence is available for *F. culmorum* (Table 1).

3.5   *F. equiseti*

Fusarium equiseti (Corda) Sacc. was described as *Selenosporium equiseti* Corda in 1838 from *Equisetum* plants in eastern Europe. It was transferred to *Fusarium* in 1886. *Fusarium equiseti* is characterized by long (5- to 7-sepate), strongly curved macroconidia with elongated apical cells, an absence of microconidia, and abundant chlamydospores in chains and clusters. In culture on PDA, it produces abundant mycelium that turns from white to beige or brown. Sporodochia are orange to brown and may form in concentric rings. The colony undersurface may be peach-colored but it usually brown to dark brown. *Fusarium scirpi* is very similar and has been considered conspecific with *F. equiseti* by some authors. The teleomorph of *F. equiseti* was described as *Gibberella intricans* Wollenweber. *Fusarium equiseti* is widely distributed and often recovered from soil, seeds, or senescent plant tissue in both temperate and tropical areas. It usually is considered to be a secondary colonizer of plants, but it has been reported to cause disease on numerous plant species. It has been reported as a minor component of the Fusarium ear rot complex of maize in Europe [117] and the Fusarium head blight complex in both Europe and North America [1]. Other reports of pathogenicity for *F. equiseti* include cucurbits [26], pine seedlings [143], rocket [144], and soybean [136]. According to Leslie and Summerell [28], "… records of *F. equiseti* as a pathogen should be treated cautiously." Regardless of its pathogenic capabilities, the presence of *F. equiseti* in grains used for human food and livestock feed requires attention due to its toxigenic properties. The species has been reported to produce a wide range of toxic metabolites and mycotoxin production varies widely among strains of *F. equiseti*. Strains of the fungus can produce several trichothecenes, zearalenone, beauvericin, fusarochromanone, equisetin, moniliformin, and other compounds [1, 28]. Diacetoxyscirpenol was first isolated from *F. equiseti* [145]. Toxicity of *F. equiseti* has been reported toward several livestock species and experimental animals, but in many cases the specific toxin involved was not identified [2, 28]. It was reported to be associated with Degnala disease of water buffalo in India, but both *F. equiseti* and *F. semitectum* were recovered from moldy rice straw connected with the outbreaks, so the role of each species is not clear. *F. equiseti* also has been reported as a human pathogen [146].

3.6   F. fujikuroi

The perithecial form of *Fusarium fujikuroi* Nirenberg was first described in 1917 as the causal agent of bakanae disease of rice, as *Lisea fujikuroi*. In 1931, the name was changed to *Gibberella fujikuroi* (Sawada) Wollenw., but it was suggested that this fungus was identical to *G. moniliformis*. Controversy over the distinction between *G. fujikuroi* and *G. moniliformis* persisted. In 1976, Nirenberg adopted the combinations *G. fujikuroi/F. fujikuroi* for the rice pathogen and *G. moniliformis/F. verticillioides* for the maize pathogen [147]. In addition, several researchers began to apply a biological species concept in this group of related fungi, and the rice pathogen *F. fujikuroi* was designated as mating population C of the *G. fujikuroi* species complex [148, 149]. At least nine of the mating populations, A through I, subsequently were recognized as *Fusarium* species [1], maintaining the combinations adopted by Nirenberg in 1976 for *F. fujikuroi/G. fujikuroi* and *F. verticillioides/G. moniliformis*. Perithecia of *F. fujikuroi* occur in the field [150]. *Fusarium fujikuroi* morphology is practically indistinguishable from *F. proliferatum*, and these two species are partially interfertile. *Fusarium fujikuroi* produces oval to clavate aseptate (sometimes 1-septate) microconidia in short chains and false heads from monophialides and polyphialides in the aerial mycelium. Macroconidia are formed in sporodochia and are narrow, thin-walled, and relatively straight, usually 3- to 5-septate. Chlamydospores are not formed, and cultures on PDA are white to pale gray or violet, with an undersurface that varies from colorless to violet gray to dark magenta [28].

Fusarium fujikuroi is primarily a global rice pathogen that causes seedling disease and bakanae disease, characterized by abnormally elongated plants due to gibberellin production by the fungus [151]. It has been reported from a few other plant species including sorghum grain and native grass species in Kansas, USA. The most important mycotoxin produced by *F. fujikuroi* is moniliformin; strains also can produce beauvericin and fusaric acid but not fusaproliferin. A small percentage of strains produce fumonisins at low levels, but most *F. fujikuroi* strains do not, in spite of the presence of key fumonisin biosynthesis genes [152]. Whole genome sequences are available for several strains of *F. fujikuroi* (Table 1).

3.7 F. graminearum Species Complex (FGSC)

Fusarium graminearum Schwabe is the most economically important toxigenic species in the genus, due to its broad host range, the significant diseases it causes on small grain crops and maize, the diversity of mycotoxins it can produce, and its association with human and animal toxicoses. Although numerous species have now been described within the *F. graminearum* "species complex" (FGSC), *F. graminearum sensu stricto* remains the most important cause of Fusarium head blight of cereals and Gibberella ear rot of maize throughout most of the world. *Fusarium graminearum* was first described in 1838 from maize. The teleomorph of *F. graminearum* was described as *Gibberella zeae* (Schwein.) Petch, a homothallic sexual stage that commonly produces perithecia and ascospores in the field and in culture. For a long period of time, *F. graminearum* was split into two cryptic species known as Group 1 and Group 2, but in 1999, Group 1 was described as *F. pseudograminearum* [28]. Some confusion in the literature also arises from the use of the name *F. roseum* Link, which for a time was applied by some authors to *F. graminearum* and other species with similar morphology. Snyder and Hansen [24] proposed combining numerous species into *F. roseum* and designating strains that were pathogenic to cereals as *F. roseum* f. *cerealis*. This nomenclature was not widely adopted; other variations proposed by the same authors were *F. roseum* cv. "Graminearum" [153] and *F. roseum* var. *graminearum*. More recently, phylogenetic analyses revealed nine or more distinct lineages within *F. graminearum*, and species names were assigned to these lineages [154, 155]. Species in the FGSC subsequently have grown to at least sixteen [142], although there is cross-fertility among the members of the species complex, and some authors maintain that the FGSC comprises a single biological species [28]. *Fusarium graminearum* is characterized by long, narrow macroconidia that are 5- to 6-septate and straight or slightly curved; no microconidia are produced, and chlamydospores occur in the mycelium or in conidia, but they are not common. Colonies on PDA usually produce a red pigment in the undersurface, and aerial mycelium is white to tan or pink. Blue-black perithecia may form, with 3-septate ascospores that are important as primary inoculum for diseases caused by *F. graminearum*. Morphology is very similar and practically indistinguishable throughout the FGSC.

Diseases caused by members of the FGSC have major global significance. Head blight of wheat, barley, and oats, and maize ear rot are the most important diseases, but infections of rice, sorghum, potatoes, coffee, and legumes also cause significant economic damage. Recently, *F. graminearum* also has emerged as an important seedling and root rot pathogen of common bean and soybean [156–158]. The species has a broad host range, including the model plant species, *Arabidopsis thaliana*, which will likely lead to important further discoveries regarding host-pathogen interactions.

Fusarium graminearum is well known as a producer of Type B trichothecenes and other secondary metabolites (Table 2). Deoxynivalenol was first characterized from *F. graminearum* [159, 160], and *F. graminearum* has been an important model organism for elucidation of the trichothecene biosynthesis pathway. Whole genome sequences are available for strains of *F. graminearum* (Table 1), and this has led to important strides in the understanding of biosynthesis of numerous secondary metabolites and other physiological processes in this fungus. Members of the FGSC can be characterized into "chemotypes" depending on their predominant trichothecene type. Some strains primarily produce DON and 15-ADON. These are the predominant types in the Americas, Europe, and South Africa [142]. Strains that primarily produce DON and 3-ADON are common in parts of Asia, while strains that primarily produce NIV can be found at low frequencies on all continents and predominate in some parts of Asia. The 3-ADON chemotype is reported to be increasing in frequency in Canada and possibly in the USA [142]. Chemotypes may differ in other traits, such as overall DON production, growth rate, and pathogenic aggressiveness [161]. Little research has been conducted on prevalence of chemotypes on hosts other than small grains and maize, but soybean isolates from N. America and Argentina appear to primarily be 15-ADON chemotypes of *F. graminearum* s.s. [162, 163]. Evidence suggests that several fairly recent introductions have led to changes in the global distribution of the chemotypes [142]. *Fusarium graminearum sensu stricto* includes strains of all three chemotypes, but there are associations between chemotype and species for some members of the FGSC (see below). Recently, new Type A trichothecenes, designated NX-2 and NX-3 were reported to be produced by strains of *F. graminearum* found in a limited area of the North Central United States [164, 165]. This is a potentially important discovery, since the compounds have toxicity similar to that of DON, and Type A trichothecene production is unusual in *F. graminearum*. *Fusarium graminearum* is the most important producer of zearalenone, and also has been reported to produce fusarins, culmorins, and other toxins but not beauvericin or moniliformin.

Species described within the FGSC include:

- *F. acaciae-mearnsii* O'Donnell, T. Aoki, Kistler et Geiser—represented by strains from *Acacia* or soil in South Africa or Australia. The strains were highly aggressive, causing Fusarium head blight symptoms when inoculated to wheat heads, and produced small amounts of DON, 3-ADON, and NIV, relative to *F. graminearum* sensu stricto [52, 53].

- *F. asiaticum* O'Donnell, T. Aoki, Kistler et Geiser—this species is an important cause of Fusarium head blight in parts of Asia, and appears to have some degree of host preference for rice.

In southern Japan, it is the predominant Fusarium head blight species. Similarly, in southern China. *F. asiaticum* was reported as the predominant species associated with head blight of wheat and barley [142]. In China, *F. asiaticum* is more common in warmer areas and where wheat and barley are rotated with rice; whereas *F. graminearum* sensu stricto is associated with cooler areas and those where small grains are rotated with maize. In South Korea, *F. asiaticum* is the predominant species on rice [142, 166]. In the USA, *F. asiaticum* has been reported from wheat in Louisiana, in areas associated with rice production [167]. Nivalenol-producing isolates of *F. asiaticum* also have been reported from wheat in Uruguay; again, in areas of the country with significant rice production [142]. *F. asiaticum* also is common on rice in Brazil, and has been reported as a pathogen of maize in Japan [168] and asparagus in China [169]. Isolates of *F. asiaticum* have been reported to produce DON, 3-ADON, and NIV [52, 53]; NIV-producing isolates are predominant in many areas [170].

- *F. austroamericanum* T. Aoki, Kistler, Geiser et O'Donnell—represented by three strains from Brazil or Venezuela. The strains caused head blight symptoms when inoculated onto wheat heads and are reported to produce DON, 3-ADON, and NIV [52, 53]. *Fusarium austroamericanum* has been reported at low frequencies from wheat and barley in Brazil [171].

- *F. boothii* O'Donnell, T. Aoki, Kistler et Geiser—this species has been reported from Argentina, China, Guatemala, Hungary, S. Korea, Mexico, Nepal, and South Africa [154, 166, 172–175], primarily from maize. Although it causes head blight symptoms when inoculated to wheat heads [52, 53], *F. boothii* has been reported to have a host preference for maize. In South Africa [172], it was the predominant species associated with Gibberella ear rot symptoms, but only about 8% of FGSC isolates from wheat were *F. boothii*. This species has been reported to produce DON and 15-ADON [52, 53].

- *F. brasilicum* T. Aoki, Kistler, Geiser et O'Donnell—described from oats in Brazil, this species also has been reported from wheat in Uruguay [176].

- *F. cortaderiae* O'Donnell, T. Aoki, Kistler et Geiser—this species was described from several strains from pampas grass or other hosts from Argentina, Australia, Brazil, and New Zealand. *Fusarium cortaderiae* constituted 14.6% of FGSC isolates from rice seeds [177] and 2.5% of FGSC isolates from wheat spikes [171] in Brazil. In the latter study, *F. cortaderia* was very common in maize stubble in at least one location, but was a minor component of the FGSC infection of wheat spikes. Two isolates of *F. cortaderiae* were highly aggressive when inoculated onto wheat heads, and produced small amounts of NIV [52, 53]. *F. cortaderia* also has been reported from soybean in Argentina [156], wheat in Italy [178] and Uruguay [176], and maize in France [179]. NIV-producing *F. cortaderia* strains were reported from wheat in New Zealand [180]. Most *F. cortaderia* isolates are reported to be of the NIV genotype [171].

- *F. gerlachii*—represented by three strains from wheat or giant reed in the USA. The strains caused head blight symptoms when inoculated on wheat, and produced NIV [155].

- *F. meridionale* T. Aoki, Kistler, Geiser et O'Donnell—This species appears to be distributed primarily in the southern hemisphere, and appears to have a degree of host preference for maize [142]. However, it also occurs as a head blight pathogen in South American wheat [171, 181, 182]. *F. meridionale* is common in South American maize, and most isolates are NIV producers. Del Ponte et al. [171] found that *F. meridionale* was predominant on maize stubble in areas where *F. graminearum* s.s. was the predominant species in diseased wheat heads. Lamprecht et al. [183] found that *F. meridionale* was associated with diseased maize roots in South Africa but was less virulent to maize seedlings than *F. boothii* or *F. graminearum* s.s. In the northern hemisphere, *F. meridionale* has been reported as a minor component of the FGSC populations from maize and wheat in southern China and maize in S. Korea [142, 184], and occurs commonly as a maize ear rot pathogen in Nepal [173]. In inoculation studies on wheat, *F. meridionale* was moderately aggressive and produced small amounts of NIV [52, 53]. Most *F. meridionale* isolates are reported to be of the NIV genotype [171, 185].

- *F. mesoamericanum* T. Aoki, Kistler, Geiser et O'Donnell— this species was described from strains originating from Honduras and Pennsylvania, USA. In inoculation studies on wheat, *F. mesoamericanum* was moderately aggressive and produced small amounts of DON and NIV [52, 53].

- *F. vorosii*—represented by three strains from wheat in Hungary or Japan. The strains caused head blight symptoms when inoculated on wheat, and produced DON and 15-ADON [155].

Fusarium graminearum has been associated with several toxicoses in animals and humans. These include red mold disease (akakabi-byo) in Japan, swine feed refusal, and estrogenic syndrome in swine [2]. Symptoms of these toxicoses have been reproduced with pure culture material of *F. graminearum* [1].

3.8   F. langsethiae

Fusarium langsethiae Torp and Nirenberg was first observed in Norway in 1999 and described in 2004 from oats, wheat, and barley from several countries in Europe [186]. Its morphology is very similar to *F. poae*, and initially was referred to as "powdery *F. poae*" [187]. Previous reports of T-2-producing strains of *F. poae* probably are attributable to *F. langsethiae*. DNA sequence data and mycotoxin profiles indicate that *F. langsethiae* is actually more closely related to *F. sporotrichioides* than to *F. poae*. In culture on PDA, it produces aerial mycelium that is white to pale yellow or pale pink and develops a powdery appearance due to abundant microconidia. The colony undersurface is pale yellow, pink, orange, or pale violet. *Fusarium langsethiae* grows more slowly than *F. poae* and lacks the characteristic odor of *F. poae*. Microconidia are napiform to globose, usually nonseptate, and produced on monophialides that are often bent. Polyphialides can occur but very rarely. Chlamydospores are absent. At least one full genome sequence is available for *F. langsethiae* (Table 1). *Fusarium langsethiae* has been reported as a common contaminant of wheat, oats, and barley, primarily in central and northern Europe [188], but it occurs as far south as Sicily [189]. The species is more frequently associated with oats than the other hosts. Although *F. langsethiae* is frequently isolated from grains of these crops, the plants and grains are often symptomless, and it has been difficult to demonstrate head blight symptoms through inoculation with *F. langsethiae* [188]. Although severe head blight symptoms may not be associated with *F. langsethiae*, the species is a major concern because of its frequent presence in grain and its production of potent mycotoxins, primarily type A trichothecenes. *Fusarium langsethiae* is now considered to be the most important cause of T-2 and HT-2 contamination of small grains in many parts of Europe, and it also produces DAS and neosolaniol [188]. This species has not been directly associated with any toxicoses, likely because of its relatively recent description. However, as a major producer of type A trichothecenes, *F. langsethiae* should be suspected of involvement in toxicoses associated with those compounds.

3.9 *F. oxysporum* Species Complex

Fusarium oxysporum Schlechtend. emend. Snyder and Hansen was first described 1824 and emended in 1940 to include a very diverse complex of species with some similar morphological features, but differing widely in many phenotypic characteristics. Many of the members of the species cause vascular wilt diseases on a wide range of crops, but others cause seedling disease or root rot, or are nonpathogenic. Vascular wilt pathogens from different hosts are often designated as *formae speciales*, some of which are host-specific. Based on DNA sequence data, *F. oxysporum* is polyphyletic, even within some *formae speciales* [190, 191]. Significant progress still needs to be made in order to classify the many genotypes collectively known as *F. oxysporum*. A number of clades have been identified within the complex [191], based on multi-locus gene sequencing analysis; some have been described as separate species (e.g., *F. commune*) [192]. Using a genealogical concordance phylogenetic species recognition approach, Laurence et al. [193] identified two phylogenetic species within a population of *F. oxysporum* from Australia that represented the known clade diversity in the species complex. It is likely that additional phylogenetic (and formally described) species will be identified in the future. Whole genome sequences are available for several *F. oxysporum* strains (Table 1). Morphological characters that are shared within the species complex include: abundant, oval to kidney-shaped nonseptate microconidia that are produced in false heads on short monophialides in the aerial mycelium; macroconidia that are short to medium length, straight to slightly curved, thin-walled, and usually 3-septate, produced in orange sporodochia that are numerous in some isolates and nearly absent in others; and chlamydospores, usually produced singly or in pairs, abundant in most isolates but sparse in others. Cultural characteristics on PDA can be highly variable, with white to yellow or pale violet mycelium, and an undersurface that varies from white to violet or magenta.

Fusarium oxysporum is one of the most economically important pathogens in the genus, but members of this species complex are generally considered to be non-toxigenic. The main impact of *F. oxysporum* is through the numerous vascular wilt diseases caused by members of this species. Important agricultural hosts include alfalfa, asparagus, banana, cabbage, common bean, cotton, cucurbits, lettuce, onion, pea, pepper, potato, soybean, spinach, sweet potato, and tomato; more than 100 *formae speciales* and races have been named [28]. Strains of *F. oxysporum* also are important as seedling and root rot pathogens in a wide range of crops, and they are the most common *Fusarium* species isolated from soils worldwide. Some nonpathogenic strains are effective as biocontrol agents against root pathogens of several crops. Although *F. oxysporum* is not considered a major mycotoxin producing species, it is a diverse species complex and thus has been reported to produce a range of mycotoxins, including enniatins, fusaric acid, and moniliformin, but not trichothecenes, fumonisins, zearalenone, fusarins, fusarochromanone, or fusaproliferin [1]. One strain of *F. oxysporum* from Korea was reported to produce C-series, but not B-series, fumonisins, and its identification has been confirmed [152]; however, this appears to be a very unusual strain. Strains of *F. oxysporum* can commonly be isolated from small grain cereals and maize, but it does not appear to make a meaningful contribution to mycotoxin contamination of these crops. It is not associated with Fusarium head blight or maize ear rot, except as an occasional secondary colonizer. *Fusarium oxysporum* is not associated with any mycotoxicoses, but some strains can infect humans [28].

3.10 *F. poae*

Fusarium poae (Peck) Wollenw. was first described in 1902 as *Sporotrichum poae* Peck and transferred to *Fusarium* in 1912. It is an important trichothecene-producing Fusarium head blight pathogen. It is characterized by its abundant globose to lemon-shaped microconidia produced in clusters on mono-phialides. Macroconidia are rare; they are slightly curved or straight, usually 3-septate. Chlamydospores are very rare or absent. In culture on PDA, *F. poae* produces abundant white mycelium that may turn reddish brown. The colony undersurface is typically yellow or pale to deep red. Cultures have a characteristic peach-like odor, though it is not recommended to use this as a diagnostic character. *Fusarium poae* differs from *F. sporotrichioides* by the absence of polyphialides and chlamydospores, and by their differing microconidial morphology. *Fusarium poae* differs from *F. langsethiae* in growth rate (*F. poae* is faster) and relative abundance of aerial mycelium (*F. poae* produces more) and microconidia (*F. poae* produces less). *Fusarium langsethiae* also lacks the characteristic odor of *F. poae*. Full genome sequences have been reported for *F. poae* [194]. *Fusarium poae* is an important head blight pathogen on wheat, barley, and oats in many part of Europe (particularly oats in northern Europe), and to some extent in North America and Japan. This species also occurs on maize as a minor component of the Fusarium ear rot complex in Europe and North America. It also is frequently recovered from soil and from seeds and seedlings of numerous plant species, especially grasses [1].

Fusarium poae is reported to produce a range of trichothecene mycotoxins, but some reports must be questioned in consideration of the separation of *F. langsethiae* from *F. poae*. *Fusarium poae* has been reported to produce DAS, T-2, neosolaniol, and nivalenol. However, it is likely that T-2 producing strains were misidentified, and actually belong to *F. sporotrichioides* or *F. langsethiae* [2, 188]. Other reports indicate that some strains of *F. poae* can produce beauvericin, fusarin C, and fusarenon-X [28]. *Fusarium poae* has been associated with toxicoses in humans and livestock, including alimentary toxic aleukia and Kashin-Beck disease in Russia, and hemorrhagic syndrome in several livestock species in the USA [2]. However, its role is uncertain because it often occurs together with *F. sporotrichioides* and strains of *F. langsethiae* may have been identified in many cases as *F. poae*. Cultures of *F. poae* are often weakly toxic or non-toxic [2, 188]. In summary, mycotoxin production by *F. poae* has been reported to be very diverse, but questions remain about the range of mycotoxins produced by *F. poae sensu stricto*.

3.11   F. proliferatum

Fusarium proliferatum (Matsushima) Nirenberg ex Gerlach and Nirenberg was first described as *Cephalosporium proliferatum* in 1971 and transferred to *Fusarium* in 1976 [147]. Prior to 1976, *F. proliferatum* was probably very often misidentified as *F. moniliforme*; therefore reports about pathology and toxigenicity of *F. moniliforme* before 1976 (and some reports after 1976) undoubtedly relate to *F. proliferatum*, but are confounded with characteristics of other members of what came to be known as the *G. fujikuroi* species complex. Currently, *F. proliferatum* is recognized as an important maize pathogen and one of the two most important fumonisin-producing fungal species. Characteristics of *F. proliferatum* include clavate microconidia produced in chains and false heads from monophialides and polyphialides on the hyphae; narrow, thin-walled, 3- to 5-septate, mostly straight macroconidia; and absence of chlamydospores. Macroconidia are produced in orange sporodochia but are sometimes sparse; they are very similar to those of *F. verticillioides*. In culture on PDA, *F. proliferatum* produces fluffy white mycelium that turns grayish violet over time. The colony undersurface is typically some shade of violet, but can be colorless, yellowish, or pale gray. *Fusarium proliferatum* can be distinguished from *F. verticillioides* by the production of microconidia from polyphialides, and distinguished from *F. subglutinans* and *F. temperatum* by the production of microconidial chains. It is very difficult or impossible to distinguish *F. proliferatum* from *F. fujikuroi* by morphology, but they can be distinguished by PCR-based methods [195] and usually by fertility tests. In general, strains with this morphology from rice are *F. fujikuroi* strains that lack fumonisin production, while morphologically identical strains from maize, asparagus and several other crops are fumonisin-producing strains of *F. proliferatum*. The teleomorph was described as *Gibberella intermedia* (Kuhlman) Samuels, Nirenberg and Seifert, also known as mating population D of the *G. fujikuroi* complex.

Fusarium proliferatum is an important component of the Fusarium ear rot and stalk rot complex on maize worldwide [3, 117, 196]. It also causes diseases on a remarkably wide range of other plant species, including asparagus, banana, date palm, fig, mango, pine, and sorghum [1]. *Fusarium proliferatum* can cause disease in rice, but does not cause typical symptoms of bakanae disease associated with *F. fujikuroi*. *Fusarium proliferatum* is well-documented as a fumonisin-producing species, and some strains can produce large quantities of fumonisins [28]. *Fusarium proliferatum* is frequently isolated from fumonisin-contaminated maize kernels, animal feeds, and human foods, including figs and asparagus spears. It also produces other mycotoxins, including beauvericin, enniatins, fusaric acid, fusarin, fusaproliferin, and moniliformin. Unlike the closely related *F. fujikuroi*, *F. proliferatum* does not produce gibberellins, which are associated with symptoms of bakanae disease caused by *F. fujikuroi* [1]. As a major fumonisin-producing species, *Fusarium proliferatum* should be considered suspect in the human and livestock toxicoses associated with fumonisins. Its role has been obscured by its misidentification as *F. moniliforme*, and its frequent co-occurrence with *F. verticillioides*.

3.12 *F. pseudograminearum*

Fusarium pseudograminearum O'Donnell and T. Aoki was originally described in 1977 from crown rot of wheat in Australia as *F. graminearum* Group I. Species rank was assigned to Group I in

1999 based on DNA sequence analysis. This species is currently recognized primarily as a root and crown rot pathogen of small grain cereals and is commonly found in Australia and the Pacific Northwest of North America; it also has been reported from South Africa and Canada. Its morphology is very similar to *F. graminearum sensu stricto*. It produces orange sporodochia with macroconidia that are long and narrow, usually 5- to 6-septate, and slightly curved to straight. Chlamydospores may form in the mycelium or conidia but they are not common. Microconidia are absent. On PDA, mycelium is white, turning yellow to red, with red pigment in the agar, so that the colony undersurface is usually dark red. The teleomorph was described as *Gibberella coronicola* T. Aoki and O'Donnell; it is heterothallic but it has been observed in the field [28]. Although morphology is essentially indistinguishable from *F. graminearum*, these two species differ in several characteristics. *Fusarium pseudograminearum* is a heterothallic crown rot pathogen, whereas *F. graminearum* on small grain cereals is homothallic and primarily a head blight pathogen. *Fusarium pseudograminearum* is important in Australia, where it is found anywhere wheat is grown. It occasionally can be responsible for head blight symptoms in very wet years. Its toxin profile is very similar to that of *F. graminearum*, but *F. pseudograminearum* poses little toxicological risk because it usually does not contaminate grain. A whole genome sequence is available for *F. pseudograminearum* (Table 1).

3.13 *F. sambucinum* and *F. venenatum*

Fusarium sambucinum Fuckel was first described in 1869 from black elder and other trees in Germany and Italy. It is the type species of the genus *Fusarium*. Gradually, the concept of this species grew quite broad and several varieties were described based on morphology. In 1995, the various varieties were grouped into three different species: *F. sambucinum sensu stricto*, *F. torulosum*, and *F. venenatum* [197]. The teleomorph, *G. pulicaris* (Fr.) Sacc., is the type species for the genus *Gibberella*. Fertility studies have confirmed that the biological and morphological species concepts for *F. sambucinum sensu stricto* are in concordance. *Fusarium sambucinum* produces abundant orange sporodochia with macroconidia that are relatively short, usually 3- to 5-septate, and falcate with a pointed and slightly curved apical cell. Microconidia and chlamydospores are rare. On PDA, *F. sambucinum* produces fluffy white mycelium that turns yellow or grayish orange, with abundant orange sporodochia, often clustered in the center of the colony. Orange to brown sclerotia are sometimes formed. Colony margins are often lobed, and the undersurface is yellow to orange or red, sometimes with brown flecks. to brown. *Fusarium sambucinum* is well known as a dry rot pathogen of potatoes worldwide. It also is reported from a range of other plant species, including woody plants, maize, and wheat. A number of diseases were reported to be caused by *F. sambucinum* before the division into three species in 1995, and it is not clear in all cases which of the three currently recognized species were involved. Similarly, a wide range of toxins has been attributed to the species, but the scope of toxin production by *F. sambucinum sensu stricto* is not completely clear. Trichothecene production is, however, well documented within *F. sambucinum sensu stricto*, and this species has been the subject of several studies on trichothecene biosynthesis. Most strains produce DAS, and some produce T-2 toxin or neosolaniol. High levels of trichothecenes have been found in potatoes rotted by *F. sambucinum* [1]. Other toxins reported for *F. sambucinum* include beauvericin, fusarin C, fusaric acid, and sambutoxin [28]. No toxicoses have been associated with *F. sambucinum*, but its trichothecene production in combination with its presence in a human food product (potato tubers) indicate that it may pose a toxicological risk.

Fusarium venenatum Nirenberg is morphologically very similar to *F. sambucinum*, but can be distinguished by the production of chlamydospores, which are very rare in *F. sambucinum*. However, it may be very difficult to separate these two species on the basis of morphology. The reported mycotoxin profile of *F. venenatum* includes DAS and other trichothecenes, culmorin, enniatins, and fusarins, but not beauvericin or zearalenone [1]. A strain of *F. venenatum* is used to produce a mycoprotein that is sold as a human food product. This strain has been confirmed to lack trichothecene production.

3.14 F. solani Species Complex

Fusarium solani (Mart.) Sacc. was first described in 1842 as *Fusisporium solani*, isolated from rotted potatoes. It was transferred to *Fusarium* in 1881 and emended by Snyder and Hansen in 1941 to include a diverse collection of fungi that shared certain morphological characteristics. Macroconidia of *F. solani* are straight to slightly curved, usually 3- to 4-septate, relatively wide and thick walled, with a rounded apical cell, produced in sporodochia that are usually cream colored but can be bluish or greenish. Abundant microconidia are oval to reniform, 0- to 2-septate, and produced in false heads on long monophialides. On PDA, cultural characteristics are variable, but usually the mycelium is white to cream and the colony undersurface is cream to pale blue or tan. In some cases, a bright blue pigment is produced in the agar. Teleomorphs in this group have been described in several genera, but not *Gibberella*. Some strains have a *Haematonectria* (formerly *Nectria*) *haematococca* teleomorph. Several approaches have been taken to address diversity in this species complex. Strains from some hosts have been given formae speciales designation; varieties have been described, mostly based on morphological characteristics. More recently, phylogenetic analyses have resulted in the recognition of several clades within the species complex, and the identification of at least 60 phylogenetic species, some of which have been formally described [29, 198, 199]. However, Most members of this complex are still known as *F. solani* because numerous phylogenetic species have not been formally named [29].

Some members of the *F. solani* species complex are important plant pathogens, including several different legume and tree species [28]. An important disease of soybean (sudden death syndrome) is caused by several species in the *F. solani* complex, including *F. virguliforme*; foliar symptoms of this disease are caused by a phytotoxic protein secreted by these fungi. *Fusarium solani* strains can cause root rots of soybean [136] and other plants and the species complex is widely distributed as a soil saprophyte. Many plant diseases have been attributed to *F. solani* but because of the taxonomic disarray in this group of fungi, some reports must be questioned [28]. The majority of *Fusarium* infections in humans and other animals are caused by members of this complex, many of which cause infections of the skin or nails [200]. Strains of *F. solani* have been reported to produce fusaric acid, moniliformin, and naphthoquinones, but it is not considered a major mycotoxin-producing species complex. There are some reports of trichothecene production by strains of *F. solani* that were isolated from rotted potatoes and produce bright blue pigment [1]. On the other hand, many other *F. solani* strains have been shown to lack trichothecene production and also to lack the *Tri5* gene, needed for trichothecene biosynthesis [201, 202]. Overall, trichothecene production by this species complex should be considered unconfirmed [28]. Toxicity of sweet potatoes molded by *F. solani* is due to ipomeamarones, compounds that are not produced by the fungus, but are plant-synthesized molecules modified by the fungus [28]. A whole genome sequence is available for *F. solani* (Table 1).

3.15 F. sporotri chioides

Fusarium sporotrichioides Sherb. was described in 1915 from potatoes with dry rot symptoms, but is now known as an important trichothecene-producing fungus in small grains. It is characterized by the production of abundant microconidia of diverse shapes in false heads on polyphialides. Microconidia range in morphology including nonseptate oval or pyriform, 1-septate fusoid or clavate, and 3- to 4-septate spindle-shaped mesoconidia. Macroconidia are usually 3-septate, narrow and falcate. They are produced in orange sporodochia and usually are abundant. Chlamydospores are usually present and formed singly or in chains. On PDA, cultures produce dense white mycelium that usually becomes pale red, with pink or red pigment on the undersurface. *Fusarium sporotrichioides* is a well-documented contributor to maize ear rot and head blight of small grains in Europe, Canada, and the northern USA [1, 117]. The first characterization of T-2 toxin (in 1968) was from a strain of *F. sporotrichioides* isolated from maize in France, and subsequently the production of T-2 and other trichothecenes including DAS and neosolaniol has been confirmed for many *F. sporotrichioides* strains. It is considered one of the most important producers of type A trichothecenes and biosynthesis of these mycotoxins has been intensively studied in *F. sporotrichioides*. Toxicoses in both humans and livestock have been associated with *F. sporotrichioides*. Following outbreaks of human alimentary toxic aleukia in Russia in the 1940s, *F. sporotrichioides* was the species most commonly isolated from associated grain, and symptoms of the disease were reproduced in cats using *F. sporotrichioides* culture material. During the 1960s, *F. sporotrichioides* was associated with hemorrhagic syndrome in several livestock species, with symptoms similar to human alimentary toxic aleukia; again, symptoms were reproduced in feeding studies with *F. sporotrichioides* culture material [2]. Other mycotoxins have been reported from *F. sporotrichioides*, including acuminatum, aurofusarin, beauvericin, culmorins, enniatins, and moniliformin, but not fusaric acid or fumonisins. Reports of production of type B trichothecenes and zearalenone by *F. sporotrichioides* are questionable [1].

3.16 *F. subglutinans*

Fusarium subglutinans (Wollenw. and Reinking) P.E. Nelson, Toussoun, and Marasas was initially described as *F. moniliforme* var. *subglutinans* in 1925 and emended to *F. subglutinans* in 1983 [27]. It is closely related to *F. temperatum*, *F. anthophilum*, and *F. circinatum*. It is a significant toxigenic ear rot pathogen of maize, although unlike other species in the *G. fujikuroi* complex that infect maize ears, *F. subglutinans* typically does not produce fumonisins. The species is characterized by abundant microconidia produced in false heads on monophialides and polyphialides. Microconidia are variable in size and septation; oval nonseptate microconidia are commonly produced along with longer spindle-shaped microconidia that are 1-, 2-, or 3-septate. Macroconidia are produced in orange sporodochia and are typical of the *G. fujikuroi* complex; they are long and slender, thin-walled, slightly curved, and usually 3-septate. Chlamydopores are not produced. On PDA, cultures have white aerial mycelium that may turn pale violet, with a colony undersurface that is colorless to deep violet or tan-orange. The teleomorph, *G. subglutinans* (E. Edwards) P.E. Nelson, Toussoun, and Marasas, was first described from maize stalks, and also has been reported as mating population E of the *G. fujikuroi* complex.

Fusarium subglutinans is recognized as a major ear rot pathogen of maize in the USA, Europe, South America, Australia, Korea, and South Africa [3, 117, 203, 204]. In North and South America, it can be the predominant species in the Fusarium ear rot or pink ear rot complex at higher latitudes or elevations, in regions where *F. verticillioides* predominates at lower latitudes or elevations [3, 196, 205]. It also causes stalk rot and seedling disease in maize. A variety of diseases on other hosts have been attributed to

F. subglutinans, but many of these have been recognized as separate species, including *F. circinatum*, *F. sacchari*, *F. mangiferae*, and others. The host range of *F. subglutinans sensu stricto* is not clear, but it includes teosinte and native North American grasses, as well as a range of other monocots and dicots [28]. *Fusarium subglutinans* has not been directly associated with toxicoses, but cultures of the fungus are often toxic due to high levels of moniliformin. Toxicological risk associated with *F. subglutinans* is obscured by its previous inclusion as a variety of *F. moniliforme* and the more recent division of *F. temperatum* and *F. subglutinans*. In addition to moniliformin, mycotoxins reported to be produced by *F. subglutinans* include beauvericin, fusaric acid, and fusaproliferin. However, beauvericin production in *F. subglutinans* should be re-evaluated in consideration of the separation of *F. temperatum*. It is likely that beauvericin-producing strains of *F. subglutinans* are actually *F. temperatum*. In two studies involving 70 *F. subglutinans* isolates from Iowa, none produced detectable beauvericin [133, 206]. Scauflaire et al. [54] and Fumero et al. [203] also found that all strains of *F. temperatum* but none of *F. subglutinans* produced beauvericin among limited numbers of isolates from Belgium and Argentina, respectively. Fumonisins in small amounts have been reported from some strains of *F. subglutinans*, but no fumonisin biosynthesis genes have been detected and the vast majority of *F. subglutinans* strains do not produce fumonisins [1]. The synthesis of fumonisins by some strains of both *F. subglutinans* and *F. temperatum* is poorly understood, and neither species is an important producer of this group of mycotoxins.

3.17 F. temperatum

Fusarium temperatum Scauflaire and Munaut was first described in 2011 as a species closely related to *F. subglutinans* [207]. *Fusarium temperatum* is characterized by white to violet mycelium with macroconidia produced in tan to orange sporodochia (macroconidia hyaline, narrow and slightly falcate, usually 4-septate), and abundant microconidia produced singly or in false heads from monophialidic and polyphialidic conidiophores in the aerial mycelium (microconidia hyaline, ellipsoid or oval to fusiform, 0- to 2-septate). Chlamydospores are not formed. The teleomorph has been described [207] but not formally named, and there are no reports of the teleomorph outside the laboratory. A limited degree of interfertility occurs between *F. temperatum* and *F. subglutinans*. *F. temperatum* and *F. subglutinans* are very similar morphologically, but the macroconidia differ slightly. The distinction between *F. subglutinans* and *F. temperatum* first became evident during the late 1990s, when several authors began to distinguish two phenotypic and phylogenetic groups within the biological species *F. subglutinans* [208, 209]. These were designated as *F. subglutinans* Group 1 and *F. subglutinans* Group 2. Strains placed in Group 1 are consistent with *F. temperatum*, and it is very likely that they are the same entity, although additional work is needed to confirm this [207]. *Fusarium temperatum* causes diseases in maize, including seedling blight, stalk rot, and ear rot [207]. It has been reported from several European countries [210, 211], the USA [212], Argentina [203], South Africa (as *F. subglutinans* Group 1), and China [213].

Mycotoxin production by *F. temperatum* has not been extensively studied. Scauflaire et al. [54] analyzed 11 *Fusarium temperatum* strains for the production of 15 different mycotoxins. All 11 strains produced beauvericin, most produced moniliformin, and one strain produced enniatins and fumonisin B_1. Most strains of *F. temperatum* from Argentina produced beauvericin and fusaproliferin, and approximately 25% of these strains also produced fumonisin B_1 [203]. Fusaproliferin and beauvericin production also was reported for strains from the USA, originally identified as *F. subglutinans* Group 1 [206], but these strains were subsequently identified as *F. temperatum* [212]. This species has not been associated with any toxicoses, possibly due to its relatively new nomenclature.

3.18   *F. tricinctum*

Fusarium tricinctum (Corda) Sacc. was initially described as *Selenosporium tricinctum* in 1838 and transferred to *Fusarium* in 1886. Its importance as a toxigenic species has been overestimated because the name has been used to refer to several different species, some of which (e.g., *F. poae* and *F. sporotrichioides*) are important trichothecene-producing fungi [28]. The species is characterized by production of abundant 0- to 1-septate microconidia of diverse shapes ranging from lemon- or pear-shaped to oval or spindle-shaped. Microconidia are produced in false heads on monophialides. Macroconidia are produced in pale orange sporodochia and are usually 3-septate, slender, thin-walled, and curved, falcate to lunate, with tapering apical and basal cells. Chlamydospores are occasionally formed singly or in chains. On PDA, cultures of *F. tricinctum* produce fluffy white mycelium that turns pink or red with age, and the colony undersurface is usually pale to deep red. The teleomorph, *G. tricincta* El-Gholl, McRitchie, Schoulties, and Ridings, has been described from laboratory crosses. *Fusarium tricinctum* has been associated with diseases of several plants, but some reports are questionable due to the broad application of the species name. It has been isolated from cereal grains with head blight symptoms in Europe, North America, China, and Argentina, but it is considered a minor contributor to Fusarium head blight [1, 214, 215]. *Fusarium tricinctum* has recently been reported as a pathogen of soybean [216] and is considered an endophyte in some plants [217]. It can be found in soils and plant debris, widely distributed in cooler climates in the Northern hemisphere. *Fusarium tricinctum* has been reported to produce enniatins, fusarin C, and moniliformin, but not beauvericin or trichothecenes [1, 218].

Reports of trichothecene production by *F. tricinctum* are due to misidentification; the species lacks *Tri5*, a gene essential for trichothecene biosynthesis [2, 28]. *Fusarium tricinctum* is not associated with any known toxicoses.

3.19 F. verticillioides

Fusarium verticillioides (Sacc.) Nirenberg (syn. *F. moniliforme* Sheldon) is the most common toxigenic species found in maize kernels and its characteristics have been thoroughly studied. It was first described from maize in Italy in 1877 as *Oospora verticillioides* [219] but became well-known as *F. moniliforme* after it was discovered to be associated with animal toxicoses [109]. The currently accepted nomenclature was established in 1976 [147] when *F. moniliforme* and *O. verticillioides* were determined to be synonymous. The teleomorph, was described as *Gibberella moniliformis* in 1924 [220], but also has been extensively reported as mating population A of the *G. fujikuroi* complex [149]. *Fusarium verticillioides* is characterized by white to violet mycelium, tan to orange sporodochia producing 3- to 5-septate hyaline macroconidia that are narrow and straight, and abundant, hyaline, oval-to-clavate single-celled microconidia that are produced in false heads and short to long chains from monophialidic conidiophores in the aerial mycelium. Chlamydospores are not formed. The teleomorph is rarely found outside the laboratory. *Fusarium moniliforme* has been reported to have a wide host range that includes as many as 11,000 plant species [28], but the more narrowly defined *F. verticillioides* is primarily a globally distributed maize pathogen. It can infect all parts of the plant at any stage of development, causing seed rot, seedling disease, stalk rot, and ear rot, as well as causing symptomless endophytic infection [119]. Other plants reported to be infected by *Fusarium verticillioides* include sorghum, millet, and North American native grasses; associations between *F. verticillioides* (reported as *F. moniliforme*) and diseases of many other plants have yet to be confirmed [28]. Strains isolated from banana and originally identified as *F. verticillioides* [221] have been assigned to a new species, *F. musae* [222]. A whole genome sequence is available for *F. verticillioides* (Table 1).

Fumonisins are the most important mycotoxins produced by *F. verticillioides*, and it is the most common fungus that produces them [119]. Fumonisins were first described in 1988 from a strain of *F. verticillioides* from South Africa [96]. Most strains of *F. verticillioides* produce fumonisins in varying amounts. The fumonisin biosynthetic pathway was first described from *F. verticillioides* [223] and has been most thoroughly studied in this organism. Because of the broad use of the name *F. moniliforme* for several decades, some confusion exists regarding the production of other mycotoxins by *F. verticillioides*. The current belief is that some strains can produce fusaric acid, fusarins, and naphthoquinones, but fusaproliferin, beauvericin, and moniliformin are not produced by this species [1].

As the primary fumonisin-producing *Fusarium* species in maize, *F. verticillioides* is associated with all toxicoses in which fumonisins are implicated. This species first gained recognition as a possible cause of "blind staggers" in horses [109], although the etiology of that condition was not demonstrated until the late 1980s, when it was described as equine leukoencephalomalacia (ELEM), caused by ingestion of fumonisins. Fumonisins produced by *Fusarium verticillioides* are believed to be the primary cause of ELEM and porcine pulmonary edema [114]. These diseases have been reproduced experimentally using cultures of *F. verticillioides*. Human toxicoses associated with *F. verticillioides* include esophageal cancer and fetal neural tube defects [1]. Culture material of *F. verticillioides* has been shown to cause renal and hepatic cancer in rats [36]. For more details, see the section on fumonisins.

3.20   Other Species

Fusarium chlamydosporum Wollenw. and Reinking is closely related to *F. poae* and *F. sporotrichioides*. It is commonly found in soils and plant debris in warm temperate and tropical regions and has been reported from a range of plant species on several continents. Cultures of *F. chlamydosporum* can be toxic to ducklings, due to moniliformin production. It has been reported to produce chlamydosporol and other minor toxins, but it has not been associated with known toxicoses [1]. However, it has been associated with mycoses in humans [28].

F. lateritium Nees is distributed worldwide and has been reported to cause a variety of disease symptoms across a very broad host range, particularly woody plants. It is a diverse species complex in which various varieties and formae speciales have been described. *Fusarium stilboides* is considered synonymous with *F. lateritium* [28]. Its teleomorph has been described as *G. baccata*. There were reports of production of a wide range of mycotoxins by *F. lateritium*, but these were determined to be incorrect by Marasas et al. [2]. More recently, enniatin production has been reported for *F. lateritium*.

F. semitectum Berk. and Ravenel has been recovered from soils, plant debris, and many species of seeds and live plants, primarily in tropical and sub-tropical areas, but also across a wide range of other geographies. Several names have been proposed for this species, including *F. pallidoroseum* and *F. incarnatum*. It has been reported as a minor pathogen on numerous plants, including cereal grains. A wide range of mycotoxins has been reported to be produced by *F. semitectum*, including DAS and other Type A trichothecenes, equisetin, moniliformin, sambutoxin, and zearalenone [28]. It was reported to be associated with Degnala disease of water buffalo in India, but both *F. equiseti* and *F. semitectum* were recovered from moldy rice straw connected with the outbreaks, so the role of each species is not clear [1].

Aside from the economically important species *F. verticillioides*, *F. proliferatum*, *F. subglutinans*, and *F. temperatum*, numerous other species have been described within the F. fujikuroi (G. fujikuroi) **species complex**. The most important mycotoxins produced by members of this group are the fumonisins, but other toxins include beauvericin, enniatins, fusaproliferin, and moniliformin. These species include:

- *F. circinatum* Nirenberg and O'Donnell emend. Britz, Coutinho, Wingfield, and Marasas is the causal agent of a significant disease problem in pine trees, pitch canker. It is closely related to *F. subglutinans*, and was formerly called *F. subglutinans* f.sp. *pini*. It was previously included under *F. moniliforme* var. *subglutinans* and also has been known as *F. lateritium* f.sp. *pini*. It also has been referred to as *G. fujikuroi* mating population H. It has been reported to produce beauvericin and possibly fusaproliferin [1].

- *Fusarium konzum* Zeller, Summerell, and J.F. Leslie was described in 2003 from native prairie grasses found in Kansas, USA [224]. It is closely related to *F. subglutinans* and *F. anthophilum*. It also has been referred to as *G. fujikuroi* mating population I. Some strains of *F. konzum* produce beauvericin, fumonisins, or fusaproliferin [225], but the species seems to be limited to non-crop hosts and likely poses little toxicological risk to humans.

- *F. lactis* Pirotta and Riboni is the causal agent of a disease known as fig endosepsis in California. The name has been used to refer to several different *Fusarium* species, and strains from fig also have been reported as *F. moniliforme* or *F. moniliforme* var. *fici*. *Fusarium lactis* produces moniliformin, but not beauvericin, fumonisins, or fusaproliferin [28].

- *F. musae* Van Hove, Waalwijk, Munaut, Logrieco, and Moretti was described in 2011 from bananas. It had been reported as *F. verticillioides*, and is interfertile with *F. verticillioides*, but was separated based on biochemical, morphological, and phylogenetic evidence. The host range and mycotoxin profile of this species are not completely clear; it produces moniliformin but not fumonisins [222]. It has been associated with human infections [226].

- *F. nygamai* Burgess and Trimboli is closely related and similar in morphology to *F. thapsinum*. It also has been referred to as *G. fujikuroi* mating population G. *Fusarium nygamai* was first observed in association with stalk and root rot of sorghum and some strains are pathogenic to sorghum and other crops. It has been isolated from a wide range of plants and soils, mostly in hot, dry areas. Cultures of *F. nygamai* are toxic to ducklings and strains can produce beauvericin, fumonisins, and fusaric acid. Moniliformin production has been reported in a few strains, but these strains likely were a related species, *F. pseudonygamai* [28].

- *F. sacchari* (E.J. Butler) Gams is the causal agent of pokkah boeng disease of sugar cane in Asia and also has occasionally been isolated from maize and sorghum. It was originally described as *Cephalosporium sacchari* and was considered synonymous with *F. subglutinans* by some authors. Its teleomorph has been referred to as *G. sacchari* or *G. fujikuroi* mating population B. It has been reported to produce beauvericin, fusaric acid, and moniliformin, but not fumonisins or fusaproliferin [1].

- *F. thapsinum* Klittich, Leslie, Nelson and Marasas causes stalk rot, grain mold, and seedling disease of sorghum. Prior to 1997, it was recognized as *F. moniliforme*. Its morphology is similar to *F. verticillioides*, but most strains of *F. thapsinum* produce a yellow pigment in culture. The teleomorph is *G. thapsina* but it also has been referred to as *G. fujikuroi* mating population F. Although it is mainly considered a sorghum pathogen, *F. thapsinum* has been recovered from other hosts, including maize, bananas, peanuts, and native North American grasses [28]. Cultures of *F. thapsinum* can be toxic to ducklings, due to moniliformin production. It also has been reported to produce fusaric acid but not beauvericin or fusaproliferin. A few strains have been reported to produce low levels of fumonisins, but fumonisin biosynthetic genes have not been detected in strains of *F. thapsinum* [1].

Acknowledgements

The author is grateful to Dr. Robert Proctor, USDA-ARS, Peoria, IL, for contributing information regarding *Fusarium* genome sequences, and to Ms. Lauren Washington, Iowa State University, for assistance in preparing the manuscript.

References

1. Desjardins AE (2006) *Fusarium* mycotoxins: chemistry, genetics, and biology. APS Press, St Paul, MN, 260 pp
2. Marasas WFO, Nelson PE, Toussoun TA (1984) Toxigenic *Fusarium* species: Identity and mycotoxicology. Pennsylvania State University Press, University Park, PA, 328 pp
3. Munkvold GP (2003) Mycotoxins in corn—occurrence, impact, and management Ch. 23 (pp. 811–881). In: White PJ, Johnson LA (eds) Corn: chemistry and technology, 2nd edn. American Association of Cereal Chemists, St. Paul, MN, 892 pp
4. European Food Safety Authority (2013) Deoxynivalenol in food and feed: occurrence and exposure. EFSA J 11(10):3379. doi:10.2903/j.efsa.2013.3379, 56 pp
5. Waskiewicz A, Beszterda M, Golinski P (2012) Occurrence of fumonisins in food—an interdisciplinary approach to the problem. Food Control 26(2):491–499

6. Wu F, Munkvold GP (2008) Mycotoxins in ethanol co-products: modeling economic impacts on the livestock industry and management strategies. J Agric Food Chem 56(11):3900–3911

7. Logrieco AF, Haidukowski M, Susca A, Mule G, Munkvold GP, Moretti A (2014) *Aspergillus* section *Nigri* as contributor of Fumonisin B2 contamination in maize. Food Addit Contam Part A Chem Anal Control Expo Risk Assess 31(1):149–155

8. Cuomo CA, Güldener U, Xu JR, Trail F, Turgeon BG, Di Pietro A, Walton JD et al (2007) The *Fusarium graminearum* genome reveals a link between localized polymorphism and pathogen specialization. Science 317:1400–1402

9. Gardiner DM, McDonald MC, Covarelli L, Solomon PS, Rusu AG, Marshall M, Kanzan K et al (2012) Comparative pathogenomics reveals horizontally acquired novel virulence genes in fungi infecting cereal hosts. PLoS Pathog 8(9):e1002952

10. Lysoe E, Harris LJ, Walkowiak S, Subramaniam R, Divon HH, Riiser ES, Llorens C et al (2014) The genome of the generalist plant pathogen *Fusarium avenaceum* is enriched with genes involved in redox, signaling and secondary metabolism. PLoS One 9(11):e112703

11. Wiemann P, Sieber CMK, Von Bargen KW, Studt L, Niehaus EM, Espino JJ, Huss K et al (2013) Deciphering the cryptic genome: genome-wide analyses of the rice pathogen *Fusarium fujikuroi* reveal complex regulation of secondary metabolism and novel metabolites. PLoS Pathog 9(6):e1003475

12. Jeong HY, Lee SH, Choi GJ, Lee T, Yun SH, Jeong HY, Lee SH, Choi GJ, Yun SH (2013) Draft genome sequence of *Fusarium fujikuroi* B14, the causal agent of the bakanae disease of rice. Genome Announc 1(1): e00035-13

13. Ma LJ, van der Does HC, Borkovich KA, Coleman JJ, Daboussi MJ, Pietro AD, Dufresne M et al (2010) Comparative genomics reveals mobile pathogenicity chromosomes in *Fusarium*. Nature 464:367–373

14. Guo L, Han L, Yang L, Zeng H, Fan D, Zhu Y, Feng Y et al (2014) Genome and transcriptome analysis of the fungal pathogen *Fusarium oxysporum* f. sp. *cubense* causing banana vascular wilt disease. PLoS One 9(4):e95543

15. Coleman JJ, Rounsley SD, Rodriguez-Carres M, Kuo A, Wasmann CC, Grimwood J, Schmutz J et al (2009) The genome of *Nectria* haematococca: contribution of supernumerary chromosomes to gene expansion. PLoS Genet 5(8):e1000618

16. Srivastava SK, Huang X, Brar HK, Fakhoury AM, Bluhm BH, Bhattacharyya MK (2014) The genome sequence of the fungal pathogen *Fusarium virguliforme* that causes sudden death syndrome in soybean. PLoS One 9:e81832

17. Geiser DM, Aoki T, Bacon CW, Baker SE, Bhattacharyya MK, Brandt ME, Brown DW et al (2013) One fungus, one name: defining the genus *Fusarium* in a scientifically robust way that preserves longstanding use. Phytopathology 103(5):400–408

18. O'Donnell K, Rooney AP, Proctor RH, Brown DW, McCormick SP, Ward TJ, Frandsen RJN et al (2013) Phylogenetic analyses of *Rpb1* and *Rpb2* strongly support a middle cretaceous origin for a clade comprising all agriculturally and medically important Fusaria. Fungal Genet Biol 52:20–31

19. O'Donnell K, Sutton DA, Rinaldi MG, Gueidan C, Crous PW, Geiser DM (2009) Novel multilocus sequence typing scheme reveals high genetic diversity of human pathogenic members of the *Fusarium incarnatum-F. equiseti* and *F. chlamydosporum* species complexes within the United States. J Clin Microbiol 47(12):3851–3861

20. Link HF (1809) Observationes in ordines plantarum naturals. Mag Ges Naturf Freunde 3:3–42

21. Wollenweber HW, Reinking OA (1935) Die Fusarien, ihre Beschreibung, Schadwirkung, und Bekämpfung. Paul Parey, Berlin

22. Snyder WC, Hansen HN (1940) The species concept in *Fusarium*. Am J Bot 27(2):64–67

23. Snyder WC, Hansen HN (1941) The species concept in *Fusarium* with reference to Martiella. Am J Bot 28(9):738–742

24. Snyder WC, Hansen HN (1945) The species concept in *Fusarium* with reference to Discolor and other sections. Am J Bot 32(10):657–666

25. Gordon WL (1952) The occurrence of *Fusarium* species in Canada. 2. prevalence and taxonomy of *Fusarium* species in cereal seed. Can J Bot 30(2):209–251

26. Booth C (1971) The genus *Fusarium*. Commonwealth Mycological Institute, Kew, Surrey

27. Nelson PE, Toussoun TA, Marasas WFO (1983) *Fusarium* species: an illustrated manual for identification (University Park. Penn State University Press, Pennsylvania

28. Leslie JF, Summerell BA (2006) The *Fusarium* laboratory manual. Blackwell, Ames, Iowa

29. Aoki T, O'Donnell K, Geiser DM (2014) Systematics of key phytopathogenic *Fusarium* species: current status and future challenges. J Gen Plant Pathol 80(3):189–201

30. Desjardins AE (2003) Gibberella from A (*venaceae*) to Z (*eae*). Annu Rev Phytopathol 41:177–198

31. Hawksworth DL (2011) A new dawn for the naming of fungi: impacts of decisions made in Melbourne in July 2011 on the future publication and regulation of fungal names. Mycokeys 1:7–20

32. Nelson PE, Toussoun TA, Cook RJ (1981) *Fusarium:* diseases, biology, and taxonomy. Pennsylvania State University Press, University Park, Pennsylvania

33. Covey PA, Kuwitzky B, Hanson M, Webb KM (2014) Multilocus analysis using putative fungal effectors to describe a population of *Fusarium oxysporum* from sugar beet. Phytopathology 104(8):886–896

34. WHO. Food Additives Series no. 47. (2001) Safety evaluation of certain mycotoxins in food. FAO Food and Nutrition Paper, 74, 701 pp

35. WHO (2002) Evaluation of certain mycotoxins in food: fifty-sixth report of the joint FAO/WHO expert committee on food additives. WHO Technical Report Series, 906, 62 pp

36. NTP (2001) Technical report on the toxicology and carcinogenesis studies of fumonisin B1 (CAS NO. 116355-83-0) in F344/N Rats AND B6C3F1 MICE (Feed Studies). NTP TR 496. (U.S. Dept. of Health and Human Services, NIH Publication No. 01-3955)

37. Alexander NJ, Proctor RH, McCormick SP (2009) Genes, gene clusters, and biosynthesis of trichothecenes and fumonisins in *Fusarium*. Toxin Rev 28:198–215

38. Kimura M, Tokai T, Takahashi-Ando N, Ohsato S, Fujimura M (2007) Molecular and genetic studies of *Fusarium* trichothecene biosynthesis: pathways, genes, and evolution. Biosci Biotechnol Biochem 71(9):2105–2123

39. McCormick SP, Stanley AM, Stover NA, Alexander NJ (2011) Trichothecenes: from simple to complex mycotoxins. Toxins 3(7):802–814

40. Woloshuk CP, Shim WB (2013) Aflatoxins, fumonisins, and trichothecenes: a convergence of knowledge. FEMS Microbiol Rev 37(1):94–109

41. Proctor RH, Van Hove F, Susca A, Stea G, Busman M, van der Lee T, Waalwijk C, Moretti A, Ward TJ (2013) Birth, death and horizontal transfer of the fumonisin biosynthetic gene cluster during the evolutionary diversification of *Fusarium*. Mol Microbiol 90(2):290–306

42. Foroud NA, Eudes F (2009) Tricothecenes in cereal grains. Int J Mol Sci 10(1):147–173

43. Sydenham EW, Shephard GS, Thiel PG, Marasas WFO, Stockenstrom S (1991) Fumonisin contamination of commercial corn-based human foodstuffs. J Agric Food Chem 39(11):2014–2018

44. Rodrigues I, Handl J, Binder EM (2011) Mycotoxin occurrence in commodities, feeds and feed ingredients sourced in the Middle East and Africa. Food Addit Contam Part B Surveill 4(3):168–179. doi:10.1080/193932 10.2011.589034

45. Schatzmayr G, Streit E (2013) Global occurrence of mycotoxins in the food and feed chain: facts and figures. World Mycotoxin J 6(3):213–222

46. Wood GE (1992) Mycotoxins in foods and feeds in the United States. J Anim Sci 70(12):3941–3949

47. Griessler K, Rodrigues I, Handl J, Hofstetter U (2010) Occurrence of mycotoxins in Southern Europe. World Mycotoxin J 3(3):301–309

48. SCF (Scientific Committee for Food). Opinion of the scientific committee for food on *Fusarium* toxins. Part 6: Group evaluation of T-2 toxin, HT-2 Toxin, nivalenol and deoxynivalenol. http://ec.europa.eu/food/fs/sc/scf/out123_en.pdf.

49. EFSA Panel on Contaminants in the Food Chain (2011) Scientific opinion on the risks for animal and public health related to the presence of T-2 and HT-2 toxin in food and feed. EFSA J 9(12):2481. doi:10.2903/j.efsa.2011.2481

50. Jestoi M (2008) Emerging *Fusarium*-mycotoxins fusaproliferin, beauvericin, enniatins, and moniliformin—a review. Crit Rev Food Sci Nutr 48(1):21–49

51. Logrieco A, Moretti A, Castella G, Kostecki M, Golinski P, Ritieni A, Chelkowski J (1998) Beauvericin production by *Fusarium* species. Appl Environ Microbiol 64(8):3084–3088

52. Goswami RS, Kistler HC (2005) Pathogenicity and in planta mycotoxin accumulation among members of the *Fusarium graminearum* species complex on wheat and rice. Phytopathology 95(12):1397–1404

53. Toth B, Kaszonyi G, Bartok T, Varga J, Mesterhazy A (2008) Common resistance of wheat to members of the *Fusarium graminearum* species complex and *F. culmorum*. Plant Breed 127(1):1–8

54. Scauflaire J, Gourgue M, Callebaut A, Munaut F (2012) *Fusarium temperatum*, a mycotoxin-producing pathogen of maize. Eur J Plant Pathol 133(4):911–922

55. Moretti A, Mule G, Ritieni A, Logrieco A (2007) Further data on the production of beauvericin, enniatins and fusaproliferin and toxicity to *Artemia salina* by *Fusarium* spe-

cies of *Gibberella fujikuroi* species complex. Int J Food Microbiol 118(2):158–163

56. Lattanzio VMT, von Holst C, Visconti A (2013) Experimental design for in-house validation of a screening immunoassay kit. The case of a multiplex dipstick for *Fusarium* mycotoxins in cereals. Anal Bioanal Chem 405(24):7773–7782

57. EFSA Panel on Contaminants in the Food Chain (2013) Scientific opinion on risks for animal and public health related to the presence of nivalenol in food and feed. EFSA J 11(6):3262

58. Rocha O, Ansari K, Doohan FM (2005) Effects of trichothecene mycotoxins on eukaryotic cells: a review. Food Addit Contam 22(4):369–378

59. Gerez JR, Pinto P, Callu P (2015) Deoxynivalenol alone or in combination with nivalenol and zearalenone induce systemic histological changes in pigs. Exp Toxicol Pathol 67(3):89–98

60. Arunachalam C, Doohan FM (2013) Trichothecene toxicity in eukaryotes: cellular and molecular mechanisms in plants and animals. Toxicol Lett 217(2):149–158

61. Beasley VR, Swanson SP, Corley RA, Buck WB, Koritz GD, Burmeister HR (1986) Pharmacokinetics of the trichothecene mycotoxin, T-2 toxin, in swine and cattle. Toxicon 24(1):13–23

62. Eriksen GS, Pettersson H (2004) Toxicological evaluation of trichothecenes in animal feed. Anim Feed Sci Technol 114:205–239

63. Flannery BM, Clark ES, Pestka JJ (2012) Anorexia induction by the trichothecene deoxynivalenol (vomitoxin) is mediated by the release of the gut satiety hormone peptide Yy. Toxicol Sci 130(2):289–297

64. Pestka JJ (2010) Deoxynivalenol: mechanisms of action, human exposure, and toxicological relevance. Arch Toxicol 84(9):663–679

65. World Health Organization (2011) Safety evaluation of certain contaminants in food: 72nd report of the joint FAO/WHO expert committee on food additives. World Health Organization (WHO Food Additives Series 63), Geneva

66. World Health Organization (WHO) (2000) Safety evaluation of certain mycotoxins in food. Fifty-third Report of the FAO/WHO Joint Expert Committee on Food Additives. WHO Technical Report Series 896, Geneva.

67. Pitt JI, Wild CP, Baan RA, Gelderblom WCA, Miller JD, Riley RT, Wu F (2012) Improving public health through mycotoxin control. IARC Scientific Publication 158, Geneva World Organization Press, Lyon

68. Yoshizawa T (1983) Red-mold diseases and natural occurrence in Japan. In: Uedo Y (ed) Trichothecenes, chemical, biological and toxicological aspects. Kodansha, Tokyo, pp 195–209

69. Eudes F, Comeau A, Rioux S, Collin J (2000) Phytotoxicity of eight mycotoxins associated with *Fusarium* in wheat head blight. Can J Plant Pathol 22(3):286–292

70. Proctor RH, Hohn TM, McCormick SP (1997) Restoration of wild-type virulence to *Tri5* disruption mutants of *Gibberella zeae* via gene reversion and mutant complementation. Microbiology 143:2583–2591

71. Harris LJ, Desjardins AE, Plattner RD, Nicholson P, Butler G, Young JC, Weston G, Proctor RH, Hohn TM (1999) Possible role of trichothecene mycotoxins in virulence of *Fusarium graminearum* on maize. Plant Dis 83(10):954–960

72. Proctor RH, Hohn TM, McCormick SP (1995) Reduced virulence of *Gibberella zeae* caused by disruption of a trichothecene toxin biosynthetic gene. Mol Plant Microbe Interact 8(4):593–601

73. Bruns T, Wise RP, Munkvold GP (2015) Colonization of maize, wheat and soybean seedlings by mycotoxin-deficient mutants of *Fusarium graminearum* and *F. verticillioides*. 13th European *Fusarium* Seminar Martina Franca, Italy, p. 63

74. Jansen C, von Wettstein D, Schafer W, Kogel KH, Felk A, Maier FJ (2005) Infection patterns in barley and wheat spikes inoculated with wild-type and trichodiene synthase gene disrupted *Fusarium graminearum*. Proc Natl Acad Sci U S A 102(46):16892–16897

75. Bai GH, Plattner R, Desjardins A, Kolb F (2008) Resistance to *Fusarium* head blight and deoxynivalenol accumulation in wheat. Plant Breed 120(1):1–6

76. Schollenberger M, Mueller HM, Ernst K, Sondermann S, Liebscher M, Schlecker C, Wischer G et al (2012) Occurrence and distribution of 13 trichothecene toxins in naturally contaminated maize plants in Germany. Toxins 4(10):778–877

77. Council for Agricultural Science and Technology, CAST (2003) Mycotoxins: risks in plant and animal systems. Task Force Report No. 139, Ames, IA

78. Urry WH, Wehrmeister HL, Hodge EB, Hidy PH (1966) The structure of zearalenone from *Gibberella zeae*, *Fusarium graminearum*. Tetrahedron Lett 27:3109–3114

79. Hidy PH, Baldwin RS, Greasham RL, Keith CL, McMullen JR (1977) Zearalenone and some derivatives: production and biological activities. Adv Appl Microbiol 22:59–82

100   Gary P. Munkvold

80. Hurd RN (1977) Structure activity relationships in zearalenones. In: Rodricks JV, Hesseltine CW, Mehlman MA (eds) Mycotoxins in human and animal health. Pathotox, Park Forest South

81. Hagler WM Jr, Towers NR, Mirocha CJ (2001) Zearalenone: mycotoxin or mycoestrogen? In: Summerell BA, Leslie JF, Backhouse D, Bryden WL, Burgess LW (eds) Fusarium: Paul E. Nelson memorial symposium. APS Press, St. Paul

82. Prelusky DB, Rotter BA, Rotter RG (1994) Toxicology of mycotoxins. In: Jd M, Trenholmes HL (eds) Mycotoxins in grain: compounds other than aflatoxin. Eagan Press, St. Paul

83. Zinedine A, Soriano JM, Molto JC, Manes J (2007) Review on the toxicity, occurrence, metabolism, detoxification, regulations and intake of zearalenone: an oestrogenic mycotoxin. Food Chem Toxicol 45(1):1–18

84. US National Toxicology Program (1982) Carcinogenesis bioassay of Zearalenone (CAS No. 17924-92-4) In F344/N Rats and B6C3F1 Mice (Feed Study). Technical report series no 235, NIH publ. No 83-1791. Research Triangle Park

85. Martins ML, Martins HM (2002) Influence of water activity, temperature and incubation time on the simultaneous production of deoxynivalenol and zearalenone in corn (Zea mays) by Fusarium graminearum. Food Chem 79(3):315–318

86. Susca A, Moretti A, Stea G, Villani A, Haidukowski M, Logrieco A, Munkvold G (2014) Comparison of species composition and fumonisin production in Aspergillus section Nigri populations in maize kernels from USA and Italy. Int J Food Microbiol 188:75–82

87. World Health Organization (WHO) (2000) In: Fumonisin B₁. Environmental health criteria, vol 219, 150 pp

88. Smith GW, Constable PD, Foreman JH, Eppley RM, Waggoner AL, Tumbleson ME, Haschek WM (2002) Cardiovascular changes associated with intravenous administration of fumonisin B1 in horses. Am J Vet Res 63(4):538–545

89. Constable PD, Smith GW, Rottinghaus GE, Tumbleson ME, Haschek WM (2003) Fumonisin-induced blockade of ceramide synthase in sphingolipid biosynthetic pathway alters aortic input impedance spectrum of pigs. Am J Physiol Heart Circ Physiol 284(6):H2034–H2044

90. Marasas WFO (2001) Discovery and occurrence of the fumonisins: a historical perspective. Environ Health Perspect 109:239–243

91. IARC (2002) Some traditional herbal medicines, some mycotoxins, naphthalene and styrene. International Agency for Research on Cancer Monographs on the Evaluation of Carcinogenic Risks to Humans 82, IARC Press, Lyon

92. Desjardins AE, Plattner RD, Nelsen TC, Leslie JF (1995) Genetic-analysis of fumonisin production and virulence of Gibberella fujikuroi mating population A (Fusarium moniliforme) on maize (Zea mays) seedlings. Appl Environ Microbiol 61(1):79–86

93. Desjardins AE, Munkvold GP, Plattner RD, Proctor RH (2002) FUM1—a gene required for fumonisin biosynthesis but not for maize ear rot and ear infection by Gibberella moniliformis in field tests. Mol Plant Microbe Interact 15:1157–1164

94. Arias SL, Theumer MG, Mary VS, Rubinstein HR (2012) Fumonisins: probable role as effectors in the complex interaction of susceptible and resistant maize hybrids and Fusarium verticillioides. J Agric Food Chem 60(22):5667–5675

95. Glenn AE, Zitomer NC, Zimeri AM, Williams LD, Riley RT, Proctor RH (2008) Transformation-mediated complementation of a FUM gene cluster deletion in Fusarium verticillioides restores both fumonisin production and pathogenicity on maize seedlings. Mol Plant Microbe Interact 21(1):87–97

96. Bezuidenhout SC, Gelderblom WCA, Gorstallman CP, Horak RM, Marasas WFO, Spiteller G, Vleggaar R (1988) Structure elucidation of the fumonisins, mycotoxins from Fusarium moniliforme. J Chem Soc Chem Commun 11:743–745

97. Rheeder JP, Marasas WFO, Thiel PG, Sydenham EW, Shephard GS, Vanschalkwyk DJ (1992) Fusarium moniliforme and fumonisins in corn in relation to human esophageal cancer in Transkei. Phytopathology 82(3):353–357

98. Shephard GS, Van der Westhuizen L, Sewram V (2007) Biomarkers of exposure to fumonisin mycotoxins: a review. Food Addit Contam 24(10):1196–1201

99. Wang H, Wei H, Ma J, Luo X (2000) The fumonisin B1 content in corn from North China, a high-risk area of esophageal cancer. J Environ Pathol Toxicol Oncol 19:139–141

100. Sun G, Wang S, Hu X, Su J, Huang T, Yu J, Tang L, Gao W, Wang JS (2007) Fumonisin B1 contamination of home-grown corn in high-risk areas for esophageal and liver cancer in China. Food Addit Contam Part A Chem Anal Control Expo Risk Assess 24(2):181–185

101. Franceschi S, Bidoli E, Baron AE, Lavecchia C (1990) Maize and risk of cancers of the oral cavity, pharynx, and esophagus in Northeastern Italy. J Natl Cancer Inst Monogr 82(17):1407–1411

102. Alizadeh AM, Roshandel G, Roudbarmohammadi S, Roudbary M, Sohanaki H, Ghiasian SA, Taherkhani A, Semnani S, Aghasi M (2012) Fumonisin B1 contamination of cereals and risk of esophageal cancer in a high risk area in Northeastern Iran. Asian Pac J Cancer Prev 13(6): 2625–2628

103. Howard PC, Eppley RM, Stack ME, Warbritton A, Voss KA, Lorentzen RJ, Kovach RM, Bucci TJ (2001) Fumonisin B1 carcinogenicity in a two-year feeding study using F344 rats and B6c3f1 mice. Environ Health Perspect 109:277–282

104. Gelderblom WCA, Abel S, Smuts CM, Marnewick J, Marasas WFO, Lemmer ER, Ramljak D (2001) Toxicity of cultured material of *Fusarium verticilloides* strain MRC 826 to nonhuman primates. Environ Health Perspect 109(Supplement 2):291–300

105. Missmer SA, Suarez L, Felkner M, Wang E, Merrill AH, Rothman KJ, Hendricks KA (2006) Exposure to fumonisins and the occurrence of neural tube defects along the Texas-Mexico border. Environ Health Perspect 114(2):237–241

106. Marasas WFO, Riley RT, Hendricks KA, Stevens VL, Sadler TW, Gelineau-van WJ, Missmer SA et al (2004) Fumonisins disrupt sphingolipid metabolism, folate transport, and neural tube development in embryo culture and *in vivo*: a potential risk factor for human neural tube defects among populations consuming fumonisin-contaminated maize. J Nutr 134(4):711–716

107. Gelineau-van WJ, Voss KA, Stevens VL, Speer MC, Riley RT (2009) Maternal fumonisin exposure as a risk factor for neural tube defects. Adv Food Nutr Res 56:145–181

108. Marasas WFO, Kellerman TS, Gelderblom WCA, Coetzer JAW, Thiel FG, van der Lugt JJ (1998) Leucoencephalomalacia in a horse induced by fumonisin B1 isolated from *Fusarium verticillioides*. Onderstepoort J Vet Res 55:197–203

109. Sheldon JL (1904) A corn mold (*Fusarium moniliforme* n. sp.). In: Agricultural experiment station of Nebraska: 17th annual report

110. Kriek NPJ, Kellerman TS, Marasas WFO (1981) A comparative-study of the toxicity of *Fusarium verticillioides* (=*F. moniliforme*) to horses, primates, pigs, sheep and rats. Onderstepoort J Vet Res 48(2):129–131

111. Osweiler GD, Ross PF, Wilson TM, Nelson PE, Witte ST, Carson TL, Rice LG, Nelson HA (1992) Characterization of an epizootic of pulmonary-edema in swine associated with fumonisin in corn screenings. J Vet Diagn Invest 4(1):53–59

112. Harrison LR, Colvin BM, Greene JT, Newman LE, Cole JR (1990) Pulmonary oedema and hydrothorax in swine produced by fumonisin B1, a toxic metabolite of *Fusarium verticillioides*. J Vet Diagn Invest 2:217–221

113. Gumprecht LA, Beasley VR, Weigel RM, Parker HM, Tumbleson ME, Bacon CW, Meredith FI, Haschek WM (1998) Development of fumonisin-induced hepatotoxicity and pulmonary edema in orally dosed swine: morphological and biochemical alterations. Toxicol Pathol 26(6):777–788

114. Haschek WM, Gumprecht LA, Smith G, Tumbleson ME, Constable PD (2001) Fumonisin toxicosis in swine: an overview of porcine pulmonary edema and current perspectives. Environ Health Perspect 109: 251–257

115. Delgado JE, Wolt JD (2010) Fumonisin B_1 and implications in nursery swine productivity: a quantitative exposure assessment. J Anim Sci 88(11):3767–3777

116. Delgado JE, Wolt JD (2011) Fumonisin B1 toxicity in grower-finisher pigs: a comparative analysis of genetically engineered Bt corn and non-Bt corn by using quantitative dietary exposure assessment modeling. Int J Environ Res Public Health 8(8):3179–3190

117. Logrieco A, Mule G, Moretti A, Bottalico A (2002) Toxigenic *Fusarium* species and mycotoxins associated with maize ear rot in Europe. Eur J Plant Pathol 108(7):597–609

118. Gelineau van Waes J, Maddox J, Ashley-Koch A, Gregory S, Torres de Matute O, Voss KA, Riley RT (2011) Evaluating human exposure to fumonisins in Guatemala and its possible role as a contributing factor to neural tube defects. Phytopathology 101(6):S222

119. Munkvold GP, Desjardins AE (1997) Fumonisins in maize—can we reduce their occurrence? Plant Dis 81(6):556–565

120. Egmond HP, Jonker MA (2004) Current regulations governing mycotoxin limits in food. In: Magan N, Olsen M (eds) Mycotoxins in food: detection and control. CRC, Boca Raton

121. Liu J, Bell AA, Stipanovic R, Puckhaber L, Shim W (2011) A polyketide synthase gene and an aspartate kinase like gene are required for the biosynthesis of fusaric acid in *Fusarium oxysporum* f. sp. *vasinfectum*. In: Abstract of the proceedings of the Beltwide Cotton Conferences, Marriott Marquis, Atlanta, 5–7 Jan 2011

122. Brown DW, Butchko RAE, Busman M, Proctor RH (2012) Identification of gene

clusters associated with fusaric acid, fusarin, and perithecial pigment production in *Fusarium verticillioides*. Fungal Genet Biol 49(7):521–532

123. Brown DW, Lee SH, Kim LH, Ryu JG, Lee S, Seo Y, Kim YH et al (2015) Identification of a 12-gene fusaric acid biosynthetic gene cluster in *Fusarium* species through comparative and functional genomics. Mol Plant Microbe Interact 28(3):319–332

124. Bryden WL, Logrieco A, Abbas HK (2001) Other significant *Fusarium* mycotoxins. In: Summerell BA, Leslie JF, Backhouse D, Bryden WL, Burgess LW (eds) *Fusarium*: Paul E. Nelson memorial symposium. APS Press, St. Paul

125. Scarpino V, Reyneri A, Vanara F, Scopel C, Causin R, Blandino M (2015) Relationship between European corn borer injury, *Fusarium proliferatum* and *F. subglutinans* infection and moniliformin contamination in maize. Field Crops Res 183:69–78

126. Han Z, Tangni EK, Huybrechts B, Munaut F, Scauflaire J, Wu A, Callebaut A (2014) Screening survey of co-production of fusaric acid, fusarin C, and fumonisins B1, B2 and B3 by *Fusarium* strains grown in maize grains. Mycotoxin Res 30(4):231–240

127. Niehaus EM, Kleigrewe K, Wiemann P, Studt L, Sieber CMK, Connolly LR, Freitag M et al (2013) Genetic manipulation of the *Fusarium fujikuroi* fusarin gene cluster yields insight into the complex regulation and fusarin biosynthetic pathway. Chem Biol 20(8): 1055–1066

128. Kleigrewe K, Soehnel AC, Humpf HU (2011) A new high-performance liquid chromatography-tandem mass spectrometry method based on dispersive solid phase extraction for the determination of the mycotoxin fusarin C in corn ears and processed corn samples. J Agric Food Chem 59(19):10470–10476

129. Bottalico A, Logrieco A, Ritieni A, Moretti A, Randazzo G, Corda P (1995) Beauvericin and fumonisin B1 in preharvest *Fusarium moniliforme* maize ear rot in Sardinia. Food Addit Contam 12(4):599–607

130. Moretti A, Logrieco A, Bottalico A, Ritieni A, Fogliano V, Randazzo G (1996) Diversity in beauvericin and fusaproliferin production by different populations of *Gibberella fujikuroi* (*Fusarium* section Liseola). Sydowia 48(1): 44–56

131. EFSA CONTAM Panel (EFSA Panel on Contaminants in the Food Chain) (2014) Scientific opinion on the risks to human and animal health related to the presence of beauvericin and enniatins in food and feed. EFSA J 12(8):3802

132. Ritieni A, Fogliano V, Randazzo G, Scarallo A, Logrieco A, Moretti A, Mannina L, Bottalico A (1995) Isolation and characterization of fusaproliferin, a new toxic metabolite from *Fusarium proliferatum*. Nat Toxins 3(1):17–20

133. Munkvold G, Stahr HM, Logrieco A, Moretti A, Ritieni A (1998) Occurrence of fusaproliferin and beauvericin in *Fusarium*-contaminated livestock feed in Iowa. Appl Environ Microbiol 64(10):3923–3926

134. Pavlovkin J, Jaskova K, Mistrikova I, Tamas L (2011) Impact of fusaproliferin on primary roots of maize cultivars differing in their susceptibility to *Fusarium*. Biologia 66(6): 1044–10451

135. Santini A, Meca G, Uhlig S, Ritieni A (2012) Fusaproliferin, beauvericin and enniatins: occurrence in food—a review. World Mycotoxin J 5(1):71–81

136. Díaz Arias MM, Leandro LF, Munkvold GP (2013) Aggressiveness of *Fusarium* species and impact of root infection on growth and yield of soybean. Phytopathology 103:822–832

137. Marin P, Moretti A, Ritieni A, Jurado M, Vazquez C, Gonzalez-Jaen MT (2012) Phylogenetic analyses and toxigenic profiles of *Fusarium equiseti* and *Fusarium acuminatum* isolated from cereals from Southern Europe. Food Microbiol 31(2):229–237

138. Nagy R, Hornok L (1994) Electrophoretic karyotype differences between 2 subspecies of *Fusarium acuminatum*. Mycologia 86(2): 203–208

139. Nichea MJ, Cendoya E, Zachetti VGL et al (2015) Mycotoxin profile of *Fusarium armeniacum* isolated from natural grasses intended for cattle feed. World Mycotoxin J 8(4):451–457

140. Ellis ML, Arias MMD, Leandro LF, Munkvold GP (2012) First report of *Fusarium armeniacum* causing seed rot and root rot on soybean (*Glycine max*) in the United States. Plant Dis 96(11):1693

141. Miedaner T, Caixeta F, Talas F (2013) Head-blighting populations of *Fusarium culmorum* from Germany, Russia, and Syria analyzed by microsatellite markers show a recombining structure. Eur J Plant Pathol 137(4): 743–752

142. van der Lee T, Zhang H, van Diepeningen A, Waalwijk C (2015) Biogeography of *Fusarium graminearum* species complex and chemotypes: a review. Food Addit Contam Part A Chem Anal Control Expo Risk Assess 32(4): 453–460

143. Lazreg F, Belabid L, Sanchez J, Gallego E, Garrido-Cardenas JA, Elhaitoum A (2014) First report of *Fusarium equiseti* causing

damping-off disease on aleppo pine in Algeria. Plant Dis 98(9):1268–1269

144. Garibaldi A, Gilardi G, Ortu G, Gullino ML (2015) First report of leaf spot of wild rocket (*Diplotaxis tenuifolia*) caused by *Fusarium equiseti* in Italy. Plant Dis 99(8):1183–1184

145. Brian PW, Norris GLF, Hemming HG, Dawkins AW, Grove JF, Lowe D (1961) Phytotoxic compounds produced by *Fusarium equiseti*. J Exp Bot 12(1):1–12

146. Goldschmied-Reouven A, Friedman J, Block CS (1993) *Fusarium* spp. isolated from non-ocular sites: a 10 year experience at an Israeli General Hospital. J Mycol Med 3(2): 99–102

147. Nirenberg HL (1976) Untersuchungen uber die morphologische und biologische differenzierung in der *Fusarium* section Liseola. Mitteilungen aus der biologischen bundesanstalt fur land–und forstwirtschaft (Berlin-Dahlem) 169:1–117

148. Kuhlman EG (1982) Varieties of *Gibberella fujikuroi* with Anamorphs in *Fusarium* section Liseola. Mycologia 74(5):759–768

149. Leslie JF (1995) *Gibberella fujikuroi*—available populations and variable traits. Can J Bot 73:S282–S291

150. Watanabe T, Umehara Y (1977) Perfect state of causal fungus of bakanae disease of rice plants recollected at Toyama. T Mycol Soc Jpn 18(2):136–142

151. Ploetz RC (2001) Significant diseases in the tropics that are caused by species of *Fusarium*. In: Summerell BA, Leslie JF, Backhouse D, Bryden WL, Burgess LW (eds) *Fusarium*: Paul E. Nelson memorial symposium. APS Press, St. Paul

152. Proctor RH, Plattner RD, Brown DW, Seo JA, Lee YW (2004) Discontinuous distribution of fumonisin biosynthetic genes in the *Gibberella fujikuroi* species complex. Mycol Res 108:815–822

153. Snyder WC, Hansen HN, Oswald JW (1957) Cultivars of the fungus, *Fusarium*. J Madras Univ 27:185–192

154. O'Donnell K, Ward TJ, Geiser DM, Kistler HC, Aoki T (2004) Genealogical concordance between the mating type locus and seven other nuclear genes supports formal recognition of nine phylogenetically distinct species within the *Fusarium graminearum* clade. Fungal Genet Biol 41(6):600–623

155. Starkey DE, Ward TJ, Aoki T, Gale LR, Kistler HC, Geiser DM, Suga H et al (2007) Global molecular surveillance reveals novel Fusarium head blight species and trichothecene toxin diversity. Fungal Genet Biol 44(11):1191–1204

156. Barros GG, Alaniz Zanon MS, Chiotta ML, Reynoso MM, Scandiani MM, Chulze SN (2014) Pathogenicity of phylogenetic species in the *Fusarium graminearum* complex on soybean seedlings in Argentina. Eur J Plant Pathol 138(2):215–222

157. Broders KD, Lipps PE, Paul PA, Dorrance AE (2007) Evaluation of *Fusarium graminearum* associated with corn and soybean seed and seedling disease in Ohio. Plant Dis 91(9):1155–1160

158. Bilgi VN, Bradley CA, Mathew FM, Ali S, Rasmussen JB (2011) Root rot of dry edible bean caused by *Fusarium graminearum*. Plant Health Prog. doi:10.1094/PHP-2011-0425-01-RS

159. Vesonder RF, Ciegler A, Jensen AH (1973) Isolation of emetic principle from *Fusarium*-infected corn. Appl Microbiol 26(6): 1008–1010

160. Yoshizawa T, Morooka N (1973) Deoxynivalenol and its monoacetate - new mycotoxins from *Fusarium roseum* and moldy barley. Agric Biol Chem 37(12):2933–2934

161. Kelly AC, Clear RM, O'Donnell K, McCormick S, Turkington TK, Tekauz A, Gilbert J et al (2015) Diversity of *Fusarium* head blight populations and trichothecene toxin types reveals regional differences in pathogen composition and temporal dynamics. Fungal Genet Biol 82:22–31

162. Ellis ML, Munkvold GP (2014) Trichothecene genotype of *Fusarium graminearum* isolates from soybean (*Glycine max*) seedling and root diseases in the United States. Plant Dis 98(7):1012–1013

163. Barros G, Alaniz Zanon MS, Abod A, Oviedo MS, Ramirez ML, Reynoso MM, Torres A, Chulze S (2012) Natural deoxynivalenol occurrence and genotype and chemotype determination of a field population of the *Fusarium graminearum* complex associated with soybean in Argentina. Food Addit Contam Part A Chem Anal Control Expo Risk Assess 29(2):293–303

164. Varga E, Wiesenberger G, Hametner C, Ward TJ, Dong Y, Schoefbeck D, McCormick S et al (2015) New tricks of an old enemy: isolates of *Fusarium graminearum* produce a type a trichothecene mycotoxin. Environ Microbiol 17(8):2588–2600

165. Liang JM, Xayamongkhon H, Broz K, Dong Y, McCormick SP, Abramova S, Ward TJ, Ma ZH, Kistler HC (2014) Temporal dynamics and population genetic structure of *Fusarium graminearum* in the upper midwestern United States. Fungal Genet Biol 73:83–92

166. Backhouse D (2014) Global distribution of *Fusarium graminearum*, *F. asiaticum* and *F. boothii* from wheat in relation to climate. Eur J Plant Pathol 139(1):161–173

167. Gale LR, Harrison SA, Ward TJ, O'Donnell K, Milus EA, Gale SW, Kistler HC (2011) Nivalenol-type populations of *Fusarium graminearum* and *F. asiaticum* are prevalent on wheat in Southern Louisiana. Phytopathology 101(1):124–134

168. Kawakami A, Kato N, Sasaya T, Tomioka K, Inoue H, Miyasaka A, Hirayae K (2015) Gibberella ear rot of corn caused by *Fusarium asiaticum* in Japan. J Gen Plant Pathol 81(4):324–327

169. Zhu P, Wu L, Liu L, Huang L, Wang Y, Tang W, Xu L (2013) *Fusarium asiaticum*: an emerging pathogen jeopardizing postharvest asparagus spears. J Phytopathol 161(10):696–703

170. Lee T, Lee SH, Shin JY, Kim HK, Yun SH, Kim HY, Lee S, Ryu JG (2014) Comparison of trichothecene biosynthetic gene expression between *Fusarium graminearum* and *Fusarium asiaticum*. Plant Pathol J 30(1):33–42

171. Del Ponte EM, Spolti P, Ward TJ, Gomes LB, Nicolli CP, Kuhnem PR, Silva CN, Tessmann DJ (2015) Regional and field-specific factors affect the composition of *Fusarium* head blight pathogens in subtropical no-till wheat agroecosystem of Brazil. Phytopathology 105(2):246–254

172. Boutigny AL, Ward TJ, Van Coller GJ, Flett B, Lamprecht SC, O'Donnell K, Viljoen A (2011) Analysis of the *Fusarium graminearum* species complex from wheat, barley and maize in South Africa provides evidence of species-specific differences in host preference. Fungal Genet Biol 48(9):914–920

173. Desjardins AE, Proctor RH (2011) Genetic diversity and trichothecene chemotypes of the *Fusarium graminearum* clade isolated from maize in Nepal and identification of a putative new lineage. Fungal Biol 115(1):38–48

174. Sampietro DA, Diaz CG, Gonzalez V, Vattuone MA, Ploper LD, Catalan CAN, Ward TJ (2011) Species diversity and toxigenic potential of *Fusarium graminearum* complex isolates from maize fields in Northwest Argentina. Int J Food Microbiol 145(1):359–364

175. Toth B, Mesterhazy A, Horvath Z, Bartok T, Varga M, Varga J (2005) Genetic variability of central European isolates of the *Fusarium graminearum* species complex. Eur J Plant Pathol 113(1):35–45

176. Umpierrez-Failache M, Garmendia G, Pereyra S, Rodriguez-Haralambides A, Ward TJ, Vero S (2013) Regional differences in species composition and toxigenic potential among *Fusarium* head blight isolates from Uruguay indicate a risk of nivalenol contamination in new wheat production areas. Int J Food Microbiol 166(1):135–140

177. Gomes LB, Ward TJ, Badiale-Furlong E, Del Ponte EM (2015) Species composition, toxigenic potential and pathogenicity of *Fusarium graminearum* species complex isolates from Southern Brazilian rice. Plant Pathol 64(4):980–987

178. Somma S, Petruzzella AL, Logrieco AF, Meca G, Cacciola OS, Moretti A (2014) Phylogenetic analyses of *Fusarium graminearum* strains from cereals in Italy, and characterisation of their molecular and chemical chemotypes. Crop Pasture Sci 65(1):52–60

179. Boutigny AL, Ward TJ, Ballois N, Iancu G, Ioos R (2014) Diversity of the *Fusarium graminearum* species complex on french cereals. Eur J Plant Pathol 138(1):133–148

180. Monds RD, Cromey MG, Lauren DR, di Menna M, Marshall J (2005) *Fusarium graminearum*, *F. cortaderiae* and *F. pseudograminearum* in New Zealand: molecular phylogenetic analysis, mycotoxin chemotypes and co-existence of species. Mycol Res 109:410–420

181. Sampietro DA, Ficoseco MEA, Jimenez CM, Vattuone MA, Catalan CA (2012) Trichothecene genotypes and chemotypes in *Fusarium graminearum* complex strains isolated from maize fields of Northwest Argentina. Int J Food Microbiol 153:229–233

182. Spolti P, Barros NC, Gomes LB, dos Santos J, Del Ponte EM (2012) Phenotypic and pathogenic traits of two species of the *Fusarium graminearum* complex possessing either 15-ADON or NIV genotype. Eur J Plant Pathol 133(3):621–629

183. Lamprecht SC, Tewoldemedhin YT, Botha WJ, Calitz FJ (2011) *Fusarium graminearum* species complex associated with maize crowns and roots in the Kwazulu-Natal Province of South Africa. Plant Dis 95(9):1153–1158

184. Fu M, Li RJ, Guo CC, Pang MH, Liu YC, Dong JG (2015) Natural incidence of *Fusarium* species and fumonisins B1 and B2 associated with maize kernels from nine provinces in China in 2012. Food Addit Contam Part A Chem Anal Control Expo Risk Assess 32(4):503–511

185. Nicolli CP, Spolti P, Tibola CS, Fernandes JMC, Del Ponte EM (2015) *Fusarium* head blight and trichothecene production in wheat by *Fusarium graminearum* and *F. meridionale* applied alone or in mixture at postflowering. Trop Plant Pathol 40(2):134–140

186. Torp M, Nirenberg HI (2004) *Fusarium langsethiae* sp. nov. on cereals in Europe. Int J Food Microbiol 95(3):247–256

187. Torp M, Langseth W (1999) Production of T-2 toxin by a *Fusarium* resembling *Fusarium poae*. Mycopathologia 147(2):89–96
188. Imathiu SM, Ray RV, Back MI, Hare MC, Edwards SG (2013) *Fusarium langsethiae*—a HT-2 and T-2 toxins producer that needs more attention. J Phytopathol 161:1–10
189. Infantino A, Santori A, Aureli G, Belocchi A, De Felice S, Tizzani L, Lattanzio VMT, Haidukowski M, Pascale M (2015) Occurrence of *Fusarium langsethiae* strains isolated from durum wheat in Italy. J Phytopathol 163:612–619
190. Baayen RP, O'Donnell K, Bonants PJM, Cigelnik E, Kroon LP, Roebroeck EJA, Waalwijk C (2000) Gene genealogies and AFLP analyses in the *Fusarium oxysporum* complex identify monophyletic and non-monophyletic formae speciales causing wilt and rot disease. Phytopathology 90(8): 891–900
191. O'Donnell K, Kistler HC, Cigelnik E, Ploetz RC (1998) Multiple evolutionary origins of the fungus causing panama disease of banana: concordant evidence from nuclear and mitochondrial gene genealogies. Proc Natl Acad Sci U S A 95(5):2044–2049
192. Skovgaard K, Rosendahl S, O'Donnell K, Nirenberg HI (2003) *Fusarium commune* is a new species identified by morphological and molecular phylogenetic data. Mycologia 95(4):630–636
193. Laurence MH, Summerell BA, Burgess LW, Liew ECY (2014) Genealogical concordance phylogenetic species recognition in the *Fusarium oxysporum* species complex. Fungal Biol 118(4):374–384
194. Vanheule A, Audenaert K, Höfte M, Warris S, van de Geest H, Waalwijk C, Haesaert G, van der Lee T (2015) Presenting the fully assembled genome of *Fusarium poae*: repeats shed light on a cryptic sexual cycle. 13th European *Fusarium* Seminar Martina Franca, Italy, 10–14 May 2015, 39
195. Amatulli MT, Spadaro D, Gullino ML, Garibaldi A (2012) Conventional and real-time PCR for the identification of *Fusarium fujikuroi* and *Fusarium proliferatum* from diseased rice tissues and seeds. Eur J Plant Pathol 134(2):401–408
196. Chulze SN, Ramirez ML, Farnochi MC, Pascale M, Visconti A, March G (1996) *Fusarium* and fumonisin occurrence in Argentinian cool at different ear maturity stages. J Agric Food Chem 44(9): 2797–2801
197. Nirenberg HI (1995) Morphological differentiation of *Fusarium sambucinum* Fuckel *sensu stricto*, F. *torulosum* (Berk and Curt) Nirenberg comb. nov. and F. *venenatum*

Nirenberg sp. nov. Mycopathologia 129(3): 131–141
198. Chehri K, Salleh B, Zakaria L (2015) Morphological and phylogenetic analysis of *Fusarium solani* species complex in Malaysia. Microb Ecol 69(3):457–471
199. O'Donnell K, Sutton DA, Fothergill A et al (2008) Molecular phylogenetic diversity, multilocus haplotype nomenclature, and in vitro antifungal resistance within the *Fusarium solani* species complex. J Clin Microbiol 46:2477–2490
200. Short DPG, O'Donnell K, Thrane U, Nielsen KF, Zhang N, Juba JH, Geiser DM (2013) Phylogenetic relationships among members of the *Fusarium solani* species complex in human infections and the descriptions of F. *keratoplasticum* sp. nov. and F. *petroliphilum* stat. nov. Fungal Genet Biol 53:59–70
201. Lenc L, Lukanowski A, Sadowski C (2008) The use of PCR amplification in determining the toxigenic potential of *Fusarium sambucinum* and F. *solani* isolated from potato tubers with symptoms of dry rot. Phytopathol Pol 48:13–23
202. Lenc L (2011) Pathogenicity and potential capacity for producing mycotoxins by *Fusarium sambucinum* and *Fusarium solani* isolates derived from potato tubers. Plant Breed Seed Sci 64:23–34
203. Fumero M, Reynoso M, Chulze S (2015) *Fusarium temperatum* and *Fusarium subglutinans* isolated from maize in Argentina. Int J Food Microbiol 199:86–92
204. Shin JH, Han JH, Lee JK, Kim KS (2014) Characterization of the maize stalk rot pathogens *Fusarium subglutinans* and F. *temperatum* and the effect of fungicides on their mycelial growth and colony formation. Plant Pathol J 30(4):397–406
205. Bottalico A (1998) Fusarium diseases of cereals: species complex and related mycotoxin profiles, in Europe. J Plant Pathol 80(2):85–103
206. Munkvold GP, Logrieco A, Moretti A, Ferracane R, Ritieni A (2009) Dominance of group 2 and fusaproliferin production by *Fusarium subglutinans* from Iowa maize. Food Addit Contam Part A Chem Anal Control Expo Risk Assess 26(3):388–394
207. Scauflaire J, Gourgue M, Callebaut A, Munaut F (2011) *Fusarium temperatum* sp. nov. from maize: an emergent species closely related to *Fusarium subglutinans*. Mycologia 103(3):586–597
208. Desjardins AE, Plattner RD, Gordon TR (2000) *Gibberella fujikuroi* mating population a and *Fusarium subglutinans* from teosinte species and maize from Mexico and Central America. Mycol Res 104:865–872

209. Steenkamp ET, Wingfield BD, Desjardins AE, Marasas WFO, Wingfield MJ (2002) Cryptic speciation in *Fusarium subglutinans*. Mycologia 94(6):1032–1043

210. Czembor E, Stepien L, Waskiewicz A (2014) *Fusarium temperatum* as a new species causing ear rot on maize in Poland. Plant Dis 98(7):1001

211. Varela CP, Aguin CO, Chaves PM, Ferreiroa MV, Sainz Oses MJ, Scauflaire J, Munaut F, Bande Castro MJ, Mansilla Vazquez JP (2013) First report of *Fusarium temperatum* causing seedling blight and stalk rot on maize in Spain. Plant Dis 97(9):1252–1253

212. Lanza F, Mayfield D, Munkvold GP (2016) First report of *Fusarium temperatum* causing maize seedling blight and seed rot in North America. Plant Dis 100:1019

213. Zhang H, Luo W, Pan Y, Xu J, Xu JS, Chen WQ, Feng J (2014) First report of *Fusarium temperatum* causing *Fusarium* ear rot on maize in Northern China. Plant Dis 98(9):1273

214. Castanares E, Stenglein SA, Dinolfo MI, Moreno MV (2011) *Fusarium tricinctum* associated with head blight on wheat in Argentina. Plant Dis 95(4):496

215. Bottalico A, Perrone G (2002) Toxigenic *Fusarium* species and mycotoxins associated with head blight in small-grain cereals in Europe. Eur J Plant Pathol 108(7):611–624

216. Chitrampalam P, Nelson BD Jr (2014) Effect of *Fusarium tricinctum* on growth of soybean and a molecular-based method of identification. Plant Health Prog. doi:10.1094/PHP-RS-14-0014

217. Zaher AM, Makboul MA, Moharram AM, Tekwani BL, Calderon AI (2015) A new enniatin antibiotic from the endophyte *Fusarium tricinctum* Corda. J Antibiot 68(3):197–200

218. Cuomo V, Randazzo A, Meca G, Moretti A, Cascone A, Eriksson O, Novellino E, Ritieni A (2013) Production of enniatins A, A1, B, B1, B4, J1 by *Fusarium tricinctum* in solid corn culture: structural analysis and effects on mitochondrial respiration. Food Chem 140(4):784–793

219. Saccardo PA (1886) Sylloge fungorum omnium hucusque cognitorum, 4th edn. Edwards Bros, Ann Arbor

220. Wineland GO (1924) An ascigerous stage and synonomy for *Fusarium moniliforme*. J Agric Res 28:909–922

221. Moretti A, Mule G, Susca A, Gonzalez-Jaen MT, Logrieco A (2004) Toxin profile, fertility and AFLP analysis of *Fusarium verticillioides* from banana fruits. Eur J Plant Pathol 110:601–609

222. Van Hove F, Waalwijk C, Logrieco A, Munaut F, Moretti A (2011) *Gibberella musae* (Fusarium musae) sp nov., a recently discovered species from banana is sister to *F. verticillioides*. Mycologia 103(3):570–585

223. Proctor RH, Brown DW, Plattner RD, Desjardins AE (2003) Co-expression of 15 contiguous genes delineates a fumonisin biosynthetic gene cluster in *Gibberella moniliformis*. Fungal Genet Biol 38(2):237–249

224. Zeller KA, Summerell BA, Bullock S, Leslie JF (2003) *Gibberella konza* (*Fusarium konzum*) sp nov from prairie grasses, a new species in the *Gibberella fujikuroi* species complex. Mycologia 95(5):943–954

225. Leslie JF, Zeller KA, Logrieco A, Mule G, Moretti A, Ritieni A (2004) Species diversity of and toxin production by *Gibberella fujikuroi* species complex strains isolated from native prairie grasses in Kansas. Appl Environ Microbiol 70(4):2254–2262

226. Triest D, Stubbe D, De Cremer K, Pierard D, Detandt M, Hendrickx M (2015) Banana infecting fungus, *Fusarium musae*, is also an opportunistic human pathogen: are bananas potential carriers and source of fusariosis? Mycologia 107(1):46–53

Chapter 5

Penicillium Species and Their Associated Mycotoxins

Giancarlo Perrone and Antonia Susca

Abstract

Penicillium are very diverse and cosmopolite fungi, about 350 species are recognized within this genus. It is subdivided in four subgenera *Aspergilloides*, *Penicillium*, *Furcatum*, and *Biverticillium*; recently the first three has been included in *Penicillium* genus, and *Biverticillium* under *Talaromyces*. They occur worldwide and play important roles as decomposers of organic materials, cause destructive rots in the food industry where produces a wide range of mycotoxins; they are considered enzyme factories, and common indoor air irritants. In terms of human health are rarely associated as human pathogen because they hardly growth at 37°, while the main risk is related to ingestion of food contaminated by mycotoxins produced by several species of *Penicillium*. Various mycotoxins can occur in foods and feeds contaminated by *Penicillium* species, the most important are ochratoxin A and patulin; for which regulation are imposed in a number of countries, and at a less extent cyclopiazonic acid. In this chapter we summarize the main aspect of the morphology, ecology and toxigenicity of *Penicillium* foodborne mycotoxigenic species which belong mainly in subgenus *Penicillium* sections *Brevicompacta*, *Chrysogena*, *Fasciculata*, *Penicillium*, and *Roquefortorum*.

Key words *Penicillium*, Food spoilage, Ochratoxins, Patulin, Ciclopiazonic acid

1 Introduction

Penicillium genus is one of the most common fungi occurring worldwide in a diverse range of habitats, from soil to vegetation, air, indoor environments, and various food products. Its species play important and varies roles, such as production of speciality cheeses, Camembert or Roquefort [1–4], and fermented sausages [5, 6], decomposition of organic materials, causing devastating rots as pre- and postharvest pathogens on food crops [7, 8], and production of diverse range of mycotoxins [9].

The biggest claim for the genus is the production of penicillin, which impacted the fame and revolutionized medical approaches to treating bacterial diseases [10, 11]. Actually it is also screened for the production of novel enzymes [12–14] and many other extrolites directed to wide range of applications, but also harmful compounds, such us mycotoxins [9].

Antonio Moretti and Antonia Susca (eds.), *Mycotoxigenic Fungi: Methods and Protocols*, Methods in Molecular Biology, vol. 1542, DOI 10.1007/978-1-4939-6707-0_5, © Springer Science+Business Media LLC 2017

In 1809 Link [15] introduced for the genus the generic name *Penicillium*, meaning brush, and described three species as *P. candidum*, *P. glaucum*, and the generic type *P. expansum*. *P. expansum* typifies the genus, even though some of its features was showed to be variable, in fact it was showed by various taxonomists that this genus has a peculiar phenotypic plasticity with unstable micromorphological features. Thus, it is important to introduce additional characters like eco-physiological and biochemical traits to stabilize the taxonomy and recognize and discover species [7]. However only highly skilled taxonomist are able to identify *Penicillium* species occurring on food due to their complex morphological and taxonomical traits.

The modern concept of *Penicillium*, was derived from revisions of Thom [16], Raper and Thom [17], and later recognized by Pitt [18] who described four subgenera, *Aspergilloides*, *Furcatum*, *Penicillium*, and *Biverticillium*. Successively subgenus *Biverticillium* was removed from *Penicillium* as a separate genus, based on consideration of morphological and ecological factors, and anamorph-teleomorph connections [19]. The teleomorph genera historically associated with *Penicillium sensu lato* are *Talaromyces* and *Eupenicillium* (in single name nomenclature, the latter is now considered a synonym of *Penicillium sensu stricto*) [20].

Introduction of DNA-based studies in fungal phylogeny evidenced that the differences between *Penicillium sensu stricto* and *Talaromyces* were more than a matter of degree, and that there might be a significant problem with the generic concept of *Penicillium sensu lato*. *Penicillium sensu stricto* and *Talaromyces* occur as distinct clades within *Trichocomaceae*, which could be considered subfamilies [21, 22].

In this respect, the recently revision of the genus using molecular, morphological and biochemical traits data divided *Penicillium* genus in two groups/genera: *Penicillium* and *Talaromyces*. In fact, *Penicillium* subgenus Biverticillium and Talaromyces were shown to form a monophyletic clade distinct from the other subgenera of *Penicillium*, with these names recombined and given Talaromyces names [23]. The remaining *Penicillium* species formed a monophyletic clade together with species classified in *Eupenicillium*, *Eladia*, *Hemicarpenteles*, *Torulomyces*, *Thysanophora*, and *Chromocleista*. These generic names were synonymized with *Penicillium*, while its species were given *Penicillium* names [20]. In addition, three Aspergillus species, *A. paradoxus* (≡Hemicarpenteles paradoxus), *A. malodoratus*, and *A. crystallinus*, respectively, resulted to belong phylogenetically in *Penicillium* genus, and renamed *P. crystallinum*, *P. malodoratum*, and *P. paradoxum*.

Penicillium species are considered to be ubiquitous and opportunistic saprophytes, most of them primarily found in soil and decaying vegetation, but also associated with human food supplies.

In general, *Penicillium* species are strictly aerobic, nutritionally undemanding and able to grow on a wide range of physico-chemical environments, but some of them are highly specialized, as fruit pathogens (*P. expansum*, on apples, *P. digitatum* and *P. italicum* on citrus fruits), on low-water content feeds (*P. brevicompactum*, *P. chrysogenum*, *P. implicatum*), and at low oxygen tension (*P. roqueforti*). They are fast growing fungi, producing high number of exogenous dry-walled spores that are easily disseminated by air. Most of the *Penicillium* foodborne species are psychrotolerant, some are hardly able to grow at 37 °C, but they are mainly mesophilic with optimum temperature around 25 °C.

Most of *Penicillium* species are able to produce mycotoxins, but mostly under laboratory conditions. The actual knowledge of *Penicillium* foodborne species allows to predict which fungi and mycotoxins could be present on a certain food product stored in known conditions.

2   Phenotypic and Morphological Species Recognition

The standard character for identification and classification of filamentous fungi is still morphology, even though variability within morphological characters of accepted species is a still actual issue. Conventional methods for identification are based on microscopic observation of morphology, growth rates, and colors/morphology on growth media [7, 24] and also on the use of secondary metabolites, introduced by Frisvad and Samson [7]. Morphology in the past has been the main important method for taxonomy, based on observation of micro and macrocharacters, but also on other phenotypic characters, e.g., growth at different temperatures, or water activities, presence of pigment and exudate, cultivation methods [18], due to high similarity for some traits within each subgenus. Additional complexity arises because some nomenclature emphasizes the naming of sexual states (teleomorph) or asexual vegetative state (anamorph) [25].

The genus *Penicillium* was applied to fungi having a woolly covering that grow in tufts, conidiophores erect, and simple or branched with conidia collecting at the conidiophore apex. Since then, 1234 names were introduced in the genus. Many of these names are of course not recognizable today, either because descriptions were considered inadequate, names published invalidly, or species considered as synonyms of others, recently a new updated list of accepted species in *Penicillium* genus was published [26].

The teleomorphic genera with anamorphs which produce penicillin are *Byssochlamys*, *Eupenicillium*, and *Talaromyces*. A further classification in subgenus is based on the numbers of branch points. The reproductive asexual apparatus of *Penicillium* is composed of a brush-like structure called a "penicillus" that produce exogenous mitotic spores called conidia. The penicillus is composed of a conidiophore bearing phialidic conidiogenus cells directly or on former metulae and branches, determining one or several verticils. Conidiophore and cleistothecium (when produced) of *Penicillium* are characters of great taxonomic importance. Although these branching patterns do not correspond perfectly with the sections currently accepted for *Penicillium*, characterizing them accurately is still considered important.

Monoverticillate conidiophores have a terminal whorl of phialides and in some species, the terminal cell of the conidiophore is slightly swollen or vesiculate; such species could be confused with diminutive *Aspergillus* conidiophores, but they have septa in the stipes unlike species of the latter genus.

Biverticillate conidiophores have a whorl of three or more metulae between the end of the stipe and the phialides; they are one stage branching but could be arranged symmetrically or asymmetrically around the axis.

Terverticillate conidiophore two or more stage branched usually asymmetric.

In colonies of many species, especially as cultures begin to degenerate, there may be more than one branching pattern or intermediate forms, and it can be challenging to decide which pattern is typical or most developed.

Subgenus *Aspergilloides* is characterized by monoverticillate penicilli, while subgenus *Penicillium* by bi-ter or tetraverticillate penicilli.

These observations could be remarkable mainly to specialists, so observation of physical architecture through which the organism functions in and adapts to its environment can be also considered in species identification. Otherwise, physiological aspects may vary or be induced by specific cues in the immediate environment, due to difference in nutrients, lighting, temperature or humidity during fungal growth. This limit could be minimized using strictly standardized working techniques for medium preparation, inoculation technique and incubation conditions [27–29]. The method is based on observation of colony characters and diameters on specific media, among which Czapek yeast autolysate agar (CYA) and malt extract agar (MEA, Oxoid) are recommended as standard for *Penicillium*. Alternatively, Czapek's agar (CZ), yeast extract sucrose agar (YES), oatmeal agar (OA), creatine sucrose agar (CREA), dichloran 18% glycerol agar (DG18), Blakeslee's MEA, and CYA with 5% NaCl (CYAS) can also be used. The medium used in taxonomic treatments by Raper and Thom [17] and Ramírez [30] was CZ; YES is recommended for determining extrolite profiles of species; OA for sexual reproduction of strains, CREA often used for discrimination of closely related species. Therefore, in general a polyphasic approach based on observation of morphological, physiological, biochemical, and molecular biological character is optimal to characterize those fungi.

3   Molecular Identification

Molecular methods offer the advantage of measuring stable genotypic characteristics and do not be affected by culture conditions and operator interpretation. Molecular identification is based on an accurate phylogenetic reference system, which offers a big advantage over conventional phenotypic methods for species diagnosis, difficult to use by nonspecialists and responsible for frequent misidentifications.

Classification of *Penicillium* undergoes almost continual changes, in particular due to availability of nucleic acids-based analyses that allow to reveal "cryptic species," species morphologically indistinguishable from other but different in nucleic acid characters.

The DNA barcodes have chosen ITS rDNA as target for barcoding of fungi, instead of *cox1*, having lower species resolving ability in fungal taxonomy, but still allowing identification of all the species. The elongation factor 1 alpha-encoding gene (*Tef1-α*) and the calmodulin gene (*cmd*) were able to fully resolve species, whereas the beta-tubulin gene *benA* resolved all species but had some issues with paraphyly [31].

The predominant mycotoxins producers belongs to the terverticillate fungi (subgenus *Penicillium*) characterized by an high grade of plasticity in characters, requiring additional assisting species recognition method and making the taxonomy instable. In this respect DNA sequencing and phylogenetics have largely contribute to resolve taxonomic relationships, although it has resulted not particularly suited for classification and recognition [32]. The phylogenetic species are morphologically similar, but differ in combinations of colony characters, sclerotium production, conidiophore stipe roughening and branching, and conidial shape. Ecological characters and differences in geographical distribution further characterize some of the species, but increased sampling is necessary to confirm these differences. Recently, using a four-gene phylogeny, Houbraken and Samson [20] showed that all the species of *Penicillium* could be divided well identified and separated into two subgenera (*Penicillium* and *Aspergilloides*) and 25 sections. Unfortunately, characters frequently used in subgeneric and sectional classification systems, such as the branching of the *Penicillium* conidiophore and growth rates on agar media [17, 18, 30], did not correspond well with the phylogeny. Currently, it is not possible to recognize all sections without employing DNA sequence data. Ideally, a system should be formulated including morphological, phenotypical and molecular characters, which is called polyphasic approach.

However, in general, an important preliminary work of molecular taxonomy is required in mycology to define intraspecies variability and boundaries of the species, prior to suggesting diagnostic tools.

4   Subgenus Penicillium: Principal Taxa and Mycotoxins

This subgenus as described before comprises species with terverticillate conidiophore, often fasciculate with heavily sporulation, the species of this group are mainly related to animal and man nutrition, plant raw or processed materials [7]. Using a combination of micro and macro—morphological characters, physiological, extrolites, and molecular data the following classification in different sections has been set:

4.1   Sect. Brevicompacta (ex Coronata)

Includes species producing compact, multiramulate penicilli and velutinous colonies, all species produces asperphenamate, seven species belongs to this section the most important and diffuse are *P. brevicompactum*, *P. bialowiezense*, and *P. olsonii*; they are able to grow at very low water activity and low temperatures. *P. brevicompactum* and *P. bialowiezene* produced the mycotoxin mycophenolic acid [7]. This section is distributed worldwide in soil and in plants growing in greenhouse, but are also common in indoor environments.

4.2   Sect. Roquefortorum (ex Roqueforti)

The main important species are *P. roqueforti*, *P. paneum*, and *P. carneum*; these species are peculiar for their rapid growth and formation of velutinous colonies; they are unique for the high tolerance to propionic acetic and lactic acid. The production of mycotoxin roquefortine C is common among this section, and *P. carneum* and *P. paneum* are able to produce patulin. Unless *P. roqueforti* is well known for cheese production like Roquefort, Stilton and Gorgonzola, it is an important spoilage microorganism of airtight-stored grain and relevant accumulation of roquefortine C in grains [33].

4.3   Sect. Chrysogena

It is an important section for the presence of species producing Penicillin, colonies have a velvety texture and species are halotolerant and growth very well at 30 °C, this is the only group of species belonging to subgenus *Penicillium* able to grow at 37 °C. Chysogine is common in this section, but *P. chrysogenum* is also able to produce roquefortine C [34].

4.4   Sect. Fasciculata (ex Lanata-Typica or Viridicata)

It is the most important Section for extolites and mycotoxin production. Species have generally globose and rough walled conidia, fasciculate colony texture and grow rather fast except species in series *Verrucosa*. All species are psychrotolerant and could be found on various foods; *P. verrucosum* and *P. nordicum* are very important species for ochtratoxin A contamination of grains and meat products, respectively. The combination of citrinin and ochratoxin A production is present in *P. verrucosum*, while *P. crustosum*, *P. melaconidium*, and *P. tulipae* produced penitrem and roquefortine C [7].

4.5 Sect. Penicillium

Most of the species are very competitive, with smooth-walled conidia with fasciculate to corenimorfm colonies; they produces important metabolites/mycotoxins like patulin, griseofulvin, fulvic acid, roquefortine C. All species are psychrotolerant, several species are plant pathogens like *P. expansum* in pomaceous fruits, responsible of patulin contamination of apple juice; or *P. italicum* and *P. ulaiense* producing rot on citrus fruits, and *P. sclerotigenum* producing rot in yams [7].

Other minor sections not relevant for the mycotoxins production and belonging to Subgenus *Penicillium* are *Digitata*, *Paradoxa*, *Ramosa*, and *Turbata* [32].

5 Main Penicillium Mycotoxins (See Table 1)

5.1 Ochratoxin A

Ochratoxin A (OTA) is a mycotoxin produced by several species of *Aspergillus* and *Penicillium* fungi that structurally consists of a para-chlorophenolic group containing a dihydroisocoumarin moiety that is amide-linked to l-phenylalanine. OTA is a nephrotoxin, affecting all tested animal species, though effects in man have been difficult to establish unequivocally. It is listed as a probable human carcinogen (Class 2B) [35]. Links between OTA and Balkan Endemic Nephropathy have long been sought, but not established [35]. However, studies show that this molecule can have several toxicological effects such as nephrotoxic, hepatotoxic, neurotoxic, teratogenic and immunotoxic. Regarding the *Penicillium* genus *P. verrucosum* is the major producer of ochratoxin A in stored cereals [36–38], while *P. nordicum* [39] is the main OTA producer found in meat products such as salami and ham. These two species have been found on cheese also, but have only been reported to be of high occurrence on Swiss hard cheeses but erroneously recognized as *Penicillium casei*; in fact the ex type culture of *P. casei* was demonstrated to be *P. verrucosum* [39]. In general, *P. nordicum* prefers food rich in NaCl and protein, like cheeses and dry cured meats [40], while *P. verrucosum* usually contaminates cereals and occasionally was found on dry cured ham [41] and brined olives [42]. Unless the *Penicillium* species diffusely proved to be OTA producer are only *P. verrucosum* and *P. nordicum*, some reports claimed the production of OTA by other *Penicillium* species, but they need confirmation data. In particular, Chen et al. [43] reported that species like *P. chrysogenum*, *P. glycyrrhizacola*, and *P. polonicum* were able to synthesize OTA on fresh or dry liquorice; and Vega et al. [44] reported OTA production from *P. brevicompactum*, *P. crustosum*, *P. olsonii*, and *P. oxalicum* isolated as endophytes in coffee plants.

5.2 Patulin

Patulin, a genotoxic mycotoxin produced by several species of *Aspergillus*, *Penicillium* and *Byssochlamys* is the most common mycotoxin in apples and apple-derived products [45]. It contaminates

Table 1
Penicillium mycotoxins occurring on plant products and associated producing species

Mycotoxins	Agricultural products	Species[a]
Major		
Ochratoxin A	Cereals, stored grains, cured meat (ham, salami, etc), cheese, and brined olives	*P. verrucosum, P. nordicum*
Patulin	Apple, apple juice, mixed fruit juice, silage, yams	*P. expansum, P. carneum, P. paneum, P. clavigerum, P. gladioli, P. griseofulvum, P. sclerotigneum*
Some minor		
Citreoviridin	Yellow rice, cereals	*P. islandicum, P. citreonigrum*, P. smithii, P. manginii, P. miczynskii
Citrinin	Cereals, foods, feedstuffs	*P. citrinum, P. verrucosum, P. expansum*, P. odoratum, P. westlingii
Cyclopiazonic acid	Long stored cereals, pasta, meat and cheese	*P. commune, P. camamberti, P. palitans, P. dipodomyicola, P. griseofulvum*
Penicillic acid	Cereals, silage, onions, carrots, potatoes	*P. aurantiogriseum, P. cyclopium, P. melaconidium, P. viridicatum*, P. polonicum, P. radicicola
Roquefortine C	Farm silage, cheese, meat products	*P. roqueforti, P. carneum, P. chrysogenum, P. crustosum, P. expansum, P. paneum*, P. albocoremium, P. allii, P. griseofulvum, P. hordei, P. melanoconidium, , P. radicicola, P. sclerotigenum*, plus other 13 Penicillium species

[a]Species in bold represent the main occurring on the relevant product

further fruits, such as grapes, oranges, pears and peaches. Cellular effects of patulin include formation of reactive oxygen species (ROS), cell cycle arrest, and cytochrome c release from mitochondria; patulin causes DNA damage and is mutagenic, carcinogenic and teratogenic [46].

Penicillium expansum is by far the most important source of patulin; it is the major species causing spoilage of apples and pears, and is the major source of patulin in apple juice and other apple and pear products. Patulin-producing strains of *P. expansum* have been isolated from a variety of fruits including apples, apricots, black mulberries, cherries, kiwis, lingon berries, nectarines, plums, strawberries, and white mulberries [46]. Several countries in Europe and the USA have now set limits on the level of patulin in apple juice. In addition, *Penicillium griseofulvum* is a very efficient producer of high levels of patulin in pure culture, and it may potentially produce patulin in cereals, pasta and similar products.

P. carneum may produce patulin in beer, wine, meat products, and rye bread as it has been found in those substrates [7], but there are no reports yet on patulin production by this species in those foods. *P. paneum* occurs in rye-bread [7], but again actual production of patulin in this product has not been reported. *P. sclerotigenum* is common in yams and has the ability to produce patulin in laboratory cultures. Other source at laboratory level of patulin are the coprophilous fungi *P. concentricum*, *P. clavigerum*, *P. coprobium*, *P. formosanum*, *P. glandicola*, and *P. vulpinum*. *Penicillium novae-zeelandiae*, *P. marinum*, *P. melinii*, and other soil-borne fungi may produce patulin in pure culture, but are less likely to occur in any foods [34].

6   Minor Mycotoxins (Not Regulated)

6.1   Citreoviridin

Citreoviridin (CTV) was reported as a cause of acute cardiac beriberi [47]; it occurs naturally in rice and corn and is considered as a neurotoxic mycotoxin [48] and a potent inhibitor of mitochondrial ATPase [49]. It has been associated with yellow rice disease, but this disease has also been associated with *P. islandicum* and its toxic metabolites cyclic peptides cyclochlorotine and islanditoxin, and anthraquinones luteoskyrin and rugulosin [50]. *Eupenicillium cinnamopurpureum* has been found in cereals in USA and in Slovakia [51] and is an efficient producer of citreoviridin. CTV was originally isolated from *P. citreonigrum* responsible for CTV occurrence in yellowed rice. Minor producers are *P. smithii*, *P. miczynskii*, and *P. manginii* [52] found mainly in soil and only rarely in foods. In particular, CTV interferes with the metabolism of nerve and muscle tissues causing deficiency of vitamin B1, known as beriberi. The occurrence of beriberi in Japan and Asian countries is attributed to the consumption of moldy and yellow rice [53].

6.2   Citrinin

Citrinin is a nephrotoxin, but probably of less importance than ochratoxin A; toxicity studies showed its action in animals as a nephrotoxin by damaging the proximal tubules of the kidney [54], together with OTA, it was implicated as a potential causative agent in human endemic Balkan nephropathy [55]. It is produced by several filamentous fungi of the genera *Penicillium*, *Aspergillus*, and *Monascus*, which has been encountered as a natural contaminant in grains, foods, feedstuffs, as well as biological fluids. It was first isolated from filamentous fungus *Penicillium citrinum* that is an efficient and consistent producer of citrinin worldwide in foods [24]. Other important *Penicillium* species producer are *P. verrucosum* mainly in cereal-borne in Europe, *P. expansum* in fruits and other foods, and *P. radicicola* is in onions, carrots, and potatoes [56]. Finally, *P. odoratum* and *P. westlingii* have been reported as producers of citrinin, but are not likely to occur often in foods [34].

6.3 Cyclopiazonic Acid

Cyclopiazonic acid (CPA) is a potent mycotoxin that produces focal necrosis in most vertebrate inner organs in high concentrations. It was demonstrated to be responsible of severe effect on the muscle and bones of turkeys affected by the Turkey X disease in association with aflatoxin contamination of peanuts [57]. It could be produced by *Aspergillus* and *Penicillium* species, within *Penicillium* genus *P. commune*, together with its domesticated species *P. camamberti* and *P. palitans* are major producers of CPA and could be found on meat and cheese products [7]. CPA could be contaminant of long stored cereals and cereal products like pasta caused by the occurrence of *P. griseofulvum* [24].

An important minor producer is *P. dipodomyicola* reported from rice in Australia and in a chicken feed mixture in Slovakia [9].

6.4 Penicillic Acid

Penicillic acid and dehydropenicillic acid are small toxic polyketides, but they have no important toxicity, in fact their major role in mycotoxicology could be related to their possible synergistic toxic effect with OTA [58]. Penicillic acid is likely to co-occur with OTA by members of the *Penicillium* series *Viridicata* (which often co-occur with *P. verrucosum*). The main species producer of penicillic acid are *P. aurantiogriseum*, *P. cyclopium*, *P. melanoconidium*, and *P. polonicum* [7].

Penicillic acid is also produced by *P. tulipae* and *P. radicicola*, occasionally found on onions, carrots, and potatoes [56].

6.5 Roquefortine C

Roquefortine C is a mycotoxin with neurotoxic (paralytic) properties [59] but is acute toxicity of roquefortine C is not very high; then it has often been questioned as mycotoxin [34]. However it is a very widespread fungal metabolite, studies on its occurrence indicate a high frequency in farm silage contaminated by *P. roqueforti* [60]; and it is produced by a large number of species like *Penicillium albocoremium*, *P. atramentosum*, *P. allii*, *P. carneum*, *P. chrysogenum*, *P. crustosum*, *P. expansum*, *P. griseofulvum*, *P. hirsutum*, *P. hordei*, *P. melanoconidium*, *P. paneum*, *P. radicicola*, *P. roqueforti*, *P. sclerotigenum*, *P. tulipae*, and *P. venetum*. In addition, the following species are minor source of roquefortine in food: *P. concentricum*, *P. confertum*, *P. coprobium*, *P. coprophilum*, *P. flavigenum*, *P. glandicola*, *P. marinum*, *P. persicinum*, and *P. vulpinum* [9].

7 Other Minor Mycotoxins of Penicillium

Some less important metabolites produced by *Penicillium* species and reported as mycotoxins are:

Chaetoglobosins are toxic compounds that may be involved in mycotoxicosis. They are produced by common foodborne *Penicillium expansum* and *P. discolor*, and have been found to occur naturally [9].

Mycophenolic acid has low acute toxicity, but could be a very important indirect mycotoxin as it highly immunosuppressive activity. It is produced primarily by *P. brevicompactum* and could occur in foods, other producers are *P. roqueforti* and *P. carneum* [34]. It has been found to occur naturally in blue cheeses [61].

Penitrem A is a highly toxic tremorgenic indol-terpene mainly produced by *P. crustosum* [62]. It has primarily been implicated in animal mycotoxicoses [63], but has also been suspected to cause tremors in humans [64].

Rubratoxin is a potent hepatotoxin that was implicated in a severe liver damage in three Canadian boys, who drank rhubarb wine contaminated with *Penicillium crateriforme*, only known as major producer [65].

PR toxin is a mycotoxin that is acutely toxic and can damage DNA and proteins, but it is unstable on cheese and this is important because its major source is *P. roqueforti* [66].

Verrucosidin is a mycotoxin from species in *Penicillium* series *Viridicata* that has been claimed to cause mycotoxicosis in animals [67]. It is produced by *Penicillium polonicum, P. aurantiogriseum,* and *P. melanoconidium* are the major known sources of verrucosidin [7].

Xanthomegnin, Viomellein, and Vioxanthin: These toxins have been reported to cause experimental mycotoxicosis in pigs and have been found to be naturally occurring in cereals. The main producing species are *P. cyclopium, P. freii, P. melanoconidium, P. tricolor,* and *P. viridicatum*. But also some species of the genus *Aspergillus* are able to produce them [34].

References

1. Thom C (1906) Fungi in cheese ripening: camembert and Roquefort. U S D A Bur Anim Ind Bull 82:1–39

2. Nelson JH (1970) Production of Blue cheese flavor via submerged fermentation by Penicillium roqueforti. J Agric Food Chem 18:567–569

3. Karahadian C, Josephson DB, Lindsay RC (1985) Volatile compounds from Penicillium sp. contributing musty-earthy notes to Brie and Camembert cheese flavors. J Agric Food Chem 33:339–343

4. Giraud F, Giraud T, Aguileta G et al (2010) Microsatellite loci to recognize species for the cheese starter and contaminating strains associated with cheese manufacturing. Int J Food Microbiol 137:204–213

5. López-Díaz TM, Santos JA, García-López ML et al (2001) Surface mycoflora of a Spanish fermented meat sausage and toxigenicity of Penicillium isolates. Int J Food Microbiol 68:69–74

6. Ludemann V, Greco M, Rodríguez MP et al (2010) Conidial production by *Penicillium nalgiovense* for use as starter cultures in dry fermented sausages by solid state fermentation. Food Sci Technol 43:315–318

7. Frisvad J, Samson RA (2004) Polyphasic taxonomy of *Penicillium* subgenus *Penicillium*. A guide to identification of food and air-borne terverticillate Penicillia and their mycotoxins. Stud Mycol 49:1–174

8. Samson RA, Houbraken J, Thrane U et al (2010) Food and indoor fungi, CBS laboratory manual. CBS Fungal Biodiversity Centre, Utrecht

9. Frisvad JC, Smedsgaard J, Larsen TO et al (2004) Mycotoxins, drugs and other extrolites produced by species in *Penicillium* subgenus *Penicillium*. Stud Mycol 49:201–241

10. Fleming A (1929) On the antibacterial action of cultures of a *Penicillium*, with special reference to their use in the isolation of *B. influenzae*. Br J Exp Pathol 10:226–236

11. Thom C (1945) Mycology present penicillin. Mycologia 37:460–475

12. Li Y, Cui F, Liu Z et al (2007) Improvement of xylanase production by Penicillium oxali-

cum ZH-30 using response surface methodology. Enzyme Microb Technol 40:1381–1388

13. Adsul MG, Bastawde KB, Varma AJ (2007) Strain improvement of *Penicillium janthinellum* NCIM 1171 for increased cellulase production. Bioresour Technol 98:1467–1473

14. Terrasan CRF, Temer B, Duarte MCT, Carmona EC (2010) Production of xylanolytic enzymes by *Penicillium janczewskii*. Bioresour Technol 101:4139–4143

15. Link HF (1809) Observationes in Ordines plantarum naturales. *Dissertatio 1ma*. Magazin der Gesellschaft Naturforschenden Freunde Berlin 3:3–42

16. Thom C, Church MB, May OE et al (1930) The *Penicillia*. The Williams and Wilkins Company, Baltimore

17. Raper K, Thom C (1949) A manual of the *penicillia*. The Williams and Wilkins Company, Baltimore

18. Pitt JI (1979) The genus *Penicillium* and its teleomorphic states *Eupenicillium* and *Talaromyces*. Academic Press, London

19. Malloch D (1985) The *Trichocomaceae*: relationships with other Ascomycetes. In: Samson R, Pitt JI (eds) Advances in Penicillium and Aspergillus systematics. Plenum Press, New York, pp 365–382

20. Houbraken J, Samson RA (2011) Phylogeny of *Penicillium* and the segregation of Trichocomaceae into three families. Stud Mycol 70:1–51

21. Lobuglio KF, Taylor JW (1993) Molecular phylogeny of *Talaromyces* and *Penicillium* species in subgenus *Biverticillium*. In: Reyolds DR, Taylor JW (eds) The fungal holomorph: mitotic, meiotic and pleomorphic speciation in fungal systematic. C.A.B International, Surrey, pp 115–119

22. Lobuglio KF, Pitt JI, Taylor JW (1993) Phylogenetic analysis of two ribosomal DNA regions indicates multiple independent losses of a sexual *Talaromyces* state among asexual *Penicillium* species in subgenus *Biverticillium*. Mycologia 85:592–604

23. Samson RA, Yilmaz N, Houbraken J et al (2011) Phylogeny and nomenclature of the genus *Talaromyces* and taxa accommodated in *Penicillium* subgenus *Biverticillium*. Stud Mycol 70:159–183

24. Pitt JI, Hocking AD (1997) Food and fungi spoilage, 2nd edn. Blackie Academic and Professional, London

25. Guarro J, Gené J, Stchigel AM (1999) Developments in fungal taxonomy. Clin Microbiol Rev 12:454–500

26. Visagie CM, Seifert KA, Houbraken J (2014) Diversity of *Penicillium* section *Citrina* within

the fynbos biome of South Africa, including a new species from a *Protea repens* infructescence. Mycologia 106:537–552

27. Samson RA, Pitt JI (1985) General recommendations. In: Samson RA, Pitt JI (eds) Advances in *Penicillium* and *Aspergillus* systematics. Plenum Press, New York, pp 455–460

28. Okuda T (1994) Variation in colony characteristics of *Penicillium* strains resulting from minor variations in culture conditions. Mycologia 86:259–262

29. Okuda T, Klich MA, Seifert KA et al (2000) Media and incubation effect on morphological characteristics of *Penicillium* and *Aspergillus*. In: Samson RA, Pitt JI (eds) Integration of modern taxonomic methods for *Aspergillus* and *Penicillium* classification. CRC Press, Harwood Academic Publishers, Amsterdam, pp 83–99

30. Ramírez C (1982) Manual and atlas of the Penicillia. Elsevier Biomedical Press, Amsterdam

31. Rivera KG, Seifert KA (2011) A taxonomic and phylogenetic revision of the *Penicillium sclerotiorum* complex. Stud Mycol 70:139–158

32. Houbraken J, Frisvad JC, Samson RA (2011) Taxonomy of Penicillium section Citrina. Stud Mycol 70:53–138

33. Petersson S, Shnürer J (1999) Growth of *Penicillium roqueforti*, *P. carneum*, and *P. paneum* during malfunctioning airtight storage of high-moisture grain cultivars. Postharvest Biol Technol 17:47–54

34. Frisvad JC, Thrane U, Samson RA et al (2006) Important mycotoxins and the fungi which produce them. In: Pitt J, Samson RA, Thrane U (eds) Advances in food mycology. Springer, New York, pp 3–31

35. JECFA (Joint FAO/WHO Expert Committee on Food Additives) (2001) Safety evaluation of certain mycotoxins in food. Prepared by the Fifty-sixth meeting of the JECFA. FAO Food and Nutrition Paper 74, Food and Agriculture Organization of the United Nations, Rome, Italy

36. Frisvad JC (1985) Profiles of primary and secondary metabolites of value in classification of *Penicillium viridicatum* and related species. In: Samson RA, Pitt JI (eds) Advances in *Penicillium* and *Aspergillus* Systematics. Plenum Press, New York, pp 311–325

37. Pitt JI (1987) *Penicillium viridicatum, P. verrucosum*, and the production of ochratoxin A. *ppl*. Environ Microbiol 53:266–269

38. Lund F, Frisvad JC (2003) *Penicillium verrucosum* in wheat and barley indicates presence of ochratoxin A. J Appl Microbiol 95:1117–1123

39. Larsen TO, Svendsen A, Smedsgaard J (2001) Biochemical characterization of ochratoxin A-producing strains of the genus *Penicillium*. Appl Environ Microbiol 67:3630–3635

40. Schmidt-Heydt M, Graf E, Stoll D et al (2012) The biosynthesis of ochratoxin a by *Penicillium* as one mechanism for adaptation to NaCl rich foods. Food Microbiol 29:233–241

41. Comi G, Orlic S, Redzepovic S et al (2004) Moulds isolated from Istrian dried ham at the pre-ripening and ripening level. Int J Food Microbiol 96:29–34

42. Heperkan D, Dazkir GS, Kansu DZ et al (2009) Influence of temperature on citrinin accumulation by *Penicillium citrinum* and *Penicillium verrucosum* in black table olives. Toxin Rev 28:180–186

43. Chen AJ, Tang D, Zhou YQ et al (2013) Identification of ochratoxin a producing fungi associated with fresh and dry liquorice. PLoS One 8:e78285

44. Vega FE, Posada F, Peterson SW et al (2006) *Penicillium* species endophytic in coffee plants and ochratoxin A production. Mycologia 98:31–42

45. Puel O, Galtier P, Oswald IP (2010) Biosynthesis and toxicological effects of patulin. Toxins (Basel) 2:613–631

46. Reddy KRN, Spadaro D, Lore A et al (2010) Potential of patulin production by *Penicillium expansum* strains on various fruits. Mycotoxin Res 26:257–265

47. Ueno Y (1974) Citreoviridin from *Penicillium citreo-viride* Biourge. In: Purchase IFH (ed) Mycotoxins. Elsevier, Amsterdam, pp 283–302

48. Nishie K, Cole RJ, Dorner JW (1988) Toxicity of citreoviridin. Res Commun Chem Pathol Pharmacol 59:31–52

49. Hongsuk S, Huh HK, Wilcox CS (1996) Chemistry of F1, F0-ATPase inhibitor, photoleomerization of citreoviridin and isocitreoviridin an structure of isocitreoviridin. Bull Korean Chem Soc 17:104–105

50. Enomoto M, Ueno I (1974) Penicillium islandicum (toxic yellowed rice) –luteoskyrin – islanditoxin cyclochlorotine. In: Purchase IFH (ed) Mycotoxins. Elsevier, Amsterdam, pp 303–326

51. Labuda R, Tancinova D (2003) *Eupenicillium ochrosalmoneum*, a rare species isolated from a pig feed mixture in Slovakia. Biologia 58:1123–1126

52. Frisvad JC, Filtenborg O (1983) Classification of Terverticillate *Penicillia* based on profiles of mycotoxins and other secondary metabolites. Appl Environ Microbiol 46:1301–1310

53. Wicklow DT, Stubblefield RD, Horn BW et al (1988) Citreoviridin levels in *Eupenicillium ochrosalmoneum*-infested maize kernels at harvest. Appl Environ Microbiol 54:1096–1098

54. Phillips RD, Hayes AW, Berndt WO et al (1980) Effects of citrinin on renal function and structure. Toxicology 16:123–127

55. IARC (1986) Some naturally occurring and synthetic food components, coumarins ultra-violet radiation. In: Monographs of the evaluation of the carcinogenic risk of chemical to human, vol. 40. Lyon, France, pp. 83–98

56. Overy DP, Frisvad JC (2003) New *Penicillium* species associated with bulbs and root vegetables. Syst Appl Microbiol 26:631–639

57. Jand SK, Kaur P, Sharma NS (2005) Mycoses and mycotoxicosis in poultry. Indian J Anim Sci 75:465–476

58. Stoev SD, Vitanov S, Anguelov G et al (2001) Experimental mycotoxic nephropathy in pigs provoked by a diet containing ochratoxin A and penicillic acid. Vet Res Commun 25:205–223

59. Wagener RE, Davis ND, Diener UL (1980) Penitrem A and roquefortine production by *Penicillium commune*. Appl Environ Microbiol 39:882–887

60. Auerbach H, Oldenburg E, Weissbach F (1998) Incidence of *Penicillium roqueforti* and Roquefortine C in Silages. J Sci Food Agric 76:565–572

61. Lafont O, Debeaupuis JP, Gaillardin M et al (1979) Production of mycophenolic acid by *Penicillium roqueforti* strains. Appl Environ Microbiol 37:365–368

62. Sonjak S, Frisvad JC, Gunde-Cimerman N (2005) Comparison of secondary metabolite production by *Penicillium crustosum* strains, isolated from Arctic and other various ecological niches. FEMS Microbiol Ecol 53:51–60

63. Rundberget T, Wilkins AL (2002) Thomitrems A and E, two indole-alkaloid isoprenoids from *Penicillium crustosum* Thom. Phytochemistry 61:979–985

64. Lewis PR, Donoghue MB, Hocking AD et al (2005) Tremor syndrome associated with a fungal toxin: sequelae of food contamination. Med J Aust 82:582–584

65. Richer L, Sigalet D, Kneteman N et al (1997) Fulminant hepatic failure following ingestion of moldy homemade rhubarb wine. Gastroenterology 112:A1366

66. Arnold DL, Scott PM, McGuire PF et al (1987) Acute toxicity studies on roquefortine C and PR-toxin, metabolites of *Penicillium roqueforti* in the mouse. Food Cosmet Toxicol 16:369–371

67. Burka LT, Ganguli M, Wilson BJ (1983) Verrucosidin, a tremorgen from *Penicillium verrucosum* var. *cyclopium*. J Chem Soc Chem Commun 9:544–545

Part II

Polymerase Chain Reaction (PCR)-Based Methods for Detection and Identification of Mycotoxigenic Fungi

Chapter 6

Targeting Conserved Genes in *Alternaria* Species

Miguel Ángel Pavón, Inés María López-Calleja, Isabel González, Rosario Martín, and Teresa García

Abstract

Real-time polymerase chain reaction (PCR) is a molecular biology technique based on the detection of the fluorescence produced by a reporter molecule, which increases as the reaction proceeds proportionally to the accumulation of the PCR product within each amplification cycle. The fluorescent reporter molecules include dyes that bind to the double-stranded DNA (i.e., SYBR® Green) or sequence-specific probes (i.e., Molecular Beacons or TaqMan® Probes). Real-time PCR provides a tool for accurate and sensitive quantification of target fungal DNA. Here, we describe a TaqMan real-time PCR method for specific detection and quantification of *Alternaria* spp. The method uses *Alternaria*-specific primers and probe, targeting the internal transcribed spacer regions ITS1 and ITS2 of the rRNA gene, and a positive amplification control based on 18S rRNA gene.

Key words TaqMan real-time PCR, Internal transcribed spacer, *Alternaria* spp.

1 Introduction

Alternaria is a cosmopolitan fungal genus that includes saprophytic, endophytic, and pathogenic species, widely distributed in soil and organic matter in decomposition [1, 2]. There are about 300 accepted *Alternaria* species, many of which are also capable of producing mycotoxins, and they can contaminate raw or manufactured plant products like juices, sauces, and preserves, constituting a potential health hazard for humans [3]. Exposure to *Alternaria* toxins has been related to a range of pathologies, from hematological disorders to esophageal cancer [2–6].

Culture-based identification and quantification of *Alternaria* spp. require a lot of expertise in morphology-based taxonomy, and it is a tedious and time-consuming procedure, whereby it takes at least days to weeks to obtain a diagnostic result [5]. In contrast, DNA methods, mainly those based on PCR, are fast and sensitive alternative approaches for the detection of food spoilage and pathogenic microorganisms.

Antonio Moretti and Antonia Susca (eds.), *Mycotoxigenic Fungi: Methods and Protocols*, Methods in Molecular Biology, vol. 1542, DOI 10.1007/978-1-4939-6707-0_6, © Springer Science+Business Media LLC 2017

A crucial requirement for successfully detecting specific microorganisms with a PCR assay is to choose adequate genetic markers during the primer design process that allow a high degree of specificity [7]. The genetic markers used with this purpose are the internal transcribed spacers, ITS1 and ITS2 [8–10], intergenic spacer region (IGS) [11], mitochondrial small subunit (mt SSU) rDNA [12], protein-coding genes such as glyceraldehyde-3-phosphate dehydrogenase (*gpd*) [13], endopolygalacturonase (*endoPG*) [14], translation elongation factor-1α (*tef*) [15], or β-tubulin (*benA*) [16].

The internal transcribed spacer (ITS) regions are highly divergent between fungi (in sequence and in length), and they are often sufficiently different to classify fungi at the species level. Thus, these regions have been used to study the phylogeny of *Alternaria* species [8, 17–20]. DNA-based methods were previously described for the detection of *Alternaria* spp. [20, 21]. However, fungal accurate quantification using these protocols is not possible. As an alternative, real-time PCR provides a tool for sensitive detection of target fungal DNA in foodstuffs. Furthermore, because real-time PCR does not require electrophoresis, it is less laborious than conventional PCR, and it is therefore suitable for automation and high-throughput testing [22].

In this chapter, we describe a TaqMan real-time PCR method for the detection and quantification of *Alternaria* spp. in foodstuffs. The method uses *Alternaria*-specific primers and probe targeting the ITS regions ITS1 and ITS2 of the rRNA gene, and a positive amplification control based on 18S rRNA gene.

2 Materials

Prepare all solutions using ultrapure water (prepared by purifying deionized water to attain a resistivity of 18 M MΩ cm at 25 °C) and analytical grade reagents. Prepare and store all reagents at room temperature (unless indicated otherwise). Dispose of all used reagents as well as any other contaminated disposable materials in accordance with any applicable regulations.

2.1 *Alternaria spp.* Culture

1. Potato dextrose agar (PDA): 4 g/L Potato extract, 20 g/L dextrose, and 15 g/L agar, pH 5.6 ± 0.2.

2. Malt extract agar (MEA): 30 g/L Malt extract, 5 g/L mycological peptone, and 15 g/L agar, pH 5.4 ± 0.2.

3. Potato carrot agar (PCA): 250 g/L Potato extract, 250 g/L carrot extract, and 15 g/L agar, pH 6.5 ± 0.2.

4. Malt extract broth: 17 g/L Malt extract, and 3 g/L mycological peptone, pH 5.4 ± 0.2.

5. Sabouraud-chloramphenicol agar (Sabouraud-CAF): 40 g/L Dextrose, 10 g/L mycological peptone, 0.05 g/L chloramphenicol, and 15 g/L agar, pH 5.6 ± 0.2.

2.2 DNA Extraction Components

1. Extraction buffer: 10 mM Tris, 150 mM NaCl, 2 mM EDTA, and 1 % SDS, pH 8.0. Weigh 1.21 g Tris and add about 400 mL water to a 1 L glass beaker. Weight 8.77 g NaCl and transfer to the glass beaker. Weight 0.74 g 2,2′,2″,2‴-(ethane-1,2-diyldinitrilo)tetraacetic acid (EDTA) and transfer to the glass beaker. Weight 10 g sodium dodecyl sulfate (SDS) and add to the glass beaker (*see* **Note 1**). Add water to a volume of 900 mL. Mix and adjust pH with HCl (*see* **Note 2**). Make up to 1 L with water.

2. Sodium acetate: 3 M Sodium acetate. Weight 4.9 g sodium acetate. Add about 15 mL water to a glass beaker. Mix until complete dissolution. Make up to 20 mL water.

3. Proteinase K: 2 mg/mL Proteinase K. Dissolve 3 mg proteinase K in 1.5 mL water. Homogenize tube contents using a mixer such as vortex (*see* **Note 3**).

4. Sodium chloride: 1.5 M NaCl. Weight 4.9 g NaCl. Add about 40 mL water to a glass beaker. Mix until complete dissolution. Make up to 50 mL water.

5. Phenol/chloroform/isoamyl alcohol: Phenol/chloroform/isoamyl alcohol (25:24:1, v/v), pH 8.5.

6. Isopropanol.

7. Ethanol 70 %: 70 % Solution in water (*see* **Note 4**).

8. TE buffer:10 mM Tris, 1 mM EDTA, pH 8.0: Weigh 1.21 g Tris and add about 400 mL water to a 1 L glass beaker. Weight 0.37 g EDTA and transfer to the glass beaker. Add water to a volume of 900 mL. Mix and adjust pH with HCl (*see* **Note 2**). Make up to 1 L with water.

2.3 Real-Time PCR Components

1. *Alternaria*-specific primer pair: Oligonucleotides Dir1ITSAlt and Inv1ITSAlt (*see* **Note 5**).

2. AltTMTaqMan probe (*see* **Note 6**).

3. Universal primers (18Sfweu/18Srveu) and 18STM probe (*see* **Note 7**).

4. TaqMan Master Mix.

3 Methods

Carry out all procedures at room temperature unless otherwise specified.

3.1 Alternaria spp. Culture Conditions

Inoculate the *Alternaria* strains selected on PDA medium, MEA, and potato carrot agar. Incubate the inoculated plates for 7 days at 25 °C.

3.2 DNA Extraction

DNA was extracted from the fungal cultures grown in MEA, PDA, potato carrot agar, and malt extract broth.

1. Transfer 200 mg of the mycelial mat harvested from the surface of the agar with a sterile disposable loop into a 2 mL microcentrifuge tube.

2. Resuspend the mycelium in 500 μL of extraction buffer and 300 μL sodium acetate.

   Add 100 mg of 150–212 μm glass beads acid-washed, and agitate vigorously using a mixer such as vortex for 10 min.

3. Treat the resultant lysate with 50 μL of 2 mg/mL proteinase K solution. Incubate for 30 min at 37 °C in a water bath.

4. Add 100 μL of 1.5 M NaCl and mix (*see* **Note 8**). Keep the mixture for 5 min at room temperature.

5. Centrifuge the mixture at $16,000 \times g$ (*see* **Note 9**).

6. Transfer the supernatant into a new 2 mL microcentrifuge tube and add an equal volume of phenol/chloroform/isoamyl alcohol (25:24:1) (*see* **Note 10**). Mix the tube gently.

7. Centrifuge the mixture at $13,000 \times g$.

8. Collect the supernatant and add an equal volume of phenol/chloroform/isoamyl alcohol (25:24:1). Mix the tube gently.

9. Centrifuge the mixture at $13,000 \times g$.

10. Transfer the supernatant in a new tube and precipitate the DNA adding 0.6 volume of isopropanol.

11. Incubate the DNA solution at −20 °C for 1 h.

12. Centrifuge at $13,000 \times g$ for 10 min.

13. Discard the supernatant and wash the pellet with 70 % ethanol. Allow to air-dry, and finally resuspend in 100 μL of TE buffer (*see* **Note 11**).

3.3 Real-Time PCR

1. Add 2 μL TaqMan Master Mix, 200 nM of each TaqMan probe, and 300 nM forward primers and 900 nM reverse primers to a microcentrifuge tube (*see* **Note 12**).

2. Mix gently and add 100 ng of DNA from commercial food samples or 2 μL DNA from *A. alternata* culture dilutions and inoculated tomato samples. Make up to 10 μL water PCR grade (*see* **Note 13**).

3. Transfer the 10 μL to a glass capillary tube.

4. Run the real-time amplification reactions on the real-time PCR thermal cycler with the following program: 10 min at 95 °C (denaturation and Taq polymerase activation), an amplification program of 45 cycles at 95 °C for 10 s, 60 °C for 60 s, and 72 °C for 1 s. Samples were then cooled to 40 °C for 30 s (*see* **Note 14**).

4 Notes

1. Having water at the bottom of the cylinder helps to dissolve Tris and EDTA relatively easily, allowing the magnetic stir bar to go to work immediately. If using a glass beaker, Tris and EDTA can be dissolved faster provided that the water is warmed to about 37 °C. However, the downside is that care should be taken to bring the solution to room temperature before adjusting pH. Wear a mask when weighing SDS. The extraction buffer and guanidine hydrochloride can be prepared in large batches.

2. Concentrated HCl (12 N) can be used at first to narrow the gap from the starting pH to the required pH. From then on it would be better to use a series of HCl (e.g., 6 and 1 N) with lower ionic strengths to avoid a sudden drop in pH below the required pH.

3. Wear a mask when weighing the proteinase K. We recommend to prepare the proteinase K fresh each time.

4. We find that it is best to prepare this fresh each time.

5. The primers Dir1ITSAlt and Inv1ITSAlt were designed by Pavón et al. [20] on the basis of rDNA sequences from various fungal, animal, and plant species available in the NCBI (National Center for Biotechnology Information) database. These primers hybridize on the ITS regions (ITS1 and ITS2) of *Alternaria* spp., and delimit a DNA fragment of approximately 370 bp in all the *Alternaria* spp. analyzed.

6. Alt TM TaqMan probe (5′ FAM-AACACCAAGCAA AGCTTGAGGGTACAAAT-TMR 3′) was designed by Pavón et al. [23] based on alignment and comparison of the fragment generated by amplification of the corresponding *Alternaria* target, and was labeled on the 5′ end with the fluorescent reporter dye 6-carboxyfluorescein (FAM), and on the 3′ end with 6-carboxy-tetramethyl-rhodamine (TAMRA).

7. Universal primers (18Sfweu/18Srveu) and probe (18STM) designed by Martín et al. [24] were used as positive amplification control of the assay. These primers were expected to amplify a conserved region of 99 bp of the 18S rRNA gene in all the eukaryotic species.

8. Agitate the tube vigorously until a homogenous mix is observed.

9. The centrifugation condition is very important in order to separate out insoluble cell debris.

10. Use a fume hood and wear a mask when adding the phenol/chloroform/isoamyl alcohol.

11. DNA concentration should be measured with a spectrophotometer.

12. The optimum PCR concentrations of primers yielding the highest endpoint fluorescence and the lowest crossing point value (Cp) should be experimentally determined for each set of primers: in both cases, the optimum concentrations were 300 nM for forward primers and 900 nM for reverse primers.

13. Sensitivity assays were carried out on *A. alternata* culture dilutions and artificially inoculated tomato pulp. To analyze the influence of thermal treatments on the assay performance, 20 mL of the *A. Alternata* culture, grown in malt extract broth after incubation at 25 °C for 4 days, was heated at 90 °C for 5 min. Tomato pulp samples (0.9 mL) were inoculated with 0.1 mL of 10^6 to 10^{-1} CFU/mL (either viable or heat treated), corresponding to theoretical DNA concentrations ranging from 3×10^3 to 3×10^{-4} ng per reaction. DNA concentrations were estimated from DNA measurements of the first two culture dilutions ($1 \times 10^6 – 1 \times 10^5$ CFU/mL), assuming that the DNA isolation yield was the same for each extraction. After DNA isolation of the *A. alternata* culture dilutions and inoculated tomato samples, the PCR reactions were carried out in triplicate in three independent experiments. One milliliter noninoculated tomato pulp was included as a negative control.

Standard curves should be built in each run for detection/amplification of *Alternaria* spp. DNA. Standard curves were plotted as the Cp values against the logarithm of the counts of *A. alternata* dilutions plated in Sabouraud-CAF.

In case to analyze commercial food samples, add 100 ng of DNA extracted with the same protocol described to extract DNA from *Alternaria* species growth on the prepared plated media.

14. The same program was used to amplify the *Alternaria* specific PCR system, and the positive amplification control. Real-time PCR reactions were carried out in triplicate for each DNA extract. The crossing point value (Cp), which refers to the cycle number where the sample's fluorescence significantly increases above the background level, was calculated automatically by the real-time PCR thermal cycler software as the first maximum of the second derivative of the curve.

Acknowledgments

This work was supported by Grant No. AGL 2006-07659 from the Ministerio de Educación y Ciencia of Spain and the Programa de Vigilancia Sanitaria 2009/AGR-1489 from the Comunidad de Madrid (Spain).

References

1. Barkai-Golan R, Paster N (2008) Mouldy fruits and vegetables as a source of mycotoxins: part 1. World Mycotoxin J 1:147–159

2. Ostry V (2008) *Alternaria* mycotoxins: an overview of chemical characterization, producers, toxicity, analysis and occurrence in foodstuffs. World Mycotoxin J 1:175–188

3. Logrieco A, Moretti A, Solfrizzo M (2009) *Alternaria* toxins and plant diseases: an overview of origin, occurrence and risks. World Mycotoxin J 2:129–140

4. Liu GT, Qian YZ, Zhang P et al (1992) Etiological role of *Alternaria alternata* in human esophageal cancer. Chin Med J 105:390–400

5. Simmons EG (2007) *Alternaria*. An identification manual. CBS Fungal Biodiversity Centre, Utrecht

6. European Food Safety Authority (EFSA) (2011) Scientific opinion on the risks for animal and public health related to the presence of *Alternaria* toxins in feed and food. EFSA J 9:2407

7. Scheu P, Berghof MK, Stahl U (1998) Detection of pathogenic and spoilage microorganisms in food with the polymerase chain reaction. Food Microbiol 15:13–31

8. Chou H, Wu W (2002) Phylogenetic analysis of internal transcribed spacer regions of the genus *Alternaria*, and the significance of filament-beaked conidia. Mycol Res 106:164–169

9. Konstantinova P, Bonants P, van Gent-Pelzer M (2002) Development of specific primers for detection and identification of *Alternaria* spp. in carrot material by PCR and comparison with blotter and plating assays. Mycol Res 106:23–33

10. Zur G, Shimoni EM, Hallerman E, Kashi Y (2002) Detection of *Alternaria* fungal contamination in cereal grains by a polymerase chain reaction-based assay. J Food Prot 65:1433–1440

11. Hong SG, Liu D, Pryor BM (2005) Restriction mapping of the IGS region in *Alternaria* spp. reveals variable and conserved domains. Mycol Res 109:87–95

12. Morgenstern I, Powlowski J, Ishmael N et al (2012) A molecular phylogeny of thermophilic fungi. Fungal Biol 116:489–502

13. Nishikawa J, Nakashima C (2015) Morphological variation and experimental host range of *Alternaria cinerariae*. Mycoscience 56:141–149

14. Ntasiou P, Myresiotis C, Konstantinou S et al (2015) Identification, characterization and mycotoxigenic ability of *Alternaria* spp. causing core rot of apple fruit in Greece. Int J Food Microbiol 197:22–29

15. Thompson AH, Narayanin CD, Smith MF, Slabbert MM (2011) A disease survey of *Fusarium* wilt and *Alternaria* blight on sweet potato in South Africa. Crop Prot 30:1409–1413

16. Cramer RA, Lawrence CB (2004) Identification of *Alternaria brassicicola* genes expressed in planta during pathogenesis of *Arabidopsis thaliana*. Fungal Genet Biol 41:115–128

17. Kusaba M, Tsuge T (1995) Phylogeny of *Alternaria* fungi known to produce host-specific toxins on the basis of variation in internal transcribed spacers of ribosomal DNA. Curr Genet 28:491–498

18. Polizzotto A, Andersen B, Martini M et al (2012) A polyphasic approach for the characterization of endophytic *Alternaria* strains isolated from grapevines. J Microbiol Methods 88:162–171

19. Vujanovic V, Labrecque M (2008) Potentially pathogenic and biocontrol Ascomycota associated with green wall structures of basket willow (*Salix viminalis* L.) revealed by phenotypic characters and ITS phylogeny. Biocontrol 53:413–426

20. Pavón MA, González I, Rojas M et al (2011) PCR detection of *Alternaria* spp. in processed foods, based on the internal transcribed spacer genetic marker. J Food Prot 74:240–247

21. Pavón MA, González I, Pegels N et al (2010) PCR detection and identification of *Alternaria* species-groups in processed foods based on the genetic marker Alt a 1. Food Control 21:1745–1756

22. Gil-Serna J, González-Salgado A, González-Jaén MT et al (2009) ITS-based detection and quantification of *Aspergillus ochraceus* and *Aspergillus westerdijkiae* in grapes and green coffee beans by real-time quantitative PCR. Int J Food Microbiol 131:162–167

23. Pavón MA, González I, Martín R, García T (2012) ITS-based detection and quantification of *Alternaria* spp. in raw and processed vegetables by real-time quantitative PCR. Food Microbiol 32:165–171

24. Martín I, García T, Fajardo V et al (2009) SYBR-Green real-time PCR approach for the detection and quantification of pig DNA in feedstuffs. Meat Sci 82:252–259

Chapter 7

Targeting Conserved Genes in *Aspergillus* Species

Sándor Kocsubé and János Varga

Abstract

The genus *Aspergillus* is among the economically most important fungal genera, which contains about 350 species. They occur worldwide, and have both beneficial and harmful effects on humans, animals, and plants. Several molecular sequence-based approaches have been tested to identify *Aspergillus* isolates at the species level. In this chapter, we give an overview of the methods which proved to be most suitable in our experience.

Key words *Aspergillus*, Ribosomal RNA gene cluster, Calmodulin, β-Tubulin

1 Introduction

The *Aspergillus* genus was originally described by Micheli, and today comprises about 350 species [1]. This genus contains several species which are economically important. Some species produce various mycotoxins including aflatoxins, ochratoxins, fumonisins, patulin, gliotoxin, and others [1], while others are used in Oriental food fermentations (e.g., *A. oryzae, A. sojae, A. luchuensis*), or as sources of pharmaceutically important compounds like lovastatin (*A. terreus*) or penicillin (*A. nidulans*). Species assigned to this genus are also able to cause food spoilage in various agricultural products [1].

The traditional infrageneric classification of the genus *Aspergillus* is based on morphological characters. Raper and Fennell [2] divided the genus into 18 groups. The phenotype-based classification of subgenera and sections largely corresponds with the current published phylogenies. Recently, 4 subgenera (*Aspergillus, Circumdati, Fumigati,* and *Nidulantes*) and 20 sections have been proposed in this genus [3–5] (*see* Table 1).

Recently, a proposal to revise article 59 of the former botanical code was accepted, and the principle of "one fungus:one name" was established [6]. These new nomenclatural rules have large implications for *Aspergillus* and several options were considered by a meeting of the International Commission of *Penicillium* and *Aspergillus* (ICPA) in

Antonio Moretti and Antonia Susca (eds.), *Mycotoxigenic Fungi: Methods and Protocols*, Methods in Molecular Biology, vol. 1542,
DOI 10.1007/978-1-4939-6707-0_7, © Springer Science+Business Media LLC 2017

Table 1
Subgeneric classification of the *Aspergillus* genus

Subgenus	Section	Previous name of the sexual stage	Number of species (species described since 2006)	References
Aspergillus	*Aspergillus*	*Eurotium*	17	[10]
	Restricti	*Eurotium*	?	Work in progress
Fumigati	*Fumigati*	*Neosartorya*	33 (5)	[11]
	Clavati	*Neocarpenteles*	5 (1)	[12]
	Cervini	–	8 (4)	Work in progress
Circumdati	*Circumdati*	*Neopetromyces*	24 (4)	[13]
	Nigri	–	26 (7)	[14, 15]
	Flavi	*Petromyces*	22 (4)	[16]
	Cremei	*Chaetosartorya*	17 (6)	Work in progress
	Terrei	*Fennellia*	14 (7)	[17, 18]
	Flavipedes	*Fennellia*	7 (7)	[19]
	Jani	–	2 (1)	[19]
	Candidi	–	5 (1)	[20]
Nidulantes	*Nidulantes*	*Emericella*	?	Work in progress
	Versicolores	–	14 (9)	[21]
	Aenei	*Emericella*	9 (2)	[22]
	Usti	*Emericella*	21 (5)	[23]
	Sparsi	–	9 (1)	[24]
	Raperi	–	2	[4, 25]
	Silvati	–	1	[4, 25]
	Ochraceorosei	–	2	[4, 25]
	Bispori	–	1	[4, 25]

Section *Ornati* was excluded from the *Aspergillus* genus recently [26]

April 2012. At this meeting, the option to keep the name *Aspergillus* and treat other (teleomorph) names like *Neosartorya, Emericella, Eurotium, Fennellia* or *Petromyces* as synonyms of *Aspergillus* has been accepted (http://www.aspergilluspenicillium.org/images/download/minutes.pdf). This option is supported by the results of phylogenetic analyses indicating that the genus is monophyletic with minor modifications to the classical concept needed [2, 5].

The internal transcribed spacer region (ITS1-5.8S-ITS2) of the ribosomal RNA gene cluster was chosen as the official DNA barcoding region for fungi [7], because it is the most frequently sequenced marker in fungi and has primers that work universally. However, the ITS region sometimes does not contain enough variation for distinguishing among all species [7]. According to previous studies [8], ITS sequences can be used in many cases only to assign isolates to species complexes. For correct species identification, a secondary DNA barcode is needed to identify an *Aspergillus* isolate at the species level. Recently, two genes, calmodulin (*CaM*) and β-tubulin (*BenA*), were chosen as possible targets [2]. *BenA* is easy to amplify, but has been reported to vary in the number of introns, and PCR sometimes results in the amplification of paralogous genes [4, 9]. On the other hand, *CaM* is easy to amplify, and distinguishes among almost all *Aspergillus*. Besides, calmodulin sequences are available for almost all *Aspergillus* species [2]. Primers and thermal cycle protocols commonly used for PCR amplification of these genes are summarized in Tables 2 and 3 (*see* **Note 1**).

2 Materials

Prepare all solutions using ultrapure water and analytical grade reagents. Prepare and store all reagents at room temperature (unless indicated otherwise).

Table 2
The sequences of the most frequently used primers in phylogenetic studies of Aspergilli

Locus	Primer (reference)	Primer sequence (5′–3′)
Internal transcribed	ITS1 [27]	TCC GTA GGT GAA CCT GCG G
Spacer (ITS)	ITS4 [27]	TCC TCC GCT TAT TGA TAT GC
β-Tubulin (*BenA*)	Bt2a [28]	GGT AAC CAA ATC GGT GCT GCT TTC
	Bt2b [28]	ACC CTC AGT GTA GTG ACC CTT GGC
Calmodulin (*CaM*)	CMD5 [29]	CCG AGT ACA AGG ARG CCT TC
	CMD6 [29]	CCG ATR GAG GTC ATR ACG TGG
	CF1 [4]	GCC GAC TCT TTG ACY GAR GAR
	CF4 [4]	TTT YTG CAT CAT RAG YTG GAC

Table 3
The components of the PCR reaction mixture

Component	20 µL Reaction	Final concentration
10× Standard *Taq* reaction buffer	2 µL	1×
1 mM dNTPs	4 µL	200 µM
1 µM Forward primer	4 µL	0.2 µM
1 µM Reverse primer	4 µL	0.2 µM
Template DNA	Variable	<1.000 ng
Taq DNA polymerase	0.2 µL	1 unit/20 µL PCR
25 mM MgCl2 (if the polymerase supplied with separate MgCl2)	2 µL	2.5 mM
Nuclease-free water	to 20 µL	

2.1 Isolation and Cultivation

For the isolation of *Aspergilli* from different habitats several media can be used. In our laboratory we use dichloran rose bengal chloramphenicol (DRBC) agar for isolation of molds from food matrices, and dichloran 18 % glycerol agar for the isolation from indoor environments [1].

1. Rose bengal: 5 % Solution in bidistilled water.

2. Dichloran: 0.02 % Solution in ethanol.

3. Chloramphenicol solution: 1 % Solution in bidistilled water.

4. DRBC agar (m/V): 2 % Agar, 0.5 % bacteriological peptone, 0.1 % KH_2PO_4, 0.05 % $MgSO_4$ $7H_2O$, 0.5 mL of rose bengal solution, 1 mL of dichloran solution.

   Add all ingredients to approximately 800 mL of bidistilled water. Heat to dissolve agar, then make up to 990 mL with bidistilled water, and sterilize by autoclaving. After sterilization cool down the medium and complete it with 10 mL of chloramphenicol solution. The prepared medium should be protected from light, to prevent the formation of toxic photoproducts of rose bengal, which can inhibit fungal growth.

5. Dichloran 18 % glycerol agar (DG18): 2 % Agar, 0.5 % bacteriological peptone, 1 % dextrose, 0.1 % KH_2PO_4, 0.05 % $MgSO_4$ $7H_2O$, 1 mL of dichloran solution.

   Add all components to approximately 800 mL of bidistilled water, heat to dissolve agar, and then bring the volume to 990 mL with bidistilled water. Add 220 g glycerol and sterilize by autoclaving. After sterilization cool down the medium and add 10 mL of chloramphenicol solution.

6. Malt extract agar (MEA): 2 % Agar, 3 % malt extract, 0.5 % bacteriological peptone, which can be used for maintaining strains.

7. Yeast extract peptone dextrose (YPD) broth: 1 % Yeast extract, 2 % bacteriological peptone, 2 % dextrose, used for cultivation for DNA isolation.

2.2 Amplification

Recently, the ITS region of the nuclear rRNA locus was chosen as the official barcode for fungi [7]. This region is suitable for the identification of fungi on the genus level, but in several cases it does not contain enough information to assign an isolate at species level. Among *Aspergilli*, several other loci were tested for their suitability in species delineation including part of the β-tubulin (*BenA*), calmodulin (*CaM*), putative ribosome biogenesis protein (*Tsr1*), putative chaperonin complex component TCP-1 (*Cct8*), and RNA polymerase II genes (*RPB2*) [5, 26]. Based on the criteria for barcoding sequences the calmodulin and/or β-tubulin sequences can be the choice as a secondary barcode region (*see* Table 2, Fig. 1).

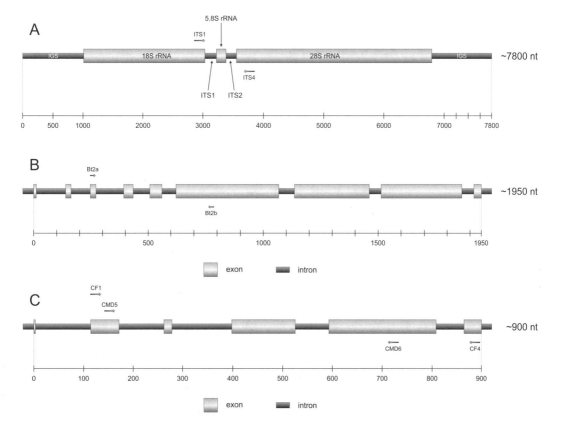

Fig. 1 Schematic representation of the loci and primer locations used routinely for the amplification and sequencing of the ITS (**a**), *benA* (**b**), and *CaM* (**c**) loci in Aspergilli

2.2.1 Preparing Stock Solutions of Primers	Dissolve the primers using sterile ultrapure water. Add the same amount of water in µL as the primer concentration in nmol (e.g., for a primer with a concentration of 25 nmol, add 25 µL water). This will give 1 mM stock solutions.
2.2.2 Preparing the Working Solution	Make a 1000 times dilution by adding 1 µL of the stock solution to 999 µL of ultrapure water. The final concentration of the working solution will be 1 µM. For the PCR reaction apply 4 µL from this solution.
2.2.3 Preparing the dNTP Working Solution	The packaging and the concentration of deoxynucleotides can vary depending on the supplier. The concentration of the working solution depends on the users, but we prefer to use 1 mM solutions of each of dATP, dCTP, dGTP, and dTTP, in sterile ultrapure water (e.g., 10 µL from a 100 mM solution of each dNTP diluted in 960 µL ultrapure water).

3 Methods

3.1 Cultivation	Inoculate the isolates into a microcentrifuge tube containing 700 µL of YPD broth and incubate at 25 °C in dark for 2 days (*see* **Note 2**). To increase the amount of mycelia, the tubes can be tilted slightly (*see* **Note 3**).
3.2 DNA Extraction	Several DNA extraction kits are available commercially for the extraction of DNA from different fungi. In our laboratory we routinely use the MasterPure™ Yeast DNA Purification Kit (Epicentre) for the extraction of fungal DNA with slight modifications of the protocol recommended by the manufacturer. The steps of the extraction protocol are the following:

1. After the incubation period the broth can be easily removed by pipetting it from below the mycelial disk. If small mycelial balls are present instead of a compact pad, the culture must be filtered using a vacuum filtration apparatus to remove the culture medium.

2. In the next step wash the mycelium twice with original culture volume of 0.1 M $MgCl_2$. Remove the $MgCl_2$ solution by pipetting or by vacuum filtration.

3. Add a small volume of sterile washed sand and 300 µL of yeast cell lysis solution (*see* **Note 4**) and grind the mycelia with a micropestle.

4. Incubate the samples at 70 °C for 20 min.

5. Cool down the samples on ice for 5 min.

6. Add 150 µL of MPC protein precipitation reagent and vortex the samples for 10 s.

7. Pellet cellular debris by centrifugation in a microcentrifuge for 10 min at ca. $10,000 \times g$.

8. Transfer the supernatant to a sterile microcentrifuge tube, add 500 μL of isopropanol, and mix thoroughly by inversion.

9. Pellet the DNA by centrifugation in a microcentrifuge for 10 min at ca. $10,000 \times g$ and remove the supernatant by pipetting and discard.

10. Wash the pellet containing the DNA with 500 μL of 70 % ethanol. Centrifuge the samples for 10 min at ca. $10,000 \times g$ and remove ethanol by pipetting. Optionally repeat this step one more time.

11. Carefully remove the ethanol by pipetting and discard. Briefly centrifuge the DNA pellet and remove any remaining ethanol by using a vacuum concentrator.

12. Resuspend the DNA pellet in 35 μL of sterile bidistilled water and store the samples at −20 °C.

3.3 Amplification

Several premade PCR mixtures are available commercially which only need the addition of primers and the template DNA. In most cases they perform well; however if problems occur during the PCR we prepare the reaction mixture by using the above-mentioned protocol (*see* Table 3).

3.3.1 Amplification Conditions for the ITS Region

	Step	Temperature	Time	Number of cycles
1.	Initial denaturation	95 °C	2 min	1
2.	Initial denaturation	95 °C	20 s	
3.	Annealing	52–55 °C (*see* **Note 5**)	20 s	
4.	Extension	72 °C	50 s	Back to **step 2**. 35×
5.	Final extension	72 °C	2 min	1

3.3.2 Amplification Conditions for the CaM Gene

	Step	Temperature	Time	Number of cycles
1.	Initial denaturation	95 °C	2 min	1
2.	Initial denaturation	95 °C	20 s	
3.	Annealing	56 °C	20 s	
4.	Extension	72 °C	40 s	Back to **step 2**. 35×
5.	Final extension	72 °C	2 min	1

3.3.3 *Amplification Conditions for the benA Gene*

	Step	Temperature	Time	Number of cycles
1.	Initial denaturation	95 °C	2 min	1
2.	Initial denaturation	95 °C	20 s	
3.	Annealing	58 °C (*see* **Note 6**)	20 s	
4.	Extension	72 °C	40 s	Back to **step 2**. 35×
5.	Final extension	72 °C	2 min	1

The amplified DNA fragments are usually sequenced by a company providing such services.

3.4 Species Identification

Sequence analyses are usually performed by nucleotide-nucleotide BLAST similarity search at the website of the National Center for Biotechnology Information (http://www.ncbi.nlm.nih.gov/BLAST), and sequences are also compared with our own sequence database. Species identification is usually carried out from the lowest expected value of the BLAST output. A good reference database containing ITS, *BenA*, and *CaM* sequences for most of the valid species from the genus *Aspergillus* has been published recently [3] and it is also available on the ICPA of the International Union of Microbiological Societies (IUMS) website (http://www.aspergilluspenicillium.org/).

4 Notes

1. Other targets have also been used for species identification in Aspergilli. The studies by Houbraken and Samson [26] and Houbraken et al. [5] used a four-gene phylogeny (*RPB1*, *RPB2*, *Tsr1*, and *Cct8*). Other gene sequences like those of actin [23], *RPB2* [4], or mitochondrial cytochrome b [30, 31] have also been used for phylogenetic studies in this genus.

2. Aspergilli can be isolated from various agricultural products after surface sterilization using DRBC medium [1], from indoor air using DG18 [1], or from clinical samples (e.g., Sabouraud agar). Purification of the isolates is usually made on MEA or CYA media [1].

3. In general 2 days are enough to get sufficient amount of mycelia, but in special cases (e.g., *Aspergillus inflatus, A. candidus*, xerophilic species assigned to *Aspergillus* section *Aspergillus*) more time is needed to have enough mycelia. If the incubation period is longer than 2 days, check the cultures regularly to prevent heavy conidiogenesis, because components released from conidia during DNA extraction can act as strong inhibitors in the PCR reaction [32, 33].

4. The MPC protein precipitation reagent can be replaced with 5 M sodium acetate solution adjusted to pH 5.5 with glacial acetic acid, while the yeast cell lysis solution can be replaced by 10% of *N*-lauroylsarcosine sodium salt solution. Optionally 0.5 µL of 2-mercaptoethanol can be added to the sample to increase the efficacy of the DNA extraction. Pay attention on not to transfer the whole supernatant, because the lower layer of the supernatant can contain small cellular debris which can affect the subsequent PCR. Approximately 300 µL is safe to use.

5. If no product is present after amplification, decrease the annealing temperature to 48 °C.

6. If unspecific bands can be detected during gel electrophoresis the annealing temperature can be increased to 65 °C.

Acknowledgments

Part of the work was supported by OTKA grant No. 84077.

References

1. Samson RA, Houbraken J, Thrane U et al (2010) Food and indoor fungi. CBS KNAW Biodiversity Center, Utrecht

2. Raper KB, Fennell DI (1965) The genus *Aspergillus*. Williams & Wilkins, Baltimore

3. Samson RA, Visagie CM, Houbraken J et al (2014) Phylogeny, identification and nomenclature of the genus *Aspergillus*. Stud Mycol 78:141–173

4. Peterson SW (2008) Phylogenetic analyses of *Aspergillus* species using DNA sequences from four loci. Mycologia 100:205–226

5. Houbraken J, de Vries RP, Samson RA (2014) Modern taxonomy of biotechnologically important *Aspergillus* and *Penicillium* species. Adv Appl Microbiol 86:199–249

6. Norvell LL (2011) Fungal nomenclature. 1. Melbourne approves a new code. Mycotaxon 116:481–490

7. Schoch CL, Seifert KA, Huhndorf S, Robert V et al (2012) Nuclear ribosomal internal transcribed spacer (ITS) region as a universal DNA barcode marker for Fungi. Proc Natl Acad Sci U S A 109:6241–6246

8. Balajee SA, Houbraken J, Verweij PE et al (2007) *Aspergillus* species identification in the clinical setting. Stud Mycol 59:39–46

9. Hubka V, Kolarik M (2012) β-tubulin paralogue *tubC* is frequently misidentified as the *benA* gene in *Aspergillus* section *Nigri* taxonomy: primer specificity testing and taxonomic consequences. Persoonia 29:1–10

10. Hubka V, Kolarik M, Kubatova A, Peterson SW (2013) Taxonomic revision of *Eurotium* and transfer of species to *Aspergillus*. Mycologia 105:912–937

11. Samson RA, Hong S, Peterson SW et al (2007) Polyphasic taxonomy of *Aspergillus* section *Fumigati* and its teleomorph *Neosartorya*. Stud Mycol 59:147–203

12. Varga J, Due M, Frisvad JC, Samson RA (2007) Taxonomic revision of *Aspergillus* section *Clavati* based on molecular, morphological and physiological data. Stud Mycol 59:89–106

13. Visagie CM, Varga J, Houbraken J et al (2014) Ochratoxin production and taxonomy of the yellow aspergilli (*Aspergillus* section *Circumdati*). Stud Mycol 78:1–61

14. Varga J, Frisvad JC, Kocsubé S et al (2011) New and revisited species in *Aspergillus* section *Nigri*. Stud Mycol 69:1–17

15. Jurjević Z, Peterson SW, Stea G et al (2012) Two novel species of *Aspergillus* section *Nigri* from indoor air. IMA Fungus 3:159–173

16. Varga J, Frisvad JC, Samson RA (2011) Two new aflatoxin producing species, and an overview of *Aspergillus* section *Flavi*. Stud Mycol 69:57–80

17. Samson RA, Peterson SW, Frisvad JC, Varga J (2011) New species in *Aspergillus* section *Terrei*. Stud Mycol 69:39–55

18. Balajee SA, Baddley JW, Peterson SW et al (2009) *Aspergillus alabamensis*, a new clinically relevant species in the section *Terrei*. Eukaryot Cell 58:713–722

19. Hubka V, Novakova A, Kolarík A et al (2014) Revision of *Aspergillus* section *Flavipedes*: seven new species and proposal of section *Jani* sect. nov. Mycologia 107:169–208

20. Varga J, Frisvad JC, Samson RA (2007) Polyphasic taxonomy of *Aspergillus* section *Candidi* based on molecular, morphological and physiological data. Stud Mycol 59:75–88

21. Jurjevic Z, Peterson SW, Horn BW (2012) *Aspergillus* section *Versicolores*: nine new species and multilocus DNA sequence based phylogeny. IMA Fungus 3:59–79

22. Varga J, Frisvad JC, Samson RA (2010) *Aspergillus* sect. *Aeni* sect. nov., a new section of the genus for *A.karnatakaensis* sp. nov. and some allied fungi. IMA Fungus 1:197–205

23. Samson RA, Varga J, Meijer M, Frisvad JC (2011) New taxa in *Aspergillus* section *Usti*. Stud Mycol 69:81–97

24. Varga J, Frisvad JC, Samson RA (2010) Polyphasic taxonomy of *Aspergillus* section *Sparsi*. IMA Fungus 1:187–195

25. Peterson SW, Varga J, Frisvad JC et al (2008) Phylogeny and subgeneric taxonomy of *Aspergillus*. In: Varga J, Samson RA (eds) Aspergillus in the genomic era. Wageningen Academic Publishers, Wageningen, pp 33–56

26. Houbraken J, Samson RA (2011) Phylogeny of *Penicillium* and the segregation of *Trichocomaceae* into three families. Stud Mycol 70:1–51

27. White TJ, Bruns T, Lee S, Taylor J (1990) Amplification and direct sequencing of fungal ribosomal RNA genes for phylogenetics. In: Innis MA, Gelfand DH, Shinsky TJ, White TJ (eds) PCR Protocols: a guide to methods and applications. Academic Press Inc, New York, pp 315–322

28. Glass NL, Donaldson GC (1995) Development of premier sets designed for use with the PCR to amplify conserved genes from filamentous Ascomycetes. Appl Environ Microbiol 61:1323–1330

29. Hong SB, Go SJ, Shin HD, Frisvad JC, Samson RA (2005) Polyphasic taxonomy of *Aspergillus fumigatus* and related species. Mycologia 97:1316–1329

30. Yokoyama K, Wang L, Miyaji M, Nishimura K (2001) Identification, classification and phylogeny of the *Aspergillus* section *Nigri* inferred from mitochondrial cytochrome b gene. FEMS Microbiol Lett 200:241–246

31. Wang L, Yokoyama K, Takahasi H et al (2001) Identification of species in *Aspergillus* section *Flavi* based on sequencing of the mitochondrial cytochrome b gene. Int J Food Microbiol 71:75–86

32. Selma MV, Martínez-Culebras PV, Aznar R (2008) Real-time PCR based procedures for detection and quantification of *Aspergillus carbonarius* in wine grapes. Int J Food Microbiol 122:126–134

33. McDevitt JJ, Lees PSJ, Merz WG, Schwab KJ (2007) Inhibition of quantitative PCR analysis of fungal conidia associated with indoor air particulate matter. Aerobiologia 23:25–45

Chapter 8

Targeting Conserved Genes in *Fusarium* Species

Jéssica Gil-Serna, Belén Patiño, Miguel Jurado, Salvador Mirete, Covadonga Vázquez, and M. Teresa González-Jaén

Abstract

Fumonisins are important mycotoxins contaminating foods and feeds which are mainly produced by *F. verticillioides* and *F. proliferatum*. Additionally, both are pathogens of maize and other cereals. We describe two highly sensitive, rapid, and species-specific PCR protocols which enable detection and discrimination of these closely related species in cereal flour or grain samples. The specific primer pairs of these assays were based on the intergenic spacer region of the multicopy rDNA unit which highly improves the sensitivity of the PCR assay in comparison with single-copy target regions.

Key words *Fusarium verticillioides*, *Fusarium proliferatum*, Fumonisins, Polymerase chain reaction (PCR), Detection, rDNA intergenic spacer (IGS)

1 Introduction

Fusarium is a worldwide distributed fungal genus and includes different phytopathogenic species which are able to infect cereals among other crops [1]. Additionally, certain *Fusarium* species can produce fumonisins, a group of mycotoxins that contaminate food and feed products and represent a risk for human and animal health [2]. Fumonisins are the most abundant toxins in infected maize kernels, with *F. verticillioides* and *F. proliferatum* being the most important producing species [3].

Detection of fumonisin-producing fungal species by conventional methods is a labor- and time-consuming task that requires highly qualified expertise and it is particularly difficult in the case of the genus *Fusarium* [3]. These methods are increasingly being replaced by DNA-based detection assays, basically PCR. These are rapid and highly sensitive, they can even be directly applied to agrofood products, and they allow discrimination at species level. Several PCR protocols have been developed and applied successfully to detect mycotoxigenic *Fusarium* species in food products although the highest sensitivity is achieved when they are based on multicopy target

Antonio Moretti and Antonia Susca (eds.), *Mycotoxigenic Fungi: Methods and Protocols*, Methods in Molecular Biology, vol. 1542, DOI 10.1007/978-1-4939-6707-0_8, © Springer Science+Business Media LLC 2017

sequences [4–8]. In this chapter, we present a detailed PCR protocol based on multicopy sequences of a region with high variability (IGS of rDNA units) for specific detection of *F. verticillioides* and *F. proliferatum*. These assays allow rapid, highly efficient, and sensitive diagnosis and discrimination of these closely related fungal species.

2 Materials

2.1 DNA Extraction

1. DNeasy Plant Mini Kit (Qiagen, Valencia, Spain): The kit can be stored at room temperature (15–25 °C). Store RNase stock solution for up to a year also at room temperature. We recommend keeping the RNase A stock solution at 2–8 °C for longer storage or if ambient temperatures often exceed 25 °C.

2. Ethanol (96–100%): Store at room temperature.

2.2 PCR Reactions

1. Molecular biology-grade water.

2. Primer pairs VERT-1 (5′-GTCAGAATCCATGCCAGAACG -3′)/VERT-2 (5′-CACCCGCAGCAATCCATCAG-3′) [6] and PRO-F (5′-CGGCCACCAGAGGATGTG-3′)/PRO-R (5′-CAACACGAATCGCTTCCTGAC-3′) [7] (IDT, Haverhill, UK): Prepare aliquots of each primer at 20 μM in sterile molecular biology-grade water and store at −20 °C.

3. 10× PCR buffer, MgCl₂ solution (50 mM), and Taq DNA polymerase (5 U/mL) (Biotools, Madrid, Spain): Store at −20 °C.

4. dNTP mix solution (100 mM) (Biotools, Madrid, Spain): Store at −20 °C.

5. Template DNA (10–200 ng): Store at −20 °C.

2.3 Agarose Gel Electrophoresis

1. TAE 1× buffer: Tris–acetate 40 mM and EDTA 1.0 mM. Store at room temperature.

2. Agarose d-1 Low EEO (Pronadisa, Madrid, Spain): Store at room temperature.

3. Ethidium bromide solution 1% (Applichem, Darmstadt, Germany): Store at room temperature.

4. Loading buffer: 0.25% Bromophenol blue, 4% (w/v) sucrose, distilled water to adjust volume. Store at 4 °C.

5. 100 bp and 1 kb DNA molecular weight markers (Biotools, Madrid, Spain): Store at 4 °C.

3 Methods

3.1 DNA Extraction

1. These instructions assume the extraction of total DNA from fungal pure culture (*see* **Note 1**), contaminated plant material,

or food matrix (*see* **Note 2**). The DNeasy Plant Mini Kit includes RNase A enzyme, buffers, columns, and collection tubes.

2. Prepare in advance all the materials required for the protocol: sterilize all material (tips, mortar and pestle, microcentrifuge tubes …); add the appropriate amount of ethanol to buffers AW1 and AW2, as indicated on the bottle, to prepare a working solution; preheat a water bath to 65 °C; label each microcentrifuge tube and column.

3. Buffer AP1 and buffer AW1 concentrate may form precipitates upon storage. If necessary, warm to 65 °C to redissolve (before adding ethanol to buffer AW1). Do not heat buffer AW1 after ethanol has been added.

4. Grind the plant, food matrix, or fungal tissue under liquid nitrogen using a sterile mortar and pestle (*see* **Note 3**). Transfer the tissue powder (a maximum of 100 mg) to 1.5 mL microcentrifuge tube.

5. Add 400 μL buffer AP1 (lysis buffer) and 4 μL RNase A stock solution (100 mg/mL) and vortex vigorously.

6. Incubate the mixture for 10 min at 65 °C to lysate cells. Mix two or three times during incubation by inverting the tube.

7. Add 130 μL of buffer P3 (precipitation buffer, *see* **Note 4**) to the lysate, mix, and incubate for 5 min in ice to precipitate detergent, proteins, and polysaccharides present in the sample.

8. Centrifuge for 6.5 min at $16,000 \times g$.

9. Transfer the supernatant into the QIAshredder Mini spin column (lilac) placed in a 2 mL collection tube, and centrifuge for 2.5 min at $16,000 \times g$.

10. Transfer the flow-through fraction (usually about 430 μL) from **step 8** into a new tube without disturbing the cell debris pellet.

11. Add 1.5 volumes (usually about 645 μL) of buffer AW1 (binding buffer) to the cleared lysate, and mix immediately by pipetting.

12. Pipet 650 μL of the mixture from **step 10**, including any precipitate that may have formed, into the DNeasy Mini spin column placed in a 2 mL collection tube. Centrifuge for 1 min at $6000 \times g$, and discard the flow-through. Reuse the collection tube in **step 12**.

13. Repeat **step 11** with the remaining sample. Discard the flow-through and collection tube.

14. Place the DNeasy Mini spin column into a new 2 mL collection tube, add 500 μL of buffer AW2 (wash buffer), and centrifuge for 1 min at $6000 \times g$. Discard the flow-through and reuse the collection tube in **step 14**.

15. Add 500 µL buffer AW2 to the DNeasy Mini spin column, and centrifuge for 2.5 min 16,000×g to dry the membrane.

16. Transfer the DNeasy Mini spin column to a 1.5 or 2 mL microcentrifuge tube and add 100 µL of molecular degree water directly onto the DNeasy membrane. Incubate for 5 min at room temperature (15–25 °C), and then centrifuge for 1 min at 6000×g to elute.

17. Measure the amount of template DNA spectrophotometrically. The concentration and purity of DNA can be determined by measuring the absorbance at 260 nm (A260) and 280 nm (A280). Purity is determined by calculating the ratio of absorbance at 260 nm to absorbance at 280 nm. Pure DNA has an A260/A280 ratio of 1.8–2.0.

18. Store the DNA at −20 °C.

3.2 PCR Amplification

1. Mix the following components in a 0.2 mL microcentrifuge tube: 2 µL of template DNA (10–200 ng total DNA amount, *see* **Note 5**) (*see* **Note 6**), 2.5 µL of 10× PCR buffer, 1 µL of MgCl$_2$, 0.2 µL of dNTPs, and 0.15 L of Taq DNA polymerase (*see* **Note 7**), and 1 µL of each primer VERT-1 and VERT-2 for *F. verticillioides* and PRO-F and PRO-R for amplifications of *F. proliferatum*. Add molecular biology-grade water up to 25 µL.

2. It is recommended to perform control PCR assays (*see* **Note 8**).

3. The *Fusarium verticillioides* PCR amplification protocol is as follows: 1 cycle of 1 min 25 s at 94 °C; 25 cycles of 35 s at 95 °C (denaturalization), 30 s at 64 °C (annealing), and 2 min at 72 °C (extension); and finally 1 cycle of 5 min at 72 °C. *The Fusarium proliferatum PCR amplification protocol is exactly the same except for the annealing temperature that is 69 °C.*

4. Store the PCR products at 4 °C.

3.3 Agarose Gel Electrophoresis

1. These instructions assume that you are making a 10 × 15 cm 1% agarose gel using a Bio-Rad Wide Mini-Sub Cell GT gel system.

2. Weigh out 0.6 g of agarose into a 250 mL conical flask. Add 60 mL of 1× TAE. Swirl to mix.

3. Microwave for about 1 min to dissolve the agarose.

4. Leave it to cool down to about 60 °C (just too hot to keep holding in bare hands).

5. Add 1 µL of ethidium bromide (10 mg/mL) and swirl to mix (*see* **Note 9**).

6. Pour the gel slowly into the tray. Insert the comb and check that it is correctly positioned. The gel should polymerize within 20 min.

7. Pour 1× TAE buffer into the gel tank to submerge the gel to 2–5 mm depth. This is the running buffer.

8. Prepare the samples adding 5 µL of loading buffer to 15 µL of each PCR reaction into the appropriate labeled tube.

9. Once the gel has set, remove carefully the comb and load 20 µL of each sample in a well. Reserve one well for the DNA molecular weight marker.

10. Complete the assembly of the gel unit and connect to a power supply. Run the gel at 80 V during about 45 min monitoring the progress of the gel by reference to the marker dye.

11. The gel is then placed in a UV light box to visualize the PCR products. An example of the results produced is shown in Figs. 1 and 2. The presence of an amplification band of the corresponding size (700 bp for *F. verticillioides* and 200 bp for *F. proliferatum*) indicates a positive result for fungal DNA on the sample tested.

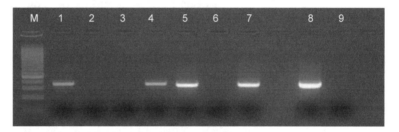

Fig. 1 PCR amplification using primers VERT-1/VERT-2 and lanes *1–7*: DNA from maize flour (samples *1*, *4*, *5*, and *7* were positive for *F. verticillioides*); lane *8*: *F. verticillioides* DNA (positive control); lane *9*: non-template control. *M*: 1 kb DNA molecular size marker

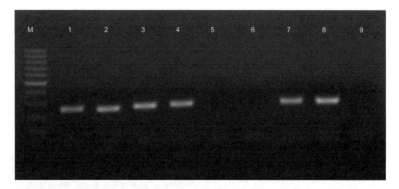

Fig. 2 PCR amplification using primers PRO-F/PRO-R and lanes *1–7*: DNA from maize flour (samples *1*, *2*, *3*, *4*, and *7* were positive for *F. proliferatum*); lane *8*: *F. proliferatum* DNA (positive control); lane *9*: Non-template control. *M*: 100 bp DNA molecular size marker

4   Notes

1. Fungal DNA can be obtained from isolates cultured in 100 mL Erlenmeyer flasks containing 20 mL liquid Sabouraud medium (Scharlau Chemie, Barcelona, Spain) and incubated at 28 °C and 150 rpm. *Fusarium* mycelia from 3-day-old cultures can be harvested by filtration through Whatman No. 1 paper and frozen in liquid nitrogen. Mycelia can be stored at −80 °C until DNA extraction.

2. In the case of maize flour or other food samples, fungal DNA can be obtained from cultures prepared in 250 mL Erlenmeyer flasks with 50 mL of Sabouraud chloramphenicol broth (Scharlau Chemie, Barcelona, Spain) and 1 g of the sample incubated at 28 °C and 150 rpm for 24 h. Cultures are subsequently harvested by filtration through Whatman No. 1 and frozen in liquid nitrogen. The sample can be stored at −80 °C until DNA extraction. In case of grains or coarse matrices, thorough grinding and homogenization are required.

3. The primary disruption of the tissue for DNA extraction is an essential step that must be performed carefully. It can be performed with another disruption method, such as commercially available homogenizers or micro beads.

4. Phenolic compounds present in some food matrices such as maize, grapes, or paprika might interfere in DNA extraction procedure or inhibit PCR assay. In these cases, polyvinylpyrrolidone (0.33%) (Sigma-Aldrich, Steinheim, Germany) can be added to P3 buffer in order to remove these compounds.

5. When the assay is performed from a fungal pure culture, DNA concentration for PCR might be between 10 and 100 ng. When the PCR assay is performed with DNA from a food matrix or a plant tissue, the total amount of DNA per reaction should be between 150 and 200 ng.

6. Positive (with 2 μL of corresponding *F. verticillioides* or *F. proliferatum* pure DNA) and negative (with 2 μL of molecular biology-grade water instead of DNA template) controls must be included in every PCR assay.

7. Use filter tips to prepare PCR reaction. Taq polymerase must be maintained on ice during its manipulation.

8. It is highly recommended to test DNA integrity using a control PCR reaction with the universal fungal primers ITS1 (5′-TCCGTAGGTGAACCTGCGG-3′) and ITS4 (5′-TCCTCCGCTTATTGATATG-3′) following the amplification program described previously [9].

9. Ethidium bromide is a potent mutagenic agent. Handle with care; always wear a suitable lab coat and disposable gloves. All ethidium bromide waste must be disposed in appropriate containers.

References

1. Bottalico A, Perrone G (2002) Toxigenic *Fusarium* species and mycotoxins associated with head blight in small-grain cereals in Europe. Eur J Plant Pathol 108:611–624
2. Creppy EE (2002) Update of survey, regulation and toxic effects of mycotoxins in Europe. Toxicol Lett 127:19–28
3. Nirenberg H, O'Donnell K (1998) New *Fusarium* species and combinations within the *Gibberella fujikuroi* species complex. Mycologia 90:434–458
4. Bluhm BH, Flaherty JE, Cousin MA, Woloshuk CP (2002) Multiplex polymerase chain reaction assay for the differential detection of trichothecene- and fumonisin-producing species of Fusarium in cornmeal. J Food Prot 65:1955–1961
5. Mulè G, Susca A, Stea G, Moretti A (2004) A species-specific PCR assay based on the calmodulin partial gene for identification of *Fusarium verticillioides*, *F. proliferatum* and *F. subglutinans*. Eur J Plant Pathol 110:495–502
6. Patiño B, Mirete S, González-Jaén MT et al (2004) PCR detection assay of fumonisin-producing *Fusarium verticilliodes* strains. J Food Prot 67:1278–1283
7. Jurado M, Vázquez C, Marín S et al (2006) PCR based strategy to detect contamination with mycotoxigenic *Fusarium* species in maize. Syst Appl Microbiol 29:681–689
8. Gil-Serna J, Mateo E, González-Jaén MT et al (2013) Contamination of barley seeds with *Fusarium* species and their toxins in Spain: an integrated approach. Food Addit Contam 30:372–380
9. Henry T, Iwen P, Hinrichs S (2000) Identification of *Aspergillus* species using internal transcribed spacer regions 1 and 2. J Clin Microbiol 38:1510–1515

Chapter 9

Targeting Conserved Genes in *Penicillium* Species

Stephen W. Peterson

Abstract

Polymerase chain reaction amplification of conserved genes and sequence analysis provides a very powerful tool for the identification of toxigenic as well as non-toxigenic *Penicillium* species. Sequences are obtained by amplification of the gene fragment, sequencing via capillary electrophoresis of dideoxynucleotide-labeled fragments or NGS. The sequences are compared to a database of validated isolates. Identification of species indicates the potential of the fungus to make particular mycotoxins.

Key words ITS, Barcode, Housekeeping genes, DNA sequencing, Identification

1 Introduction

Identification of toxigenic *Penicillium* species has historically been a difficult task. The problematic nature of identification is rooted in two major areas, species concepts and species recognition. Thom [1] defined species as organisms displaying a correlated set of morphological and physiological characters, allowing for variation of those characters in different isolates but having a majority of characters within the limits of the species. Raper and Thom [2] recognized a species as having a largely fixed morphology, culturally and microscopically, and fixed common reactions. Pitt [3] agreed with prior authors [1, 2] adding more physiological criteria to his species concept. Monographic treatments of *Penicillium* based on morphology and physiology however [1–3] resulted in very different appreciations of the number of species in the subgenus based on each author's species concepts. The phylogenetic species concept [4] has resolved many species concept issues encountered using morphological taxonomy. Polymerase chain reaction (PCR) was first described by Mullis and associates [5] and with the purification of thermostable DNA polymerases PCR has become a low-cost, powerful, and standard laboratory technique. This discovery enabled the routine addition of genetic characters to the phenotypic

Antonio Moretti and Antonia Susca (eds.), *Mycotoxigenic Fungi: Methods and Protocols*, Methods in Molecular Biology, vol. 1542, DOI 10.1007/978-1-4939-6707-0_9, © Springer Science+Business Media LLC 2017

characters previously used to identify toxigenic *Penicillium* species.

Single-locus phylogenetic studies of *Penicillium* [6, 7] provided an assessment of the genetic variability among isolates of morphologically identified species and sometimes guided taxonomic decisions in toxigenic species [8]. Multilocus studies with phylogenetic and concordance analysis provide a firm basis for defining the limits of variation within species [9] and can be used to resolve some of the disputes over the synonymy of species in subgenus *Penicillium*. Much of the sorting out of subgenus *Penicillium* was resolved by Samson and Frisvad [8].

Seifert and associates [10] tested the ability to distinguish the species in *Penicillium* subgenus *Penicillium* using sequences from the mitochondrial cytochrome oxidase gene (CO1). However, that gene fragment distinguished only 38 of 58 species in the subgenus. Schoch and associates [11] proposed the ITS region as a universal barcode for fungi, but in some *Aspergillus* and *Penicillium* species ITS does not have sufficient discriminatory power [12, 13]. Samson and Frisvad [8] found that they were able to distinguish the species using a β-tubulin gene fragment, BT2. Databases such as GenBank (http://www.ncbi.nlm.nih.gov/) provide quick and easy comparison of DNA sequences. GenBank often contains sequences from isolates that are not well characterized, or sequences are listed under species names that are invalid or synonyms of others, which can be confusing. Currently more narrow databases containing well-characterized and ex-type cultures are becoming available for identifications (e.g., **ARS Trichocomaceae database** http://199.133.98.43/; **MycoBank identification database**, http://www.mycobank.org/). These databases are not as inclusive as GenBank but are highly reliable for the taxon range they cover. To guard against sequencing failures or unexpected sequence variation in any single locus, sequencing of three loci is recommended.

2   Materials

Prepare all solutions using ultrapure water (18 MΩ cm at 25 °C) and analytical grade chemicals. Contamination of cultures, DNA samples, and reagents must be guarded against at all times [14].

2.1   Pure Cultures

1. Malt extract agar (ME3): 30 g Malt extract, 20 g agar, 1 L water. Mix components and heat sterilize for 20 min at 121 °C.

2. Antibiotic solution: 1 g Penicillin-G, 1 g streptomycin, 100 mL water. Filter sterilize and store at 4 °C.

3. Melt ME3 agar and cool to 50 °C in a water bath. Add 2.5 mL of antibiotic solution per liter molten ME3 agar prior to pouring plates.

2.2 DNA Isolation
1. CTAB buffer: 2 g Hexadecyltrimethyl ammonium bromide (CTAB), 8.18 g NaCl, 10 mL of 1 M Tris–HCl pH 8.4, 5 mL of 0.5 M EDTA pH 8.0, water to make 100 mL. Heat to 55 °C to promote dissolution of the CTAB.

2.3 Amplification
1. Hot start *Taq* polymerase with amplifying buffer.
2. Amplification tubes or 96-well amplification plates depending on the number of samples.
3. Primers (*see* Table 1).
4. Pipets capable of delivering 1–10, 10–100, and 100–1000 µL volume.
5. Pipet tips, sterile and with anti-aerosol filters.

2.4 Electrophoresis
1. Tris-acetate-EDTA (TAE −50×) buffer: 242 g Tris base, 57.1 mL glacial acetic acid, 100 mL 0.5 M EDTA, water to 1 L.
2. Agarose: 1 g Low EEO agarose, 100 mL 1× TAE, boil to dissolve agarose.
3. DNA stain: Low-toxicity alternatives to ethidium bromide are available and recommended.
4. Gel electrophoresis mechanism and power supply.
5. UV transilluminator.

2.5 Sequencing
1. Reagent kit for sequencing system being used.

Table 1
Primer pairs most commonly used for DNA-based identification of *Penicillium* spp.

Locus	Anticipated length (nt)	Primer forward direction	Primer reverse direction
β-Tubulin (BT2) [15]	400–500	BT2a—5′-GTT AAC CAA ATC GGT GCT GCT TTC-3′	BT2b—5′-ACCCTCAGTGTAGTGACCCTTGGC-3′
Calmodulin [16]	450–550	Cmd5—5′-CCG AGT ACA AGG AGG CCT TC-3′	Cmd6—5′-CCG ATA GAG GTC ATA ACG TGG-3′
Internal transcribed spacer (ITS) [17]	500–800	ITS5—5′-GCA ATGT AAA AGT CGT AAC AAG-3′	ITS4—5′-TCC TCC GCT TAT TGA TAT GC-3′

Annealing temperature for all amplifications is 51 °C. If double bands occur in the amplification, the annealing temperature can be increased by 1–2 °C; if no band appears, the annealing temperature can be reduced by 4–5 °C to account for partial primer mismatch

3   Methods

3.1   Culture Isolation and Purification

1. Place a loop of spores from a culture plate (or natural substrate) in a 1.5 mL sterile microfuge tube with 1 mL of sterile water, and vortex (*see* **Notes 1** and **2**).

2. Serially dilute 0.5 mL of suspension with 4.5 mL of sterile water to obtain 10^{-4}, 10^{-5}, and 10^{-6} dilutions of the original spore suspension.

3. Place 1 mL of 10^{-4} spore suspension in a 9 cm petri plate, add 15 mL of molten (50 °C) ME3, and swirl the contents to mix. Use ME3 with antibiotics for natural substrates. Make three replicate plates. Repeat with 10^{-5} and 10^{-6} spore dilutions. Allow the agar to solidify.

4. Incubate the plates at 25 °C and check for growth each day. Transfer individual colonies to separate ME3 plates (*see* **Note 3**).

3.2   DNA Isolation

1. Collect biomass by centrifugation from liquid cultures (e.g., 1 min at $5000 \times g$) or by scraping the surface of agar cultures.

2. Add 100–200 mg biomass to a 1.5 mL microfuge tube, freeze, and freeze-dry (ca. 24 h).

3. Pulverize the dried biomass with a sterile pipet tip.

4. Add 450 µL CTAB buffer, vortex to mix thoroughly, and use a sterile pipet tip to suspend clumps.

5. In a chemical hood add 450 µL chloroform, cap, and shake the tube vigorously to form an emulsion.

6. Centrifuge for 10 min at maximum speed in a microcentrifuge.

7. Remove 350 µL of the upper, aqueous phase to a clean microcentrifuge tube.

8. Add 350 µL isopropanol, and invert the tube several times to mix and precipitate nucleic acids.

9. Centrifuge at maximum speed for 2 min, and pour off the supernatant.

10. Rinse the pellet with 100–200 µL 70 % ethanol, centrifuge for 2 min, and gently discard the supernatant.

11. Dissolve the pellet in 100 µL TE buffer (10 mM Tris–HCl, 1 mM EDTA, pH 8.0). Dissolve by heating at 55 °C for 10–60 min with vortex mixing or stirring with a sterile pipet tip.

12. Store DNA at −20 °C.

3.3   PCR Amplification

1. Place 10 µL of hot start *Taq* reagent premix in each amplification tube or amplification plate well (*see* **Note 4**).

2. Add 1 μL of forward and 1 μL of reverse primer to each tube.

3. Add 10–50 ng DNA to each tube.

4. Negative control: Add all reagents to a tube except the DNA.

5. Bring reaction volume to 20 μL using ultrapure water.

6. Mix the reaction by gently pipetting the mixture several times.

7. Place tubes/plate in a thermal cycler with thermal profile of 2 min at 94 °C; 35 cycles of 30 s at 96 °C, 30 s at 51 °C, and 60 s at 72 °C; 5-min hold at 72 °C; and hold indefinitely at 4 °C.

3.4 Gel Electrophoresis

1. Agarose stock: 1 g Agarose, 100 mL 1× TAE buffer. Melt in microwave or autoclave, and cool to ca. 50 °C in a water bath.

2. Add low-toxicity DNA stain to agarose, pour agarose into casting mold, place the well-forming comb, and cool to solidify (*see* **Note 5**).

3. Remove comb, place gel in electrophoresis tray, flood with 1× TAE, and add 2 μL of sample to each well.

4. Attach power cords and electrophorese the sample for 10 min at 200 V.

5. Place gel on a UV transilluminator, view results, and photograph to record results.

3.5 DNA Sequencing

1. Follow the sequencing kit instructions provided for your particular DNA sequencing environment.

3.6 Data Quality Analysis

1. Sequencing center provides chromatogram files from the sequencer along with files of the base sequences in Fasta format. The chromatograms need to be examined using programs to visualize the sequences, to trim off primer sequences and remove poor-quality data that often exist at the beginning and end of reads.

2. Compare the forward and reverse reads, correcting any disagreement between the strands.

3. The corrected Fasta-format DNA sequence is submitted to a database for species identification.

3.7 Databases

3.7.1 GenBank

1. Navigate to http://blast.ncbi.nlm.nih.gov/Blast.cgi in your Internet browser.

2. Select nucleotide blast.

3. Paste the DNA sequence (Fasta format) into the box.

4. Select Others (nr etc.) from the databases.

5. Accept the default search (Megablast).

6. Check the "show results in new window" box and press Blast.

7. Blast reports sequences with high similarity, how well the query and reference agree in length, and the percent identity.

8. 99–100 % similarity with full-length coverage of the query normally indicates identification, but several different names may be returned (*see* **Note 6**).

3.7.2 MycoBank

1. Navigate to http://www.mycobank.org/BioloMICSSequences. aspx?expandparm=f&file=allin in an Internet browser.

2. Paste a sequence (Fasta format) into the alignment box.

3. Press the start alignment button.

4. A list of percent similarity is returned. The 99–100 % similarity list may contain multiple species (*see* **Note 7**).

3.7.3 ARS Trichocomaceae MLSA

1. Navigate to http://199.133.98.43 (Fig. 1) in an Internet browser.

2. Choose Trichocomaceae tab.

3. Choose sequence database link (Fig. 2).

4. Choose sequence query.

5. Place your sequences into the query box and press submit. Single or multiple sequences can be entered.

6. The database either returns the perfect match genotype and species name or, if no perfect match is found, the most similar genotype and the name associated with that genotype (Fig. 3) (*see* **Note 8**).

---

4   Notes

1. DNA isolated from natural substrates will give uninterpretable results if significant amounts of contaminating species are present. Pure cultures are crucial to correct identifications using Sanger sequencing. NGS techniques work with mixed cultures, but current NGS techniques use only the ITS sequence [18, 19].

2. Not all fungi grow on ME3, but *Penicillium* subg. *Penicillium* species grow well on this medium. Low-water-activity-loving spoilage fungi (e.g., *Aspergillus* section *Restricti* species) require high-sugar or high-salt media to thrive. Media with 20–60 % sucrose may be necessary to obtain good growth.

3. Transferring colonies early from isolation plates is important with subg. *Penicillium* isolates because they usually sporulate prolifically and early in growth. Cross-contamination of colonies is likely shortly after sporulation is apparent.

4. Hot start *Taq* polymerases are important to prevent nonspecific primer binding and amplification. Aliquoting primer solu-

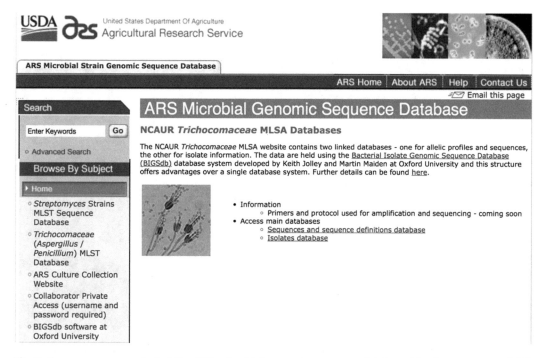

Fig. 1 Home page screenshot of the ARS microbial genomic sequence database. The "sequences ..." link leads to the query input page

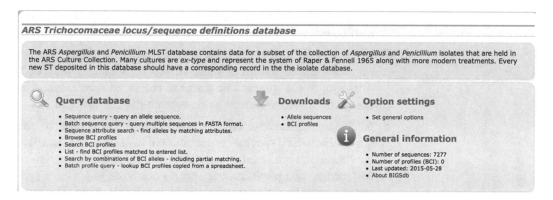

Fig. 2 Links to single sequence query, batch query, and downloads of the DNA sequences in the database

tions helps prevent cross-contamination of the solutions. Pipet tips with aerosol barrier plugs are very useful for prevention of contamination of pipets and solutions.

5. DNA stains can be incorporated into the agarose prior to autoclaving or after autoclaving, or as a post-run stain. It is most convenient to add the stain post-autoclaving.

6. The oldest and most inclusive DNA sequence database is the National Center for Biotechnology Information (GenBank). You normally get the top 100 hits against your query sequence.

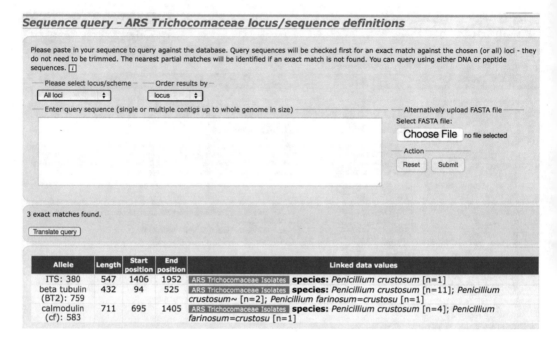

Fig. 3 DNA sequences of β-tubulin, calmodulin, and ITS region from a fungal isolate. The program identifies each allele and reports that the isolate is *Penicillium crustosum*

The list often includes several 100% matches under different names. Unless the type isolate is identified and present in the results, the taxonomic identification is in doubt.

7. MycoBank includes very good coverage in the Trichocomaceae, but sometimes produces a list of several species with 100% similarity to the query. The status of the listed species must be examined in the literature to determine the current name.

8. ARS Trichocomaceae MLSA has imperfect coverage of the family. However, species assignments of isolates in the database are based on type isolates, genealogical concordance analysis of 3–7 genes, and phylogenetic species concepts [4]. Older synonym names are not included in the results returned. Using all three databases will give the best coverage of available DNA reference sequences.

Acknowledgments

Amy E. McGovern very carefully produced high-quality sequences to support this project. The mention of firm names or trade products does not imply that they are endorsed or recommended by the US Department of Agriculture over the firms or similar products not mentioned.

References

1. Thom C (1930) The Penicillia. Williams and Wilkins, Baltimore, p 644

2. Raper KB, Thom C (1949) A manual of the Penicillia. Williams and Wilkins, Baltimore, p 875

3. Pitt JI (1980) The genus *Penicillium* and its teleomorphic states *Eupenicillium* and *Talaromyces*. Academic, New York, p 634

4. Taylor JW, Jacobson DJ, Kroken S et al (2000) Phylogenetic species recognition and species concepts in fungi. Fungal Genet Biol 31:21–32

5. Mullis K, Faloona F, Scharf S et al (1986) Specific enzymatic amplification of DNA in vitro: the polymerase chain reaction. Cold Spring Harb Symp Quant Biol 51:263–273

6. Peterson SW (2000) Phylogenetic analysis of *Penicillium* species based on ITS and LSU-rDNA nucleotide sequences (pp 163–178). In: Samson RA, Pitt JI (eds) Integration of modern taxonomic methods for *Penicillium* and *Aspergillus* classification. Harwood Academic Publishers, Amsterdam, p 510

7. Seifert KA, Louis-Seize G (2000) Phylogeny and species concepts in the *Penicillium aurantiogriseum* complex as inferred from partial B-tubulin gene DNA sequences (pp 189–198). In: Samson RA, Pitt JI (eds) Integration of modern taxonomic methods for *Penicillium* and *Aspergillus* classification. Harwood Academic Publishers, Amsterdam, p 510

8. Samson RA, Frisvad JC (2004) *Penicillium* subgenus *Penicillium*: new taxonomic schemes, mycotoxins and other extrolites. Stud Mycol 49:1–260

9. Peterson SW, Jurjevic Z, Frisvad JC (2015) Expanding the species and chemical diversity of *Penicillium* section *Cinnamopurpurea*. PLoS ONE 10(4), e0121987

10. Seifert KA, Samson RA, Dewaard JR et al (2007) Prospects for fungus identification using CO1 DNA barcodes, with *Penicillium* as a test case. Proc Natl Acad Sci U S A 104:3901–3906

11. Schoch CL, Seifert KA, Huhndorf S et al (2012) Nuclear ribosomal internal transcribed spacer (ITS) region as a universal DNA barcode marker for Fungi. Proc Natl Acad Sci U S A 109:6241–6246

12. Peterson SW (2012) *Aspergillus* and *Penicillium* identification using DNA sequences: barcode or MLST? Appl Microbiol Biotechnol 95:339–344

13. Osmundson TW, Robert VA, Schoch CL et al (2013) Filling gaps in biodiversity knowledge for macrofungi: contributions and assessment of an herbarium collection DNA barcode sequencing project. PLoS ONE 8(4), e62419

14. Seifert KA (1990) Isolation of filamentous fungi (pp 21–52). In: Labeda DP (ed) Isolation of biotechnological organisms from nature. McGraw-Hill, New York, p 322

15. Glass NL, Donaldson GC (1995) Development of primer sets designed for use with the PCR to amplify conserved genes from filamentous ascomycetes. Appl Environ Microbiol 61:1323–1330

16. Hong SB, Cho HS, Shin HD et al (2006) Novel *Neosartorya* species isolated from soil in Korea. Int J Syst Evol Microbiol 56:477–486

17. White TJ, Bruns T, Lee S, Taylor J (1990) Amplification and direct sequencing of fungal ribosomal RNA genes for phylogenetics (pp 315–322). In: Innes MJ, Gelfand DH, Sninsky JJ, White TJ (eds) PCR protocols. Academic, New York, p 482

18. Kuczynski J, Stombaugh J, Walters WA, et al (2011) Using QIIME to analyze 16S rRNA gene sequences from microbial communities. Curr Protoc Bioinformatics, UNIT 10.7. Wiley Online Library

19. Zimmerman NB, Vitousek PM (2012) Fungal endophyte communities reflect environmental structuring across a Hawaiian landscape. Proc Natl Acad Sci U S A 109:13022–13027

Chapter 10

Targeting Aflatoxin Biosynthetic Genes

Ali Y. Srour, Ahmad M. Fakhoury, and Robert L. Brown

Abstract

Chemical detoxification and physical destruction of aflatoxins in foods and feed commodities are mostly unattainable in a way that preserves the edibility of the food. Therefore, preventing mycotoxins in general and aflatoxins in particular from entering the food chain is a better approach. This requires early detection of the aflatoxin-causing organisms. Detection and quantification of aflatoxin-producing fungi has always been a challenge, especially within species of *Aspergillus* and *Penicillium*. Culture-based methods require a high level of expertise and a list of sophisticated equipment. Furthermore, even for a trained taxonomist, species that are identical in morphology, physiology, and nutritional aspects can be challenging to classify. Fungal taxonomy has changed over the past few decades; more species are being reclassified, and new species are being described due to advances in sequencing and genome assembly. These developments make the use of PCR-based approaches practical, rapid, and more reliable for the identification of fungi to the species level. This chapter presents a variety of protocols to detect and quantify aflatoxin-producing fungi using mycotoxin biosynthesis pathway genes.

Key words Mycotoxins, Fungi, Aflatoxins, Aspergillus, Penicillium, Biosynthesis, PCR, qPCR

1 Introduction

Mycotoxin contamination is a major concern for food and feed producers worldwide. According to the FAO, one-fourth of harvested crops were contaminated with mycotoxins to levels that render them unsuitable for consumption [1, 2]. Aflatoxins are some of the most harmful mycotoxins [3]. These secondary metabolites are mainly produced by filamentous fungi in the *Aspergillus* genus section Flavi, especially *Aspergillus flavus* and *Aspergillus parasiticus* [4, 5]. Other minor aflatoxin producers include species within the genera *Emericella*, *Rhizopus*, and *Penicillium* [6, 7]. The most common aflatoxins present on a wide range of commodities include aflatoxins B1, B2, G1, and G2. *A. parasiticus* produces all four toxins whereas *A. flavus* can only produce aflatoxins B1 and B2 [8]. Aflatoxin B1 can be metabolized to M1, which is excreted in

Antonio Moretti and Antonia Susca (eds.), *Mycotoxigenic Fungi: Methods and Protocols*, Methods in Molecular Biology, vol. 1542, DOI 10.1007/978-1-4939-6707-0_10, © Springer Science+Business Media LLC 2017

the milk of humans and animals following ingestion of foods contaminated with aflatoxin B1. Sterigmatocystin (ST) is a precursor of aflatoxin B1, and is a mycotoxin [9].

Aflatoxins are considered potent naturally occurring carcinogens and mutagens capable of causing acute and chronic diseases (liver cancer, hemorrhagic necrosis, jaundice, hepatitis, cirrhosis, Reye's syndrome) that threaten human and animal health. Aflatoxins have been reported on a wide range of commodities including nuts (peanuts, pistachio, and almonds), cereals (corn, wheat, sorghum, oat, rye, and barley), soybean, rice, milk, cheese, dried fruits, culinary herbs, and spices [3, 7, 10]. Given the carcinogenic potency of aflatoxins and the extent of human and animal exposure to these compounds, aflatoxins may pose a greater danger to human health than pesticide residues and food additives [11]. In addition, most aflatoxin-producing fungi infect crops prior to harvest and can cause economic losses due to reduced yield [12]. These fungi can impact food staples from preharvest to after harvest including storage, processing, and transport [13].

Aflatoxins are low-molecular-weight difuran coumarin derivatives (Fig. 1) that can be secreted in amounts in the magnitude of micrograms per kilogram (μg/kg). Hence, there is a need for reliable methods to rapidly assess food and feed stuff contamination by aflatoxins through the detection of aflatoxins and aflatoxin-producing fungi. Polymerase chain reaction (PCR)-based methods offer sensitive and rapid molecular diagnostic tools for the early detection of fungi. In fact, the accurate identification of aflatoxin-producing fungi that contaminate food and feedstuffs can be difficult, labor intensive, and time consuming when using traditional culture-based methods [14]. Moreover, detection with conventional methods is often only effective after toxins have accumulated, which may be too late to safeguard the food chain. In this chapter we focus on methods used to detect and quantify aflatoxin-producing fungi using PCR-based methods. These methods permit an early, rapid, and sensitive detection of these mycotoxin-producing species, which is crucial to prevent mycotoxins from entering the food chain [10, 15].

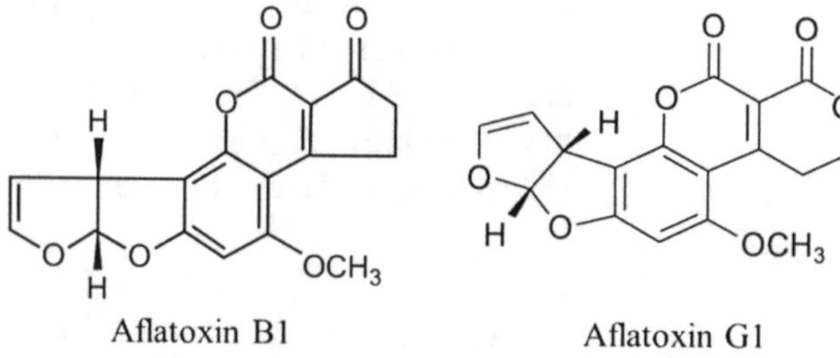

Aflatoxin B1   Aflatoxin G1

Fig. 1 Structure of major aflatoxins

The use of PCR-based assays has been bolstered in the past decades by the sequencing of numerous housekeeping genes and ribosomal DNA as well as the discovery of genes involved in the biosynthetic pathways of mycotoxins. Several PCR assays designed and successfully used to detect single or multiple species of aflatoxin-producing fungi have been described in the literature. The majority of these assays target genes involved in the aflatoxin biosynthesis pathway, whereas few are based on conserved DNA in hypervariable genomic regions. Methods using techniques such as restriction fragment length polymorphism (RFLP), denaturing gradient length polymorphism (DGGE), loop-mediated length polymorphism (LAMP), and single-strand conformational polymorphism (SSCP) are typically good for detection but less useful for the quantification of fungi; they can also be somewhat complicated to use. Hence, the focus in this chapter is on simple PCR, as well as on real-time PCR assays (qPCR) that do not require any downstream applications for the detection and quantification of aflatoxin-producing fungi. These assays are rapid and practical and do not require extensive expertise to be used.

2 Materials

2.1 DNA Extraction

1. Sterile nanopure water.
2. Filter pipette tips: 0.1–10, 10–100, 100–1000 μL (*see* **Note 1**).
3. EZNA Fungal DNA Mini Kit (Omega Bio-Tek).
4. PowerLyzer Microbial DNA kit (MoBio).
5. Filter bags BagPage.
6. 10 mg/mL Proteinase K solution.
7. 10 mg/mL Lyticase solution or zymolyase.
8. 10 mg/mL RNase A solution.
9. 1 mg/10 mL Chelex-100 resin.
10. TBE buffer: 50 mM Tris, 50 mM boric acid, 1 mM EDTA.
11. 0.5 mg/mL Gel Red dye.
12. Gel-loading mixture: 60% Glycerol, 60 mM EDTA, 0.3% bromophenol blue, 0.3% xylene cyanol, 10 mM Tris–HCl, pH 7.6.

2.2 Polymerase Chain Reaction

1. dNTP mix: 10 mM.
2. 10v× PCR buffer.
3. $MgCl_2$: 25 mM.
4. Template DNA.
5. Taq DNA polymerase: 5 u/μL.
6. Oligonucleotide primers (Table 1).

Table 1
PCR assays used for the detection of aflatoxin-producing fungi

Assay nb	Species	Assay	Targeted region	Primer pair/probe (5′ → 3′)	Sensitivity (LOD)[a]/ Amplicon size	Thermal profile	References
1 *See Note* 12	A. flavus (aflatoxin) A. parasiticus (aflatoxin) A. versicolor (sterigmatocystin)	Species-specific detection (conventional PCR)	aflD (nor1/2) aflM (ver1/2) aflP (omt1/2) aflR (aflR1/R2)	nor1, ACCGGTCACGCCGGCACTCTCGGCAC, nor2, TTGGCCGCCAGCTTCGACACTCCG ver1, GCCGGCAGGCGGCGGAGAAAGTGGT, ver2, GGGGATATACTCCCGGACACAGCC omt1, GTGGACGGACCTAGTCCGACATCAC, omt2, GTCGGCGGCCACGCACTGGGTTGGGG aflR1, TATCTCCCCCGGGCATCTCCCGG, aflR2, CCGTCAGACAGCACTGGACACGG	400 bp 600 bp 797 bp 1000 bp	Quadruplex PCR 35 cycles of [30 s at 95 °C; 30 s at 65 °C; 30 s 72 °C]	[18–20]
2 *See Note* 13	A. flavus	Species-specific detection (conventional PCR)	aflS aflR aflD aflO aflQ	AflJ-gF, GAACGCTGATTGCCAATGCC, AflJ-giR, CGGTCAGGATGTTACTAAGC, Aflj-cR, GACTGGGCGCCACCGTTGC AflR-R, CCGTCAGACAGCACTGGACACGG, AflR-F1, TGACCCACCTCTTCCCCACG Nor-F, ACGGATCACTTAGCCAGCAC, Nor-R, CTACCAGGGGAGTTGAGATCC OmtB-F, GCCTTGACATGGAAACCATC, OmtB-R, CCAAGATGGCCTGCTCTTTA Ord-gF, TTAAGGCAGCGGAATACAAG, Ord-gR, GAGCGCCAAAGCCGAACACAAA, Ord-cR, GAATATCTGACGTTTACCC3′	1256 bp 598 bp 300 bp 812 bp 1131 bp 599 bp 487 bp	QuantuplexPCR 5 cycles of [1 m at 94 °C; 1 m at 60 °C; 1 m 72 °C] followed by 30 cycles of [1 m at 94 °C; 1 m at 55 °C; 1 m 72 °C]	[22]
3 *See Note* 14	A. flavus and A. parasiticus	Species-specific detection (conventional PCR)	aflP (omt 1), aflM (ver 1), aflR	OMT-208, GGCCGCCGGTTCCTTGGCTCCTAAGC, OMT-1232, GCCCCAGTGAGACCCTTCCTCG VER-496, ATGTCGGATAATCACCGTTTA GATGGC, VER-1391, CGAAAAGCGCCACCATCCACC CCAATG AFLR-F, AATACAT GGTCTCCAAGCGG, AFLR-R, GAAGACA GGGTGCTTTGCTC	1254 bp 895 bp 352 bp LOD 10–500 pg	Single and multiplex PCR 37 cycles of [1 m at 94 °C; 2 m at 66 °C; 2 mat 72 °C]	[10, 23]
4 *See Note* 15	Aflatoxin-producing, Aspergillus spp., Penicillium spp. and Rhizopus spp.	Species-specific detection (conventional PCR)	O-methyltransferase gene (omt-1)	AFF1, CTTCGAGGATGTGCCCAGCGC AFR3, CGAACCTCGTCCACAGTGC	LOD 15 pg 381 bp	Single PCR 30 cycles of [1 m at 94 °C; 2 min at 58 °C; 1.5 m at 72 °C]	[21]

5 See Note 16	A. flavus and A. parasiticus	Species-specific detection and quantification RT-PCR (Taqman)	nor-1 gene	nortaq-1, GTCCAAGCAACAGGCCAAGT; nortaq-2, TCGTGCATGTTGGTGATGGT; norprobe, TGTCTTGATCGGCGCCCG	LOD 1 conidium 66 bp	qPCR 40 cycles of [30 s at 95 °C; 30 s at 53 °C; 20 s at 72 °C]	[3, 4]
6 See Note 17	Aflatoxigenic, Aspergillus, Penicillium, and Rhizopus	Species-specific detection and quantification RT-PCR (Taqman)	omt-1 gene	F-omt, GGCCGCCGCTTTGATCTAGG; R-omt, ACCACGACCGCCGCC; OMTprobe, CCACTGGTAGAGGAGATGT	LOD=1–2 log cfu/g 123 bp	qPCR 40 cycles of [15 s at 95 °C; 1 m at 60 °C]	[7]

[a]LOD limit of detection

7. TBE buffer: 50 mM Tris, 50 mM boric acid, 1 mM EDTA.

8. Thermal cycler, real-time PCR.

9. Gel-loading dye: 60% Glycerol, 60 mM EDTA, 0.3% bromophenol blue, 0.3% xylene cyanol, 10 mM Tris–HCl (pH 7.6).

10. 100 bp Molecular marker.

2.3 Real-Time PCR (qPCR) Quantification

1. Gene-specific oligonucleotides and/or Taqman probes (*see* **Notes 2** and **3**).

2. 2× TaqMan Universal PCR Master Mix.

3. Optical 96-well PCR plate.

4. PCR clear adhesive seal.

5. SYBR Green reagent.

6. Molecular biology-grade water.

7. Centrifuge.

8. RT-qPCR thermocycler.

9. Quantification program.

3 Methods

3.1 Nucleic Acid Extraction from Food Matrices Contaminated with Aflatoxins

Prior to performing PCR, it is crucial to have adequate and good-quality DNA to detect and/or quantify mycotoxin-producing fungi. A wide variety of methods to extract DNA from different food matrices have been reported in the literature. However, only a few have proven to be efficient and reproducible for the detection of mycotoxin-producing fungi. The components of fungal cell walls such as chitin, glucans, and other polymers make them resistant to lysis resulting in low DNA yield and inefficient DNA purification. Quality is another concern when extracting DNA from complex food matrices. In some cases, lysis results in the release of matrix-associated inhibitors that interfere with cell lysis, degrade DNA, and reduce PCR sensitivity and efficiency. There is no single nucleic acid purification method adequate for all food matrices and all mycotoxin-producing genera. We found the method developed by Rodriguez et al. [16] to be adequate for the extraction of fungal DNA in terms of quality and recovery of nucleic acid. This method can be used to prepare DNA from a wide variety of food matrices such as ripened foods (ripened cheese), nuts, and grapes. It involves a combination of thermal disruption of conidia, enzymatic treatment, incubation with Chelex-100 resin, and final extraction with the EZNA fungal DNA Mini kit.

The steps are as follows:

1. Take 5 g of contaminated foods.

2. Homogenize with 10 mL of Tris–HCl buffer (pH 8.0) in a filter bag BagPage using a homogenizer for 5 min (*see* **Note 4**).

3. Transfer the filtrate to a sterile tube.

4. Incubate at 95 °C for 10 min (*see* **Note 5**).

5. Add 50 μL of proteinase K solution.

6. Add 40 μL of lyticase solution (*see* **Note 5**).

7. Incubate at 65 °C for 15 min (*see* **Note 6**).

8. Add 20 μL of RNase A solution.

9. Incubate at 37 °C for 15 min.

10. Add Chelex-100 resin (*see* **Note 7**).

11. Incubate at 40 °C for 5 min with shaking (~$4 \times g$).

12. Centrifuge for 5 min an aliquot (1 mL) of each sample at $15,000 \times g$.

13. Transfer as much as possible of the top aqueous layer to a clean tube.

14. Proceed to use the EZNA Fungal DNA Mini kit starting by applying the sample to a HiBind DNA column (**step 10**, protocol B).

15. Elute DNA from the EZNA column with 50–100 μL of elution buffer.

16. Check quality and quantity of DNA (*see* **Notes 8, 9**, and **10**).

17. Store DNA at −20 °C.

3.2 PCR-Based Assays

For general PCR and/or qPCR, the DNA to be tested needs to be quantified and its quality assessed using a nanophotometer. Gel electrophoresis can also be used to check the quality of the DNA. For qPCR experiments DNA samples should conform to MIQE guidelines [16] (*see* **Note 11**). The general thermal cycling profile varies with the type of assay, primers used, and their annealing temperatures (*see* **Notes 12–17**).

3.2.1 Conventional PCR Setup Protocol

Prepare a master mix (for 50 μL reaction) combining the following (*see* **Notes 18** and **19**):

5 μL 10× PCR buffer, 6 μL of MgCl$_2$, 4 μL dNTP mix, 0.5 μL of Taq DNA polymerase, 10 μL of 10 mM primer pair mix, and 5 μL containing 10–20 ng template DNA. Finally add up to 50 μL sterile nanopure water.

3.2.2 Quantitative (qPCR) or Real-Time (RT) PCR Assays

This method can be used to prepare standard DNA curves using known amounts of fungal DNA in qPCR. Fungal DNA from both aflatoxin- and non-aflatoxin-producing strains of fungi need first to be extracted from pure cultures. We use a modified protocol with the PowerLyzerUltraClean Microbial DNA Isolation Kit (*see* **Note 20**).

1. Scrape mycelia off the agar and collect 250 mg of tissue for the extraction of genomic DNA.

2. Resuspend mycelia in 300 μL of Micro Bead solution and transfer to a 0.1 mm glass MicroBead tube.

3. Add 50 μL of lysis solution and heat at 70 °C for 10 min.

4. Secure tubes horizontally on a vortex adapter. Disrupt cells by vortexing at maximum speed for 10 min (*see* **Note 21**).

5. Follow the Power Lyzer DNA kit protocol and elute DNA with 50 μL of TE buffer.

6. Adjust extracted DNA to the same concentration for qPCR assays.

7. For the standard DNA extracted from the reference aflatoxin-producing fungi, make tenfold dilution series from 1/10 to 1/107 (*see* **Note 22**).

For the detection and quantification of aflatoxin producers belonging to different species and genera of fungi, the assay developed by Rodriguez et al. [7] is recommended. For species-specific detection and quantification of *A. flavus* and/or *A. parasiticus*, assays developed by Passone et al. [3] are a good choice (Table 1). Prepare the qPCR assay for the samples to be analyzed in triplicates of 25 μL reactions. The no-template control (NTC) (just water), negative control (aflatoxin non-producer), and standard curve (aflatoxin-producing fungus) reactions are also prepared in triplicate. The following qPCR protocol can be used with Taqman probes:

1. Prepare a TaqMan master mix (20 μL reaction) which contains 1× TaqMan Universal PCR Master Mix, primer, and probe mix (*see* **Note 23**). Vortex and spin down the tube.

2. Add 20 μL aliquot into each well of the 96-well reaction plate.

3. Add 5 μL of DNA samples, standard DNA (tenfold dilution series), NTC, and negative control to the appropriate wells (*see* **Note 24**).

4. Seal the plate and centrifuge briefly.

5. Load the plate in the real-time PCR machine.

6. Edit the plate by entering the names of each sample and standard curve including the concentrations of each standard.

7. Select the right wavelength for the fluorophore of the probe.

8. Run the thermal cycling which includes an initial denaturation step for 3 min at 95 °C, followed by 40 cycles of 95 °C for 15 s and 60 °C for 1 min (*see* **Notes 15** and **17**).

9. Use default parameters for Ct determinations (*see* **Notes 25** and **26**).

10. Export the quantification data in Excel or csv format.

3.2.3 Target Genes for Specific Detection and Quantification of Aflatoxigenic Strains

Mycotoxin biosynthesis genes offer a great advantage for the detection of mycotoxin-producing fungi since many of these genes are conserved and present within gene clusters that appear to have undergone horizontal transfer between several species producing the same toxin. Thus, if the major concern is to identify the producers of a specific mycotoxin, regardless of species identity, these genes would be of great use to design primers and probes specific to the particular mycotoxin. Many PCR and q-PCR assays were developed to target aflatoxin-producing fungi and were proven to be specific and efficient (Table 1). Most assays use a multiplex PCR in which 3–4 biosynthetic genes are amplified in order to differentiate between aflatoxin- and sterigmatocystin-producing fungi [17]. Geisen et al. [18] developed an assay (assay 1 in Table 1) targeting the norsolorinic acid reductase (*nor-1*)-encoding gene, the versicolor in *A dehydrogenase* (*ver-1*) encoding gene, the sterigmatocystin *O*-methyltransferase (*omt-1*) encoding gene, and the aflatoxin regulatory (*aflR*) gene. The *omt-1* gene converts sterigmatocystin to *O*-methylsterigmatocystin, and its amplification differentiates aflatoxin producers from aflatoxin non-producing species. Similar assays (assays 1, 2, 3) were developed by Shapira et al. [10], Criseo et al. [19], and Chen et al. [20]. These assays were successful in detecting *A. flavus* and *A. parasiticus* from other fungi. Recently Luque et al. [21] developed a uniplex assay (Assay 4) targeting the *omt-1* gene for the detection of aflatoxin-producing *Aspergillus* spp., *Penicillium* spp., and *Rhizopus* spp. The first real-time PCR assay (assay 5) to detect aflatoxin-producing fungi was developed by Mayer et al. [4]. However, this protocol has not been tested against non-aflatoxin producers. A more reliable qPCR assay (assay 6) has recently been reported by Rodriguez et al. [7] based on the *omt-1* gene, which is an adequate target for the differentiation between sterigmatocystin and aflatoxin producers.

4 Notes

1. Filter tips are recommended to be used to avoid cross contamination.

2. Primers and probes should preferably be dissolved in 10 mM Tris–HCl, pH 7.5 or 0.1× TE.

3. Fluorescent probes are light sensitive and tend to degrade with repeated freeze-thawing cycles. It is a good practice to make aliquots of your probe in smaller volume (50–100 μL).

4. Homogenizing in Tris buffer helps release the fungal cells from food matrices and results in higher fungal DNA recovery, whereas in methods grinding the total food matrix (in liquid nitrogen, or dried specimens) the fungal DNA will be underrepresented by DNA coming from food matrix.

5. Incubation at 95 °C helps weaken cell walls and disrupts conidia. Enzymatic treatment with lyticase is crucial to degrade fungal cell walls by cleaving the glycosidic bonds of the water-insoluble cell wall polymer. In addition, proteinase K which is a serine protease is used to digest contaminating proteins mainly nucleases which attack nucleic acids.

6. Incubation at 65 °C is essential in part to help dissolve the DNA; on the other hand it is required to inactivate lyticase and proteinase K.

7. Chelex resin is made from polystyrene-divinylbenzenei minodiacetate material and is used to chelate heavy metals which inhibit polymerase activity in PCR.

8. DNA needs to be quality and quantity assessed using a nano-drop or a nanophotometer. The ratio of absorbance at 260 and 280 nm is used to assess the purity of DNA. A ratio of ~1.8 is generally regarded as pure for DNA. A lower ratio may indicate the presence of protein, phenol, or other contaminants.

9. Sheared DNA will give higher readings than intact DNA. It is good to check DNA integrity on a gel to make sure that yields obtained on nanophotometer (or a spec) are accurate. DNA can be checked using a standard agarose gel (1 % agarose in 1× TAE gel) stained with GelRed dye and run at 70 V.

10. Low-yield DNA may be due to extensive shearing or degradation. Therefore, avoid using contaminated solutions and aliquot nanopure water in 1.5 mL tubes instead of using large containers.

11. It is recommended to test for PCR inhibition after DNA extraction since inhibitors affect amplification efficiency and may result in false negatives or inaccurate measurements of analyzed samples. To test for inhibition, samples can be spiked with standard fungal DNA (positive control). PCR efficiency of samples with and without added fungal DNA can be compared in order to deduce the presence of inhibitors. In case of significant reduction in efficiency there may be no need to re-extract or repurify DNA. An alternative would be to dilute samples to reduce the effect of inhibitors.

12. Assay 1 is a multiplex conventional PCR using a set of four primer pairs targeting aflR, aflD, aflM, and Aflp genes of the biosynthetic pathway in order to differentiate between aflatoxin-producing and non-producing *Aspergillus flavus*. Afla+ strains of *A. flavus* give a quadruplet banding pattern marking the presence of all four genes of the aflatoxin biosynthesis pathway.

13. Assay 2 is a multiplex PCR that was applied to *A. flavus* isolated from corn grains. The assay targets five genes of the aflatoxin pathway; when used on both genomic DNA and cDNA

it can provide a correlation between genes and their expression. PCR conditions are the same for genomic DNA; ratio of aflR and aflS is recommended at 1:10 with respect to aflD, aflO, and AflQ primer sets. Ord-cR and AflJ-cR primers selectively amplify cDNA only. This is the only assay that might require RNA extraction using TRIZOL (sigma) and cDNA synthesis.

14. Assay 3 was designed to detect Afla+ strains of *A. flavus* and *A. parasiticus* on maize kernels; the assay targets three genes in the aflatoxin biosynthesis cluster (omt1, ver1, and AflR genes). A complete banding pattern is a marker of Afla+ strains. For higher sensitivity single PCR can be used to increase the detection threshold.

15. Assay 4 is a simple, sensitive, and reliable PCR assay for detecting Afla+ of various species contaminating various food matrices. The assay targets O-methyltransferase gene (*omt-1*) of the aflatoxin biosynthetic pathway.

16. Assay 5 is a high-throughput real-time PCR (qPCR) assay targeting the nor-1 gene of the aflatoxin pathway. It is a sensitive and rapid assay which was initially developed for stored peanuts but can be used for other food matrices. Recommended tPrimer/probe ratio is 2:1 for optimal amplification.

17. Assay 6 is a real-time PCR (qPCR) protocol designed to detect and quantify Afla+ contaminating a variety of food matrices. The assay is highly sensitive and specific enough to discriminate Afla+ from Afla− strains. The assay can be run with Sybr green as well by using just the primers.

18. Thaw PCR reagents and DNA prep and keep on ice all the time.

19. When setting up a PCR reaction, always start pipetting with the negative control followed by your samples to analyze and finish with the positive control to avoid amplicon carryover.

20. PowerLyzerUltraClean Microbial DNA Isolation Kit from MoBio is suitable for the extraction and purification of fungal DNA and can yield high-quality DNA by removing PCR inhibitors. While other methods and kits are widely used and available (such as CTAB), this kit provides a rapid, clean, and high-yield DNA

21. It is important not to exceed 250 mg of tissue used as it will not result in more DNA yield. This method combines physical, thermal, and chemical disruption by using glass beads (with vortexing), heating at 70 °C, and using a lysis buffer. Vortex uses less force in comparison to high-powered bead-beating instruments; therefore it yields high DNA quality with less shearing.

22. Tenfold serial dilution of the reference aflatoxigenic producer can be prepared from 1 ng/μL stock to produce the following concentrations: 1 ng/μL, 100 pg/μL, 10 pg/μL, 1 pg/μL, 100 fg/μL, 10 fg/μL, and 1 fg/μL.

23. Optimum primer/probe ratio in qPCR assays varies with type of fluorophore and quencher used; therefore it is recommended to test a matrix of combinations to find the best working ratio.

24. Make sure to run all reactions in triplicates including the standard curve, unknowns, and controls.

25. The threshold value can be set manually within the exponential phase of amplification to maximize the efficiency and R^2. A perfect assay will have a slope of -3.3, meaning that every tenfold dilution is 3.3 cycles higher. This is 100% doubling in each cycle.

26. PCR efficiency is calculated based on the standard curve; efficiencies in the range of 85–115% are considered acceptable, but 90–110% is ideal. In case PCR efficiency falls below these values consider re-optimizing primer/probe ratios, changing reaction conditions, and running a secondary cleaning of your DNA.

References

1. Konietzny U, Greiner R (2003) The application of PCR in the detection of mycotoxigenic fungi in foods. Braz J Microbiol 34:283–300
2. Lawlor PG, Lynch PB (2005) Mycotoxin management. Afr Farm Food Process 46:12–13
3. Passone MA, Rosso LC, Ciancio A, Etcheverry M (2010) Detection and quantification of Aspergillus section Flavi spp. in stored peanuts by real-time PCR of nor-1 gene, and effects of storage conditions on aflatoxin production. Int J Food Microbiol 138:276–281
4. Mayer Z, Bagnara A, Färber P, Geisen R (2003) Quantification of the copy number of nor-1, a gene of the aflatoxin biosynthetic pathway by real-time PCR, and its correlation to the cfu of Aspergillus flavus in foods. Int J Food Microbiol 82:143–151
5. Horn BW (2007) Biodiversity of Aspergillus section Flavi in the United States: a review. Food Addit Contam 24:1088–1101
6. Varga E, Sulyok M, Schuhmacher R, Krska R (2009) Validation and application of an LC–MS/MS based method for multi-mycotoxin analysis in different nuts and dried fruits'. In Poster 236 at the First International Society for Mycotoxicology Conference, Tulln, Austria, 9–11 Sep 2009. Book of abstracts, available from: http://www.ism2009.at/ISM2009_posters.pdf. Accessed 29 May 2010, p 261
7. Rodriguez A, Rodriguez M, Luque M et al (2012) Real-time PCR assays for detection and quantification of aflatoxin-producing molds in foods. Food Microbiol 31:89–99
8. Xu HX, Annis S, Linz J, Trail F (2000) Infection and colonization of peanut pods by Aspergillus parasiticus and the expression of the aflatoxin biosynthetic gene, nor-1, in infection hyphae. Physiol Mol Plant Pathol 56:185–196
9. O'Brian GR, Fakhoury AM, Payne GA (2003) Identification of genes differentially expressed during aflatoxin biosynthesis in Aspergillus flavus and Aspergillus parasiticus. Fungal Genet Biol 39:118–127
10. Shapira R, Paster N, Eyal O et al (1996) Detection of aflatoxigenic molds in grains by PCR. Appl Environ Microbiol 62:3270–3273
11. Kuiper-Goodman T (1998) Food safety: mycotoxins and phycotoxins in perspective. In: Miraglia M, van Egmond H, Brera C, Gilbert J (eds) Mycotoxins and phycotoxins–developments in chemistry. Toxicology and food safety. Proceedings of the IX IUPAC International Symposium, Fort Collins, CO, AlakenInc, pp 25–48
12. Dagnas S, Membré JM (2012) Predicting and preventing mold spoilage of food products. J Food Prot 76:538–551

13. Yu SR, Liu XJ, Wang YH, Liu J (1995) A survey of moniliformin contamination in rice and corn from Keshan disease endemic and non-KSD areas in China. Biomed Environ Sci 8:330–334

14. Chandra Nayaka S, Udaya Shankar AC, Niranjana SR et al (2010) Detection and quantification of fumonisins from *Fusarium verticillioides* in maize grown in southern India. World J Microbiol Biotechnol 26:71–78

15. Gil-Serna J, Patiño B, González-Jaén MT, Vázquez C (2009) Biocontrol of *Aspergillus ochraceus* by yeast. In: Mendez-Vilas A (ed) Current research topics in applied microbiology and microbial biotechnology. World Scientific, Singapore, pp 368–372

16. Bustin S, Benes V, Garson J et al (2009) The MIQE guidelines: minimum information for publication of quantitative real-time PCR experiments. Clin Chem 55:611

17. Levin RE (2012) PCR detection of aflatoxin producing fungi and its limitations. Int J Food Microbiol 156:1–6

18. Geisen R (1996) Multiplex polymerase chain reaction for the detection of potential aflatoxin and sterigmatocystin producing fungi. Syst Appl Microbiol 19:388–392

19. Criseo G, Bagnara A, Bisignago G (2001) Differentiation of aflatoxin-producing and non-producing strains of *Aspergillus flavus* group. Lett Appl Microbiol 33:291–295

20. Chen RS, Tsay JG, Huang YF, Chiou RY (2002) Polymerase chain reaction-mediated characterization of moulds belonging to the *Aspergillus flavus* group and detection of *Aspergillus parasiticus* in peanut kernels by a multiplex polymerase chain reaction. J Food Prot 65:840–844

21. Luque MI, Rodríguez A, Andrade MJ et al (2012) Development of a PCR protocol to detect aflatoxigenic molds in food products. J Food Prot 75:85–94

22. Degola F, Berni E, Dall'Asta C et al (2007) A multiplex RT-PCR approach to detect aflatoxigenic strains of *Aspergillus flavus*. J Appl Microbiol 103:409–417

23. Del Fiore A, Reverberi M, De Rossi P et al (2010) Polymerase chain reaction-based assay for the detection of aflatoxigenic fungi on maize kernels. Qual Assur Saf Crops Foods 2:22–27

Chapter 11

Targeting Trichothecene Biosynthetic Genes

Songhong Wei, Theo van der Lee, Els Verstappen, Marga van Gent, and Cees Waalwijk

Abstract

Biosynthesis of trichothecenes requires the involvement of at least 15 genes, most of which have been targeted for PCR. Qualitative PCRs are used to assign chemotypes to individual isolates, e.g., the capacity to produce type A and/or type B trichothecenes. Many regions in the core cluster (consisting of 12 genes) including intergenic regions have been used as targets for PCR, but the most robust assays are targeted to the *tri3* and *tri12* genes. Quantitative PCRs, that work across trichothecene-producing members of the Fusarium head blight complex, are described along with procedures to quantify the amount of fungal biomass in wheat samples. These assays are directed to the chemotype(s) present in field samples and quantify the total fungal biomass of trichothecene-producing fungi, irrespective of their genetic identity.

Key words Trichothecenes, PCR, Quantitative PCR

---

1 Introduction

Fusarium species produce two types of trichothecene mycotoxins, namely type A trichothecenes, like T-2 and HT-2, and type B trichothecenes such as deoxynivalenol (DON) and nivalenol (NIV). In the toxin profile of most Fusarium strains also acetylated versions of both DON (3ADON and 15ADON) or a di-acetylated version of NIV (4,15-diANIV) can be found. The biochemical pathways leading to these compounds have been the subject of multiple studies over several decades (e.g., [1–3]). Most enzymatic steps have been elucidated in the pathways leading to the final products, T2-toxin and HT2 toxin (type A), and deoxynivalenol and nivalenol (type B) and the corresponding genes have been identified. The genes concerned with trichothecene production are largely clustered in the genome of producing organisms. These genes and corresponding transcription factors are located at three different regions in the genome [1, 4, 5]: a *tri101* locus is situated on chromosome IV in the reference genome of *F. graminearum* strain PH-1; a two-gene locus consisting of *tri1* and *tri16* located

Antonio Moretti and Antonia Susca (eds.), *Mycotoxigenic Fungi: Methods and Protocols*, Methods in Molecular Biology, vol. 1542, DOI 10.1007/978-1-4939-6707-0_11, © Springer Science+Business Media LLC 2017

on chromosome I and a 12-gene cluster, also known as the core trichothecene cluster, that encodes most of the enzymatic steps leading to the final mycotoxins (*see* Table 1 and Fig. 1) is found on chromosome II.

Individual isolates are typically capable of producing only one of these trichothecenes, either type A or type B, and only one of the acetylated derivatives, e.g., 3ADON or 15ADON [6]. Because of the different toxicological properties of each of these mycotoxins, there is demand for assays that can identify which toxins are produced. Hence, many of these core cluster genes have been targeted for PCR-based identification of the chemotype of the fungal isolate.

Table 1
Genes of the core trichothecene cluster and their function

Cluster gene	Predicted function
TRI8	Trichothecene-3-*O*-esterase
TRI7	Trichothecene-4-*O*-acetyltransferase
TRI3	Trichothecene-15-*O*-acetyltransferase
TRI4	Trichodiene oxygenase
TRI6	Transcription factor
TRI5	Trichodiene synthase
TRI10	Regulatory gene
TRI9	Unknown
TRI11	Isotrichodermin 15-oxygenase
TRI12	Trichothecene efflux pump
TRI13	Calonectrin 4-oxygenase
TRI14	Virulence factor

Fig. 1 Structural organization of the core trichothecene cluster, containing 12 Tri genes. The predicted function of each of these genes is listed in Table 1. In NIV-producing isolates all genes are functional, but several deletions and/or insertions in *tri7* and *tri13* (indicated by the symbols Δ and ψ at these loci) lead to functional inactivation of these genes and 3ADON or 15ADON are the final productions on the mycotoxin synthesis

The genes *tri7* and *tri13* were targeted because their gene products are involved in production of DON vs. NIV [7–9]. Both genes encode functional enzymes in NIV-producing strains, while in DON producers multiple mutations and/or deletions occur in *tri7* and in *tri13*, rendering the gene products nonfunctional, which ultimately leads in a failure to produce NIV. These Tri7 and Tri13 PCRs work well in most strains of *F. graminearum*, but results from other species were less conclusive. Therefore many other genes in the trichothecene core cluster have been used as target for chemotype discrimination in Fusarium; for example Zhang et al. [10] developed a discriminatory PCR based on *tri11* sequences that could discriminate NIV, 3ADON, and 15ADON producers within the *Fusarium graminearum* species complex. However the most frequently used chemotype-specific PCRs are based on the *tri3* gene and the *tri12* gene [11, 12]. The primers in these assays always generate a PCR product with DNA from any of the type B-producing Fusarium species. The size of the product will indicate the chemotype of the isolate (*see* also Table 2). A combination of both genes will also make the assay more robust as the results from the *tri3* gene must be in agreement with the results from the *tri12* gene (as demonstrated in Fig. 2). The *tri12* gene was also targeted for qPCR based on TaqMan technology that allowed detection of several of the tested B-clade species [13]. This assay allowed for the detection/quantification of 3ADON, 15ADON, and/or NIV producers in single wheat seeds, which showed reasonable correlation with quantity of fungal biomass [13].

However, independent analyses on two different targets also add to the robustness of the assay, because both reactions should deliver the same outcome.

Table 2
Primers used for the generation of Tri3- or Tri12-specific amplicons

Primer	Sequence	Amplicon length*
3_CONS	TGGCAAAGACTGGTTCAC	n.a.
3_NIV_F	GTGCACAGAATATACGAGC	840 bp
3_15ADON_F	ACTGACCCAAGCTGCCATC	610 bp
3_3ADON_F	CGCATTGGCTAACACATG	243 bp
12_CONS	CATGAGCATGGTGATGTC	n.a.
12_NIV_F	TCTCCTCGTTGTATCTGG	840 bp
12_15ADON_F	TACAGCGGTCGCAACTTC	670 bp
12_3ADON_F	CTTTGGCAAGCCCGTGCA	410 bp

*Length of the amplicon after PCR in combination with the conserved primers 3_CONS or 12_CONS

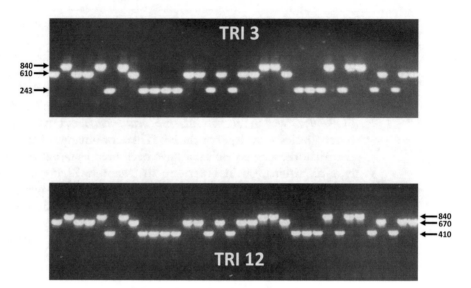

Fig. 2 Differentiation of isolates according to chemotype. PCR fragments of specific sizes are generated, depending on the chemotype of the isolate. In a Tri3-specific PCR, 3ADON, 15ADON, and NIV producer fragments of 243, 610, or 840 bp, respectively, are obtained. In a PCR directed to the Tri12 alleles, PCR fragments of, respectively, 410, 610, or 840 bp are generated (*see* **Note 10**)

---

2   Materials

All solutions should be prepared using ultrapure water, e.g., MilliQ water. Pre-PCR reagents should be handled in a clean room well separated from any post-PCR activities. To avoid cross-contamination use "PCR-grade water" when for dilution of stocks and working solutions of oligonucleotides. PCR reagents and oligonucleotides should be stored at −20 °C and buffers at 4 °C.

2.1 Fungal Cultures

Monospore cultures of *Fusarium* strains (*see* **Note 1**).

Potato dextrose agar and potato dextrose broth for cultivation of isolates.

2.2 Freeze-drying

Christ Epsilon 1–4 LSC.

2.3 Grinding

Chrome steel beads of 3.2 mm.

TissueLyser II.

TissueLyser adapter set for 96-well plates.

2.4 DNA Extraction

Home-made press to ensure that caps do not open during handlings.

Sbeadex Maxi Plant.

RNase A (8 mg/mL).

Ultrapure water.

KingFisher Flex

KingFisher Flex MicrotiterDeepwell 96 plate, V-bottom.

KingFisher 96 tip comb for DW magnets.

KingFisher 96 KF microplate (200 μL).

Sealing tape.

2.5 Seed Inoculation of Wheat Seeds with Fungal Spores

Seeds of a organically grown wheat cultivar.

Inoculum suspension of fungal spores.

2.6 DNA Extraction from Wheat

Grinding Machines

1. Peppink model 200AN milling machine with 1 mm sieve (http://peppink.com/en/grinders/peppink-an-grinders/peppink-200-an/).

2. Retsch MM200 with 25 mL stainless steel grinding jars http://www.retsch.com/products/milling/ball-mills/mixer-mill-mm-200/function-features/) and 15 mm stainless steel balls; this machine is also available from Qiagen as TissueLyser II (https://www.qiagen.com/nl/products/catalog/automated-solutions/sample-disruption/tissuelyser-ii).

2.7 Quantification and Quality Control of DNA

TecanInfinite M200PRO.

PicoGreen assay (cat. no: P7589).

Prepare 1× TE buffer (10 mM Tris–HCl, 1 mM EDTA, pH 7.5) from the 20× TE stock, which is supplied in the PicoGreen kit (to make 50 mL, add 2.5 mL of 20× TE to 47.5 mL sterile distilled DNase-free water). 50 mL is sufficient for 250 assays.

Black microplate 96 wells (cat. no: 655076 Greiner Bio-One).

2.8 Amplification

PCR machine: Applied Biosystems Verity 96-well Thermal Cycler.

qPCR machine: Applied Biosystems 7500 Real-Time PCR System (96 wells); BioRad CFX 384 (384 wells).

2.9 Primers

Primers used for the chemotype determination are listed in Table 2 and the primers for the real-time qPCR can be found in Table 3.

2.10 Gel Electrophoresis

Tris borate buffer (TBE) as running buffer 10× buffer.

2.11 Software/ Websites

Visual OMP http://www.dnasoftware.com/.

CLC main workbench http://www.clcbio.com/products/clc-main-workbench/.

NCBI http://www.ncbi.nlm.nih.gov/.

Table 3
Primers and probes for quantification of chemotypes

1	Fw_TRI3_F	AAGAACCTGAGCCCTCCAGTC
2	NIV TRI3_R	GGCCACAAGAGCGCTCG
3	NIV TRI3 probe	TGCCAAGAGTACTCACGTC
4	3ADON_TRI3_R	GCCGGAACATCAACTCACATAG
5	3ADON_TRI3probe[a]	AGTTGGACATCAGCACTCT
6	15ADON_TRI3_R	ATGAAGTCGGAATATCAACTCACATAA
7	15ADON_TRI3probe[a]	AGTTGGACGTAAGCACTCT
12	NIV_TRI12_F	TGTGCCTGATGAGATGGAACAC
13	NIV_TRI12_NIV	GGATGACAGCGACTGCCTC
14	NIV_TRI12probe[a]	CCCAGAATCATCACACCAGT
15	3ADON_TRI12_F	CCACAGAGCCCCGACGA
16	3ADON_TRI12_R	ATGGATGACAGCAACTGCCTC
17	3ADON_TRI12probe[a]	AGATGAAACACGTTGCC
18	15ADON_TRI12_F	GCAAGTATTTGCCAGCGGATAT
19	15ADON_TRI12_R	ATGACAGCGACTGCCGC
20	15ADON_TRI12probe[a]	ATGGAACAAGTTTCCCAGAAC

[a]Probes have minor groove binding properties (MGB probes) and are labeled with both FAM dye and non-fluorescent quencher

3   Methods

3.1   DNA Extraction from Fungal Tissue

Mycelium is grown in liquid medium and harvested on sterile filter paper.

Small amounts of mycelium are put in the Qiagen strips and freeze-dried.

Put 10 mg freeze-dried mycelium per collection microtube and add two chrome steel beads. Close the tubes with the caps. Place the rack (without the lid) in the TissueLyser adapter, balance a second rack, and bead beat both racks during 20 s at a frequency of 30 strokes/s. Change the orientation of the racks and bead beat again. Spin the powder in the racks down in the centrifuge.

3.1.1   Mini Method Sbeadex Kit for DNA Extraction in KingFisher

Add 300 µL lysis buffer with 0.5 µL RNase (2 mg/mL) using a dispenser and close the caps.

Bead beat for 30 s at a frequency of 30/s. Then change the orientation of the plate and bead beat another time. Close the tubes with the caps. Place the rack (without the lid) in the TissueLyser adapter, balance a second rack, and bead beat both racks during 20 s at a frequency of 30/s. Change the orientation of the racks and bead beat again (if necessary, spin the tubes in the centrifuge (1 min at $1650 \times g$)).

Place the tubes, rack in the "flower" press, and tighten the nuts to prevent popping off the caps during incubation at 65 °C in a water bath for 30 min (*see* **Note 2**).

Fill the plates as follows:

Plate 1	Deepwell plate	120 µL Binding buffer
Plate 2	Deepwell plate	200 µL Wash buffer 1
Plate 3	Deepwell plate	200 µL Wash buffer 2
Plate 4	Deepwell plate	200 µL MQ water
Plate 5	96-Well plate	70 µL Elution buffer

After 30-min incubation at 65 °C in a water bath spin the tubes in the centrifuge (5 min, $1650 \times g$).

Then take 50 µL supernatant from the tubes and pipette it to plate 1.

Finally add 10 µL Sbeadex particle suspension in plate 1. Make sure that the particles are very well suspended before pipetting.

Switch on the KingFisher machine.

Select the second tab (person icon) in the screen, OK, select DNA/RNA (protocol selection), OK, and select the mini protocol. Press START (several times) and the machine tells you which plate you have to place on each position. When all plates are in their right position the machine starts automatically with the DNA isolation. It takes about 30 min.

3.1.2 DNA Extraction from Wheat

Freeze-dry the kernels in Christ Epsilon.

Grind kernels using sieve mesh 1 mm (200AN mill) or Retsch MM200 with 2 balls/jar.

Put 20 mg milled seeds per collection tube and add two chrome steel beads.

Add 10–20 mg freeze-dried wheat material into the Qiagen tube.

Put the 96 collection microtubes in the blue rack.

3.1.3 Mini Method Sbeadex Kit for DNA Extraction in KingFisher

Add 300 µL lysis buffer with 0.5 µL RNase (2 mg/mL) using a dispenser and close the caps.

Bead beat for 30 s at a frequency of 30 strokes/s. Then change the orientation of the plate and bead beat another time. Close the tubes with the caps. Place the rack (without the lid) in the

TissueLyser adapter, balance a second rack, and bead beat both racks during 20 s at a frequency of 30 strokes/s. Change the orientation of the racks and bead beat again (if necessary, spin the tubes in the centrifuge (1 min $1560 \times g$)).

Place the tubes and rack in the "flower" press and tighten the nuts to prevent popping off the caps during incubation at 65 °C in a water bath for 30 min (*see* **Note 2**).

Fill the plates as follows:

Plate 1	Deepwell plate	120 μL Binding buffer
Plate 2	Deepwell plate	200 μL Wash buffer 1
Plate 3	Deepwell plate	200 μL Wash buffer 2
Plate 4	Deepwell plate	200 μL MQ water
Plate 5	96-Well plate	70 μL Elution buffer

After 30-min incubation at 65 °C in a water bath spin the tubes in the centrifuge (5 min, $1650 \times g$).

Then take 50 μL supernatant from the tubes and pipette it to plate 1. Finally add 10 μL Sbeadex particle suspension in plate 1. Make sure that the particles are very well suspended before pipetting.

Switch on the KingFisher machine.

Select the **second tab (with the person)** in the screen, OK, select DNA/RNA (protocol selection), OK, and select the mini protocol. Press **START** (several times) and the machine tells you which plate you have to place on each position. When all plates are in their right position the machine starts automatically with the DNA isolation. It takes about 30 min.

3.2 Quantification and Quality Control of DNA

Turn on the fluorescence plate reader at least 10 min before reading results. Use the following settings to read the PicoGreen results:

Wavelength/bandwidth	
Excitation	~480 nm
Emission	~520 nm

Dilute DNA standards from 100 to 0.1 μg/mL with 1× TE.

Prepare the standard curves in [part of] a microtiter plate as shown in the scheme below:

	Standards	Unknown samples
	A	B
1	100	Sample A
2	10	Sample B
3	1	Sample C
4	0.1	Sample D
5	Water	Sample E
6	–	Sample F
7	–	Sample G
8	–	Water

For each unknown, add 1 μL of sample to 99 μL of 1× TE in the black microplate well. Mix by pipetting up and down.

Prepare a 1:200 dilution of the PicoGreen reagent in 1× TE. For each standard and each unknown sample, a volume of 100 μL will be needed. For examples, two standard curves with eight points each will require 1.6 mL. To calculate the total volume of diluted PicoGreen reagent needed, determine the total number of samples and unknowns you will be testing and multiply this number by 100 μL (if using a multichannel pipet, make extra reagent). The PicoGreen reagent is light sensitive and should be kept wrapped in foil while thawing and in the diluted state. Vortex well.

Add 100 μL of diluted PicoGreen to every standard and sample. Mix by pipetting up and down.

Cover the microtiter plate with foil and allow to incubate at room temperature for 2–5 min.

Read the plate (without cover) in the Tecan (Infinite M200 pro).

Generate a standard curve using the values of the standards and determine the concentrations of DNA in the unknown samples.

3.3 Chemotype PCR on Tri3 (and Tri12)

Prepare PCR mixes for either Tri3 or Tri12 amplification. These mixes contain 1 μL each of all for Tri3 primers, dNTPs, and Taq polymerase.

Pipetting Schedule

PCR reaction (25 μL): Master mix 23 and 2 μL target (10 ng/μL).

Mastermix per reaction	1 reaction (μL)	Final conc.
Roche PCR buffer (2×)	2.5	1×
dNTPs (5 mM)	0.25	50 μM
Common primer (3CON or 12 CON)	1.0	200 nM

3ADON primer (3D3A or 12D3F)	1.0	200 nM
15ADON primer (3D15A OR 12D15F)	1.0	200 nM
NIV primer (3NA OR 12NF)	1.0	200 nM
Roche Taq polymerase 5U/μL	0.15	0.75U
Water	16.1	
Subtotal	23	
DNA solution	2	
Total	25	

PCR conditions

Stage/step	Temp. (°C)	Time	Repeat
1	95	2:00	
2/1	95	0:30	25×
2/2	52	0:30	
2/3	72	1:00	
3	10	∞	

Run 5 μL (or 20% of the sample) of the amplification mix on agarose gel (1%, buffer) along with size markers, or amplicons of isolates with known chemotype (*see* Fig. 2).

3.4 Primer Design

1. Retrieve DNA sequences from NCBI http://www.ncbi.nlm.nih.gov/ by using the advanced search option using **trichothecene gene cluster[Title]** as query (on April 21st, 2015 this resulted in 41 hits). These accessions contain the majority of the genes in the core trichothecene gene cluster (*see* Fig. 1).

2. Import the sequences in CLC main workbench http://www.dnasoftware.com/ and trim the sequences for *tri3* according to the annotations in the NCBI accessions.

3. Group the sequences according to the chemotype of the corresponding strain (*see* **Note 3**).

4. Scan the sequences for SNPs that are present in **ALL15ADON**-producing isolates **AND** absent in both **ALL 3ADON** and **ALL NIV** producers (*see* Fig. 3).

5. Among 15ADON producers the sequence variation is very low, 3ADON producers show some variation in the *tri3* sequence, and NIV producers show the highest degree of variation (*see* **Note 4**).

6. Primers and probes should be designed on the basis of the observed SNPs.

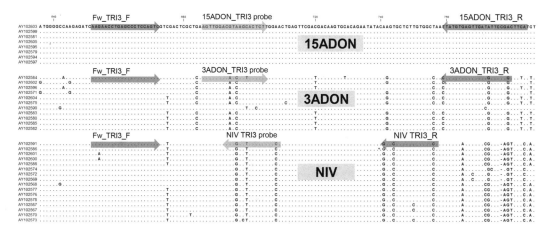

Fig. 3 Alignment of [part of] the *tri3* gene of accessions AY102567-AY102605 from NCBI (http://www.ncbi.nlm.nih.gov/) arranged according to the chemotype of these isolates from the *Fusarium graminearum* species complex (*see* **Note 9**)

7. Several additional criteria must be considered for primer and probe design:

   (a) Limited spacing between forward and reverse primer (preferably less than 100 bp).

   (b) (Absence of) secondary structure formation.

   (c) (Absence of) cross-hybridization with non-targets (*see* **Note 5**). Generally this will mean that primers and probe should be designed to work optimally at 60 °C (*see* **Note 6**).

8. Forward primer, 5′-AAGAACCCTCCAGT-3′, is common to all three chemotypes and can be used for all trichothecene-producing Fusarium isolates (*see* Fig. 3).

9. Design reverse primers to accommodate specificity between 3ADON, 15ADON, and NIV producers, e.g., 5′-ACTGACCCAAGCTGCCATC-3′ for 15ADON producers, 5′-CGCATTGGCTAACACATG-3′ for 3ADON producers, and 5′-GTGCACAGAATATACGAGC-3′ for NIV producers (*see* **Note 7**).

10. Use OMP software http://www.dnasoftware.com/ to scan primer and probe combinations for melting temperature and hybridization efficiency.

11. Table 4 illustrates that the probe p_TRI3_NIV has almost full hybridization at 60 °C with NIV target, while the same probe shows no hybridization with the 3ADON target and very limited hybridization with 15ADON targets.

3.5 Protocol for Real-Time (TaqMan) PCR

1. Select isolates from the *Fusarium graminearum* species complex, FGSC, with known (*see* **Notes 3** and **4**) or unknown chemotype.

Table 4
Hybridization properties of primers and probe as predicted by Visual OMP

Probe	Tri3 target					
	NIV		3ADON		15ADON	
	Tm (°C)	Hyb (%)	Tm (°C)	Hyb (%)	Tm (°C)	Hyb (%)
p_ TRI3_NIV	68	98.3	15	0	33	1
p_ TRI3_3ADON	27	1	70	97.3	55	12
p_ TRI3_15ADON	23	2	59	48.2	68	97.3

2. Prepare reaction mixes for 96-well format or 384-well format (*see* **Note 8**).

3. To detect *Fusarium* use 1 μL DNA in 96-well format. AB7500 Machine.

Pipetting Schedule

Real-time (TaqMan) PCR reaction (25 μL): Mastermix 24 and 1 μL target.

Mastermix per reaction	1 reaction	Final conc.
TaKaRa mastermix (2×)	12.5 μL	1×
ROX II	0.25 μL	
Forward primer (5 μM)	1.5 μL	300 nM
Reverse primer (5 μM)	1.5 μL	300 nM
Probe (5 μM)	1.5 μL	100 nM
Water	7.75 μL	
Subtotal	24 μL	
DNA solution	1 μL	
Total	25 μL	
ROX II is included for normalization (*see* **Note 10**)		

PCR conditions:

Stage/step	Temp. (°C)	Time	Repeat
1	95	5:00	
2/1	95	0:15	40×
2/2	60	1:00	

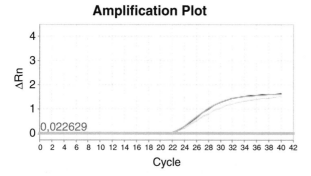

Amplification Plot

TRI3 3ADON	Ct
16D1(NIV)	ND
31F1(3ADON)	20.92
68D2(15ADON)	ND
bfb0082_1(3ADON)	20.56
bfb0982_1(NIV)	ND
CH024b(15ADON)	ND
SVP 8906(3ADON)	20.63

Fig. 4 Amplification curve of 3ADON-producing *F. graminearum*, *F. asiaticum*, and *F. culmorum* isolate, using a 3ADON-specific primer/probe combination based on the *tri3* gene (*see* **Notes 11** and **12**)

Start Applied Biosystems 7500 software, click Experiment Properties, and enter the experiment name. Click Plate Setup, and assign targets (dye) and samples. Click Run Method, and edit reaction volume and thermal profile. Load the reaction plate into the instrument and start the run after saving.

Analyze the data by opening the 7500 software and click Analysis. The analysis is performed automatically using the threshold (*see* Fig. 4).

To detect Fusarium in plant material use 5 μL in 384-well format. CFX384 Machine (BioRad) Replace ROX II with additional water when working with equipment from BioRad (*see* **Note 10**).

Pipetting Schedule

Real-time (TaqMan) PCR reaction (15 μL): Mastermix 10 and 5 μL target.

Mastermix per reaction	1 reaction (μL)	Final conc.
TaKaRa mastermix (2×)	7.5	1×
ROX II	0.15	
Forward primer (5 μM)	0.9	300 nM
Reverse primer (5 μM)	0.9	300 nM
Probe (5 μM)	0.3	100 nM
Water	0.25	
Subtotal	10	
DNA solution	5	
Total	15	

PCR conditions

Stage/step	Temp. (°C)	Time	Repeat
1	95	2:00	
2/1	95	0:10	40×
2/2	60	0:30	

Open Bio-Rad CFX Manager software, select required protocol, and make necessary changes for volume, cycling steps, and plate read. Choose the scan mode, and quick plate 384 SYBR/FAM only or all channels. Save plate file and start run. Analyze the data automatically using Cq Determination mode Regression starting from cycle 5 (*see* **Note 13**).

3.6 Preparation of Mung Bean Medium

40 g of mung beans are added to a beaker with 1 L of water, pre-heated to 100 °C. Continue heating this mixture for an additional period until the first skins are released from the beans. Pour the mixture over a funnel coated with cheesecloth to collect the filtrate. Aliquot the liquid into smaller vessels and autoclave for 15 min at 121 °C.

3.7 Preparation of Spore Suspensions

Inoculated mycelium plugs into mung bean medium and incubate for several days at 25 °C in a rotary shaker at 120 RPM. Add spores to the wheat kernels in a Erlenmeyer flask and rotate flask to mix spores and kernels well. Continue incubation for 2–4 days at 25 °C and rotate flask to mix kernels daily.

3.8 Protocol for Real-Time (TaqMan) PCR on Wheat Samples

Grind wheat samples (artificially infected with *Fusarium* isolates of different chemotypes in Peppink milling machine with a 1 mm mesh sieve).

Also grind another batch of seeds by bead beating for 3 min at maximal intensity using 2 balls of 15 mm.

Both methods will result in powders with different levels of coarseness, the Peppink milling results in a fine powder, while the ball beating leads to a very fine powder.

Prepare qPCR mixes for 16 reactions as above, using DNA extracted from 8 × 20 mg of each powder, and perform reactions as above in TaqMan machine.

With the fine powder from the Peppink machine milling amplification curves will be obtained that show some variation in the CT, while the very fine powders from the bead beating show almost no variation in CT (*see* Fig. 5).

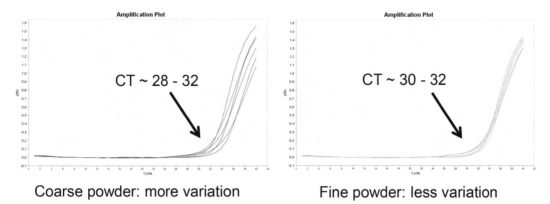

Fig. 5 Reproducibility of amplification curves obtained with DNA extracted from powders generated by milling in a Retch milling machine resulting in coarse powders (*left*) or by bead beating at full speed with two beads (3 mm) for 3 min that generates fine powders

---

4   Notes

1. Test isolates used in this study are listed in Table 5.

2. Mark the position of the collection tubes in the rack on the tubes so that you will never be puzzled about the original order of the samples.

3. PCR costs can be calculated via worksheets available at http://splice-bio.com/costs-of-your-next-qpcr-experiment/.

4. NRRL strains can be obtained at ARS Culture collection http://nrrl.ncaur.usda.gov/cgi-bin/usda/process.html?mv_doit=return&mv_nextpage=mold%2fnrrl&mv_click=nrrl&query_type=nrrl or at the CBS-KNAW Fungal Biodiversity Centre (http://www.cbs.knaw.nl/Collections/Biolomics.aspx?Table=CBS%20strain%20database).

5. Chemotype of NRRL isolates can be found in O'Donnell et al. [14].

6. The high sequence variation found in NIV producers suggests that NIV producers form the ancestral chemotype. This is in agreement with the notion that NIV-producing isolates form the endogenous population in China, while DON producers (both 3ADON and 15ADON) are displacing this original population [10, 15].

7. Note that the reverse primers for 3ADON, 15ADON, and NIV are designed to detect these three genotypes, irrespective of the species that harbors these alleles.

Table 5
Chemotypes of the isolates used in this study

Isolate code	Species	Toxin[a]	TRI3[b]	TRI12[b]	Origin	Year	Host
16D1	*F. graminearum*	NIV	NIV	NIV	The Netherlands	2000	Wheat
31F1	*F. graminearum*	3ADON	3ADON	3ADON	The Netherlands	2000	Wheat
68D2	*F. graminearum*	15ADON	15ADON	15ADON	The Netherlands	2001	Wheat
bfb0082_1	*F. asiaticum*	DON	3ADON	3ADON	China	2005	Barley
bfb0982_1	*F. asiaticum*	NIV	NIV	NIV	China	2005	Barley
CH024b	*F. asiaticum*	15ADON	15ADON	15ADON	China	2002	Wheat
SVP 8906	*F. culmorum*	Unknown	3ADON	3ADON	Switzerland	Unknown	Wheat

[a]Chemotype based on chemical analysis
[b]Chemotype according to PCR on *tri3* and *tri12* genes

8. Compatibility with other qPCRs is recommended, because this decreases handling time in screening large numbers of samples.

9. The sequence variation of the NIV alleles is so high that it was only possible to design a reverse primer in a target region showing an SNP among NIV producers. The same is true for the probes for 3ADON and NIV.

10. ROX Reference Dye is included in real-time quantitative PCR or RT-PCR equipment from Life Technologies. It normalizes for non-PCR-related fluctuations in fluorescence and provides a stable baseline for multiplex quantitative PCR and RT-PCR. See TaqMan® Protocol (https://www.lifetechnologies.com/order/catalog/product/12223012).

11. The patterns for Tri3 (upper panel) and Tri12 (lower panel) are nearly identical. The only difference between both gels is the distance between the different bands.

12. Positive or negative result is based on the *cycle threshold* (CT) value, e.g., the number of cycles needed for the sample generate a signal that exceeds the threshold. The CT value is calculated by an algorithm that is integrated in the software of the PCR machine.

13. A positive or negative result is based on the regression starting at cycle 5. The Cq value is calculated by an algorithm that is integrated in the software of the PCR machine.

Acknowledgments

This work was supported by the MycoRed project (FP7 Food Quality and Safety Priority—Large Collaborative Project—GA 222690) and the Dutch Main Board for Arable Products.

References

1. Alexander NJ, Proctor RH, McCormick SP (2009) Genes, gene clusters, and biosynthesis of trichothecenes and fumonisins in *Fusarium*. Toxin Rev 28:198–215

2. Kimura M, Tokai T, Takahashi-Ando N et al (2007) Molecular and genetic studies of fusarium trichothecene biosynthesis: pathways, genes, and evolution. Biosci Biotechnol Biochem 71:2105–2123

3. McCormick SP, Stanley AM, Stover NA et al (2011) Trichothecenes: from simple to complex mycotoxins. Toxins 3:802–814

4. Proctor RH, McCormick SP, Alexander NJ et al (2009) Evidence that a secondary metabolic biosynthetic gene cluster has grown by gene relocation during evolution of the filamentous fungus Fusarium. Mol Microbiol 74:1128–1142

5. Rep M, Kistler HC (2010) The genomic organization of plant pathogenicity in Fusarium species. Curr Opin Plant Biol 13:420–426

6. Alexander NJ, McCormick SP, Waalwijk C et al (2011) The Genetic Basis for 3-ADON and 15-ADON Trichothecene Chemotypes in Fusarium. Fungal Genet Biol 48:485–495

7. Brown DW, McCormick SP, Alexander NJ et al (2001) A Genetic and Biochemical Approach to Study Trichothecene Diversity in *Fusarium sporotrichioides* and *Fusarium graminearum*. Fungal Genet Biol 32:121–133

8. Brown DW, McCormick SP, Alexander NJ et al (2002) Inactivation of a cytochrome P-450 is a determinant of trichothecene diversity in Fusarium species. Fungal Genet Biol 36:224–233

9. Lee T, Oh DW, Kim HS et al (2001) Identification of deoxynivalenol- and nivalenol-producing chemotypes of *Gibberella zeae* by using PCR. Appl Environ Microbiol 67:2966–2972

10. Zhang H, Zhang Z, van der Lee T et al (2010) Population genetic analyses of *Fusarium asiaticum* populations from barley suggest a recent shift favoring 3ADON producers in southern China. Phytopathology 100:328–336

11. Ward TJ, Bielawski JP, Kistler HC et al (2002) Ancestral polymorphism and adaptive evolution in the trichothecene mycotoxin gene cluster of phytopathogenic *Fusarium*. Proc Natl Acad Sci U S A 99:9278–9283

12. Waalwijk C, Köhl J, De Vries I et al (2009) Fusarium in winter tarwe (in 2007 en 2008) (in Dutch). Wageningen, The Netherlands, Plant Research International. Report 272

13. Kulik T (2011) Development of TaqMan assays for 3ADON, 15ADON and NIV *Fusarium* genotypes based on *Tri12* gene. Cereal Res Commun 39:200–214

14. O'Donnell K, Kistler HC, Tacke BK et al (2000) Gene genealogies reveal global phylogeographic structure and reproductive isolation among lineages of *Fusarium graminearum*, the fungus causing wheat scab. Proc Natl Acad Sci U S A 97:7905–7910

15. Yang LJ, van der Lee TAJ, Yang XJ et al (2008) Fusarium Populations on Chinese Barley Show a Dramatic Gradient in Mycotoxin Profiles. Phytopathology 98:719–727

Chapter 12

Targeting Ochratoxin Biosynthetic Genes

Antonia Gallo and Giancarlo Perrone

Abstract

The pathway of ochratoxin A (OTA) biosynthesis has not yet been completely elucidated. Essentially, two kind of genes have been demonstrated to be involved in the biosynthesis of OTA. One of them is the *nrps* gene encoding a non-ribosomal peptide synthetase (NRPS) which catalyzes the ligation between the iso-coumarin group, constituting the polyketide group of OTA molecule, and the amino acid phenylalanine.

Here we describe a conventional PCR method developed for the detection of OTA-producing molds belonging to *Penicillium* and *Aspergillus* genera by Luque et al. (Food Control 29:270–278, 2013). This method is based on the OTA *nrps* gene of *Penicillium nordicum*. It produces a specific amplicon of 459 bp and its functionality in naturally infected samples was also demonstrated.

Key words Ochratoxin A, PCR detection, Biosynthetic gene, *nrps*, Food commodities, OTA-producing molds, *Aspergillus* and *Penicillium* genera

1 Introduction

Polymerase chain reaction has been widely applied in diagnosis of OTA-producing molds in food raw material and in processed food and feed. A great number of PCR primers have been developed in conventional and real-time PCR applications for detection of OTA producers. Most of them have been targeted to anonymous genomic markers, such as AFLP and RAPD markers, or genetically defined sequences, such as ribosomal RNA, calmodulin, and β-tubulin genes [1–3].

For a complete quality control and a quantitative estimation of contamination in order to rate the hazard associated to a food sample, the availability of molecular assays developed on OTA biosynthetic pathway genes seems to be more appropriate and convenient for their direct relation to the hazard. Diagnostic assays based on the presence of mycotoxin biosynthetic genes support the risk assessment linked to the presence of fungi responsible of mycotoxin production. For this purpose, a detailed knowledge of the molecular aspect of biosynthesis pathway is necessary.

Antonio Moretti and Antonia Susca (eds.), *Mycotoxigenic Fungi: Methods and Protocols*, Methods in Molecular Biology, vol. 1542, DOI 10.1007/978-1-4939-6707-0_12, © Springer Science+Business Media LLC 2017

Until recently, the OTA biosynthetic pathway was almost completely unknown. Initially, a polyketide synthase (PKS) was postulated to be involved in the biosynthesis of OTA at the first step of the pathway for the formation of isocoumarin portion as pentaketide moiety of toxin molecule. Firstly, the gene encoding the PKS responsible of this biosynthetic step was identified in *P. nordicum* [4] and *A. ochraceus* [5]. Thereafter, OTA PKS have been characterized also in other producing fungi such as *A. westerdijkiae* [6], *A. niger* [7], *P. verrucosum* [8], and *A. carbonarius* [9]. With the identification of this key gene, there has been a rapid development of PCR assays based on primers designed on *pks* sequences. As most of the secondary metabolite biosynthetic genes in fungi are organized in cluster, for example aflatoxins, fumonisins, and trichothecenes, also OTA genes are grouped in defined genomic region which have been identified in *A. carbonarius* and *A. niger*, following their genome sequencing. Also in *P. nordicum* the group of Geisen [4, 10] has identified a partial cluster containing genes coding proteins likely involved in the biosynthesis process. Another important key enzyme is the peptide synthetase catalyzing the bond between the polyketide residue and the phenylalanine to result in OTA as the toxic end product. Non-ribosomal peptide synthetases (NRPS) involved in the biosynthesis of OTA have been identified in *P. nordicum* [4] and *A. carbonarius* [11]. At the moment, OTA *pks* and *nrps* are the most important genes on which investigation has been focused because of their strategic roles in the biosynthetic mechanism that is not yet completely clarified. Other genes are likely to be involved such as an oxidase, a chloroperoxidase, maybe genes coding transporters, other than regulatory genes and fungal transcription activators; all of them are under investigation or are to be yet identified.

Polyketide synthases and NRPSs are multidomain enzymes consisting of characteristic functional domains, which are usually present in the structure of the enzymes of different origins. Most of the filamentous fungi present a great number of *pks* and *nrps* genes that are involved in the biosynthesis of several fungal secondary metabolites, as has become clear from the recent genome sequencing studies. Because of this, primers targeting functional domains always present in the protein structures of these enzymes, like β-ketosynthase (KS) and acyl-transferase (AT) domains in PKS and adenylation (A) and condensation (C) domains in NRPS, require special attention in the design in order to avoid amplification of unspecific PKSs and NRPSs. These could be related to the production of other secondary metabolites different from the toxin under investigation.

Since the first elucidation of molecular aspects of OTA biosynthesis, several PCR diagnostic assays targeting OTA biosynthetic genes have been developed. Most of them were species specific or able to detect species belonging to the same genus, such as *Penicillium* in the case of systems described by Geisen et al. [12] and Bogs et al. [13] based on the OTA *pks* and/or *nrps* of *P. nordicum*, or *A. niger* aggregate in the real-time PCR system described by Castella and Cabanes [14] based on the OTA *pks* of *A. niger*. A method capable of detecting OTA-producing organisms regardless of their genus and species could be more helpful and convenient for the assessment of contamination risk in the field of food safety. In this regard, Rodriguez and coworkers [15] developed two protocols of real-time qPCR based on SYBR Green and TaqMan, whose primers and probes were designed from *OTAnpsPN*, the *nrps* gene involved in OTA biosynthesis in *P. nordicum*. They allowed detection of all the tested OTA-producing strains from genera *Penicillium*, *Aspergillus*, and *Emericella*. The same research group realized a simple, specific, and sensitive method based on conventional PCR able to detect ochratoxigenic molds in foods [16], with an approach more rapid and less laborious than other molecular methods. In particular, by using a primer pair designed on the sequence of *OTAnpsPN* by Bogs et al. [13], amplified products were obtained in some OTA-producing strains (*Aspergillus* and *Penicillium*). The sequences of these amplicons showed a similarity upper than 99 % with the *OTAnpsPN* sequence. A primer pair (F1OT/R1OT) designed on the conserved regions of the aligned sequences was able to produce a single specific amplicon of the expected size of 459 bp in all the tested OTA-producing strains, while none of the non-ochratoxigenic reference strains gave a positive result with this primer pair. Sensitivity of this protocol in the detection of OTA-producing molds was established in pure culture (25 pg of mold DNA), and on artificially inoculated food matrices with a detection limit ranging between 10^2 and 10^4 cfu/g, depending on the tested food matrix, with the lowest value of sensitivity found in almond and walnut. The capacity of the designed PCR protocol for detecting ochratoxigenic molds in naturally infected food sample (dry-cured ham, paprika, and wheat semolina) was also evaluated. An assay about the inhibition from the food components was carried out showing that they had a low influence on the method sensitivity. This chapter describes the abovementioned PCR method using primer pair F1OT/R1OT; this method also contemplates a parallel amplification of the universal fungal β-*tubulin* gene with primers pairs Bt2a and Bt2b [17] to test the presence of mold DNA. The visualization of results is produced by electrophoresis analysis on agarose gel which reveals the presence of the amplified product, that is, the presence of OTA-producing mold in the extracted DNA from the tested sample, as a single band of 459 bp in length.

2 Materials

2.1 PCR Reagents for Amplification of OTAnps and β-Tubulin Genes

1. Taq DNA polymerase: 5 U/μL.
2. A specific PCR buffer usually supplied by the manufacturer of the DNA polymerase in a concentrated solution (10× or 5×).
3. MgCl$_2$ if it is not included in the supplied PCR buffer.
4. 10 mM Deoxynucleotides (dNTPs).
5. DNA template extracted from samples to be analyzed.
6. The two primer sets F1OT/R1OT and Bt2a/Bt2b (*see* Table 1).

Concentrations of stock reagents may vary depending on the supplier or the use in laboratory.

2.2 PCR Reactions

The final concentrations of reagents for a final reaction volume of 50 μL as described in Luque et al. [16] (*see* **Note 2**):

1. 1× PCR buffer.
2. 200 μM dNTPs.
3. 2 mM MgCl$_2$ for the reaction with F1OT/R1OT, while the final concentration of MgCl$_2$ in the reaction for amplification of *β-tubulin* is 1.5 mM as described in Glass and Donalsdon [17]. Add only if it is not present in the 10× buffer at the requested concentration.
4. 0.4 mM (20 pmol) of both the two primer pairs.
5. Around 10 ng of DNA template.
6. 0.5–2.5 units of DNA polymerase per 50 μL reaction (see manufacturers recommendations).
7. Sterile distilled water Q.S. (*see* **Note 3**) to obtain a 50 μL final volume per reaction as predetermined in the table of reagents.

When setting up several PCR reactions using the same reagents, you can scale appropriately and combine reagents together in a master mixture (Master Mix). This step can be done in a sterile 1.5/2 mL microcentrifuge tube (*see* **Note 4**).

Table 1
Primers used in this PCR protocol

Primer name	Sequence nucleotides (5′–3′)	Expected PCR product	Targeted gene	References
F1OT	GCCCAACGACAACCGCT	459 bp	*OTAnps*	[16]
R1OT	GCCATCTCCAAACTCAAGCGTG			
Bt2a	GGTAACCAAATCGGTGCTGCTTTC	Approx. 453 bp (*see* **Note 1**)	*β-tubulin*	[17]
Bt2b	ACCCTCAGTGTAGTGACCCTTGGC			

2.3 Agarose Gel Electrophoresis Components

To analyze the amplicons resulting from PCR experiment, the agarose gel electrophoresis requires TAE (Tris-acetate-EDTA) buffer 1×: 40 mM Tris, 20 mM acetic acid, 1 mM EDTA.

This buffer is used as running buffer but also for preparation of agarose gel. It is commonly prepared as a 50× stock solution for laboratory use. A 50× stock solution can be prepared by dissolving 242 g Tris base in water, adding 57.1 mL glacial acetic acid, and 100 mL of 500 mM EDTA (pH 8.0) solution, and bringing the final volume up to 1 L. This stock solution can be diluted 50:1 with water to make a 1× working solution.

3 Methods

3.1 PCR Reactions

When setting up PCR experiments, wear gloves to avoid contaminating the reaction mixture or reagents.

Determine the number of samples to be analyzed. False positives may occur as a consequence of carryover from another PCR reaction which would be visualized as multiple undesired products on an agarose gel after electrophoresis. Therefore, it is prudent to use proper technique, including negative and positive controls. For the negative control, add all the reagents with the exception of template DNA which is replaced by the same volume of water; an additional negative control from DNA extracted from a non-OTA-producing strains is also advisable in this case. A positive control is constituted by DNA template of a known OTA-producing strain.

Arrange all reagents needed for PCR experiment in a freshly filled ice bucket, and let them thaw completely before setting up reactions. Keep the reagents on ice throughout the experiment.

Reaction volumes will vary depending on the concentrations of the stock reagents.

1. Start by making a table of reagents that will be added to the reaction mixture (*see* Table 2).

2. Next, label PCR tubes with the ethanol-resistant marker.

3. Place a 96-well plate into the ice bucket as a holder for the 0.2 mL thin-walled PCR tubes. Allowing PCR reagents to be added into cold 0.2 mL thin-walled PCR tubes will help prevent nuclease activity and nonspecific priming.

4. Pipette the following PCR reagents in the following order into a 1.5/2 mL tube: Sterile water (*see* **Note 9**), 10× PCR buffer, dNTPs, $MgCl_2$, primers, and template DNA (*see* Table 2, in which the stock concentrations used by authors of the protocol are reported to give an example).

5. Set the 100 μL pipettor to deliver a volume of 45.0 μL. Transfer 45.0 μL from each reaction cocktail into each PCR reaction tube (to give a final reaction volume of 50 μL with 5 μL DNA

Table 2
Table of reagents for PCR reaction mixture

Reagents	Concentration of stock solution	Volume	13× Master mix (*see* Note 5)	Final concentration
Sterile H$_2$O		32.5/33 µL (*see* Note 6)	422.5/429 µL	
PCR buffer	10×	5 µL	65 µL	1×
dNTPs	10 mM	1 µL	13 µL	200 µM
MgCl$_2$	50 mM	2 µL/1.5 µL (*see* Note 6)	26 µL/19.5 µL (*see* Note 6)	2.0 mM/1.5 mM
Forward primer	10 mM	2 µL	26 µL	20 pmol
Reverse primer	10 mM	2 µL	26 µL	20 pmol
Template DNA	2 ng/µL	5 µL (*see* Note 7)		10 ng
Taq DNA polymerase (*see* Note 8)	2 Units/µL	0.5 µL	6.5 µL	1 Unit
				50 µL/reaction

sample or water). Lock the caps across the tops of the tubes and place the finished reactions on ice.

6. Set the 20 µL pipettor to deliver 5 µL. Place a sterile pipette tip on the end of the pipettor. Transfer 5 µL of the diluted DNA sample (2 ng/µL) (or water) for a final amount of 10 ng to each 200 µL PCR tube. Change pipette tips between DNA samples to avoid cross-contamination of sample DNAs.

7. Spin down the contents of the tubes in the microcentrifuge and transfer them in the thermocycler. Once the lid of the thermal cycler is firmly closed start the programs as described in Tables 3 and 4.

8. When the PCR program has finished, the 0.2 mL tubes may be removed and stored on ice or at 4 °C. There should be ~50 µL of reaction in each tube. Gently open the lids of the tubes and use an aliquot (5–50 µL) of each reaction to check PCR product amplification by electrophoresis analysis on agarose gel.

3.2 Agarose Gel Electrophoresis

This step allows to determine whether PCR amplifications were successful, whether the resulting products are of the correct size, and whether other unspecific products were amplified as well.

To visualize PCR products obtained by using the current protocol, DNA samples are separated in a gel made with 1.5–2 % (w/v) agarose and 1× TAE buffer containing 5 µL/100 mL of a stock solution (10 mg/mL) of ethidium bromide (EtBr) (*see* Note 10).

Table 3
PCR cycling for OTAnps gene

Cycle step	Temperature (°C)	Time	Number of cycles
Initial denaturation	94	5 min	1
Denaturation	94	30 s	34
Annealing	57	40 s	
Extension	72	1 min	
Final extension	72	5 min	1
Hold	4		1

Table 4
PCR cycling for β-tubulin gene

Cycle step	Temperature (°C)	Time (min)	Number of cycles
Initial denaturation	94	5	1
Denaturation	95	1	33
Annealing	68	1	
Extension	72	1	
Final extension	72	5	1
Hold	4		1

3.2.1 Mixing, Melting, and Pouring the Gel (See Note 11)

1. Add 1.5/2 g of agarose powder to Erlenmeyer flask.
2. Add 100 mL of 1× TAE to Erlenmeyer flask.
3. Swirl vigorously to thoroughly mix agarose.
4. Put agarose and 1× TAE slurry into microwave.
5. Heat on HIGH for 30–45 s at a time, remove from microwave and swirl, and repeat heating and swirling; if the mixture begins to boil, stop and remove the agarose. Swirl until the mixture is clear (*see* **Note 12**).
6. Let agarose solution cool down and add fluorescent dye (Sybr Green, ethidium bromide) in the concentration normally used in your lab (*see* **Note 13**).
7. Pour the cooled melted agarose solution in the tray cast for electrophoresis gel, equipped with the casting for the wells, according to the number of samples to analyze.
8. Let the agarose gel to polymerize until it has completely solidified and appears opaque.

3.2.2 Loading Samples and Running the Gel

1. Carefully remove combs by pulling them upwards firmly and smoothly in a continuous motion. The remaining depressions are the wells into which your samples will be loaded (*see* **Note 14**).

2. Transfer the gel in the electrophoresis box in which the gel will be run. Add sufficient 1 × TAE buffer to fill the reservoirs at both ends of the gel box and to cover the gel completely.

3. Add loading buffer to each sample (*see* **Note 15**).

4. Load the samples into the well by using a pipette, carefully to avoid the cross loading (*see* **Note 16**).

5. Load a molecular size standard with reference into one well before or after the samples to be analyzed.

6. Put gel box cover into place, and hook up the electrodes for each gel. DNA molecules are negatively charged at neutral pH and will migrate to the positive electrode. Turn the power supply on and adjust to 80–150 V. Let the gel run for 45 min to an hour until the dye line is approximately 75–80 % of the way down the gel. The electric field will move the negatively charged DNA toward the positive electrode.

3.2.3 Visualization of the Gel

1. Turn off the power, unhook the electrodes, and with gloved hands gently lift the gel and gel mold out of the gel box.

2. Place the gel and gel mold (made of UV transparent plastic) on a UV light transilluminator. Turn off the room lights, put the plexiglass shield down to protect yourself from the UV radiation, and turn on the blue background. Pictures of the gel may be taken at this time to record the results of the experiment.

4   Notes

1. The length of *β-tubulin* amplicon varies depending on the fungal species tested.

2. In the original paper of Glass and Donaldson [17], *β-tubulin* PCR reaction is performed in a volume of 100 μL, whereas here we consider the volume of reagents for a PCR reaction in a 50 μL final volume.

3. Q.S. is a Latin abbreviation for quantum satis meaning the amount that is needed.

4. The master mix should be prepared in excess. A surplus of master mix volume (10 %) is suggested. For instance you can prepare a master mix for 11 (10 + 1) reactions. The reagents in the master mix are mixed thoroughly by gently pumping the plunger of a micropipettor up and down. Aliquot the master mix in each PCR tube using the volume including all reagents except the DNA template, which will be added later separately.

5. To give an example, the master mix depicted in Table 2 is calculated for 11 reactions plus 2 extra reactions to accommodate pipette transfer loss ensuring that there is enough to aliquot to each reaction tube.

6. The first number is referred to the F1OT/R1OT reaction, and the second number is referred to the *β-tubulin* reaction.

7. DNA template is different for each PCR tube, so it is added as last reagent in each single PCR tube after that master mix has been dispensed. After isolation, the concentration of fungal DNA from pure cultures or from contaminated sample was determined and the genomic DNA extract diluted in sterile double-distilled H_2O (if needed) to a working concentration of 2 ng/μL, in this case.

8. Taq DNA polymerase is typically stored in a 50 % glycerol solution and complete dispersal in the reaction mix requires gentle mixing of the PCR reagents by pipetting up and down. The micropipettor should be set to about half the reaction volume of the master mix when mixing, and care should be taken to avoid introducing bubbles.

9. Water is added first but requires initially making a table of reagents and determining the volumes of all other reagents added to the reaction.

10. While ethidium bromide is the most common stain for nucleic acids, there are several safer and less toxic alternatives including methylene blue, crystal violet, SYBR Safe, and Gel Red.

11. Always wear glass. This is essential to limit the spread of PCR products around the lab as well as protect against intercalating dyes.

12. Watch carefully to ensure that agarose mixture DOES NOT boil over in the microwave.

13. Ethidium bromide is a known mutagen; wear a lab coat, eye protection, and gloves when working with this chemical.

14. When pouring the gel, avoid creating bubbles as this will prevent current from flowing through the gel. Do not pull comb out too quickly as wells will form holes, resulting in the loss of samples.

15. The loading buffer is a 50:50 (v:v) glycerol:H_2O mixture containing a tracking dye (xylene cyanol, cresol red, bromophenol blue) which helps with gel loading and assesses DNA migration during electrophoresis, and glycerol which makes the sample denser than the running buffer so that when loaded the sample will settle to the bottom of the well instead of diffusing.

16. Insert pipette tip into well at an angle to avoid putting a hole in the bottom of the well.

References

1. Perrone G, Susca A, Stea G, Mulè G (2004) PCR assay for identification of *Aspergillus carbonarius* and *Aspergillus japonicus*. Eur J Plant Pathol 110(5/6):641–649

2. Patiño B, González-Salgado A, González-Jaén MT, Vázquez C (2005) PCR detection assays for the ochratoxin-producing *Aspergillus carbonarius* and *Aspergillus ochraceus* species. Int J Food Microbiol 104(2):207–214

3. Morello LG, Sartori D, de Oliveira Martinez AL et al (2007) Detection and quantification of *Aspergillus westerdijkiae* in coffee beans based on selective amplification of beta-tubulin gene by using real-time PCR. Int J Food Microbiol 119(3):270–276

4. Karolewiez A, Geisen R (2005) Cloning a part of the ochratoxin A biosynthetic gene cluster of *Penicillium nordicum* and characterization of the ochratoxin polyketide synthase gene. Syst Appl Microbiol 28:588–595

5. O'Callaghan J, Caddick MX, Dobson ADW (2003) A polyketide synthase gene required for ochratoxin A biosynthesis in *Aspergillus ochraceus*. Microbiology 149:3485–3491

6. Bacha N, Atoui A, Mathieu F et al (2009) *Aspergillus westerdijkiae* polyketide synthase gene "aoks1" is involved in the biosynthesis of ochratoxin A. Fungal Genet Biol 46:77–84

7. Pel HJ, de Winde JH, Archer DB et al (2007) Genome sequencing and analysis of the versatile cell factory *Aspergillus niger* CBS 513.88. Nat Biotechnol 25:221–231

8. Abbas A, Coghlan A, O'Callaghan J et al (2013) Functional characterization of the polyketide synthase gene required for ochratoxin A biosynthesis in *Penicillium verrucosum*. Int J Food Microbiol 161:172–181

9. Gallo A, Knox BP, Bruno KS et al (2014) Identification and characterization of the polyketide synthase involved in ochratoxin A biosynthesis in *Aspergillus carbonarius*. Int J Food Microbiol 179:10–17

10. Geisen R, Schmidt-Heydt M, Karolewiez A (2006) A gene cluster of the ochratoxin A biosynthetic genes in *Penicillium*. Mycotoxin Res 22:134–141

11. Gallo A, Bruno KS, Solfrizzo M et al (2012) New insight into the ochratoxin A biosynthetic pathway through deletion of a nonribosomal peptide synthetase gene in *Aspergillus carbonarius*. Appl Environ Microbiol 78:8208–8218

12. Geisen R, Mayer Z, Karolewiez A, Farber P (2004) Development of a real time PCR system detection of *Penicillium nordicum* and for monitoring Ochratoxin A production in foods by targeting the ochratoxin polyketide synthase gene. Syst Appl Microbiol 27: 501–507

13. Bogs C, Battilani P, Geisen R (2006) Development of a molecular detection and differentiation system for ochratoxin A producing *Penicillium* species and its application to analyse the occurrence of *Penicillium nordicum* in cured meats. Int J Food Microbiol 107(1):39–47

14. Castellá G, Cabañes FJ (2011) Development of a real time PCR system for detection of ochratoxin A-producing strains of the *Aspergillus niger* aggregate. Food Control 22(8): 1367–1372

15. Rodríguez A, Rodríguez M, Luque MI et al (2011) Quantification of ochratoxin A-producing molds in food products by SYBR Green and TaqMan real-time PCR methods. Int J Food Microbiol 149:226–235

16. Luque MI, Córdoba JJ, Rodríguez A et al (2013) Development of a PCR protocol to detect ochratoxin A producing moulds in food products. Food Control 29:270–278

17. Glass NL, Donaldson GC (1995) Development of primer sets designed for use with the PCR to amplify conserved genes from filamentous Ascomycetes. Appl Environ Microbiol 61: 1323–1330

Chapter 13

Targeting Fumonisin Biosynthetic Genes

Robert H. Proctor and Martha M. Vaughan

Abstract

The fungus *Fusarium* is an agricultural problem because it can cause disease on most crop plants and can contaminate crops with mycotoxins. There is considerable variation in the presence/absence and genomic location of gene clusters responsible for synthesis of mycotoxins and other secondary metabolites among species of *Fusarium*. Here, we describe a quantitative real-time PCR (qPCR) method for distinguishing between and estimating the biomass of two closely related species, *F. proliferatum* and *F. verticillioides*, that are pathogens of maize. The qPCR assay is based on differences in the two species with respect to the genomic location of the gene cluster responsible for synthesis of fumonisins, a family of carcinogenic mycotoxins. Species-specific qPCR primers were designed from unique sequences that flank one end of the cluster in each species. The primers were used in qPCR to estimate the biomass of each *Fusarium* species using DNA isolated from pure cultures and from maize seedlings resulting from seeds inoculated with *F. proliferatum* alone, *F. verticillioides* alone, or a 1:1 mixture of the two species. Biomass estimations from seedlings were expressed as the amount of DNA of each *Fusarium* species per amount of maize DNA, as determined using maize-specific qPCR primers designed from the ribosomal gene L17. Analyses of qPCR experiments using the primers indicated that the assay could distinguish between and quantify the biomass of the two *Fusarium* species. This finding indicates that genetic diversity resulting from variation in the presence/absence and genomic location of SM biosynthetic gene clusters can be a valuable resource for development of qPCR assays for distinguishing between and quantifying fungi in plants.

Key words *Fusarium proliferatum*, *Fusarium verticillioides*, Fumonisin, Secondary metabolite, Maize, Quantitative PCR

1 Introduction

Fusarium is a species-rich genus of fungi that is of concern to agriculture because it can cause disease on many crops and can produce mycotoxins, including three (fumonisins, trichothecenes, and zearalenone) that are among the mycotoxins of greatest concern to food and feed safety [1, 2]. Some crops are affected by multiple *Fusarium* species. For example, the closely related species *F. proliferatum*, *F. subglutinans*, and *F. verticillioides* and the more distantly related species *F. graminearum* occur on maize and can cause seedling blight as well as ear rot [2]. Although these diseases are well documented, the degree to which the *Fusarium* species

Antonio Moretti and Antonia Susca (eds.), *Mycotoxigenic Fungi: Methods and Protocols*, Methods in Molecular Biology, vol. 1542, DOI 10.1007/978-1-4939-6707-0_13, © Springer Science+Business Media LLC 2017

interact with one another in maize tissue is not well understood. Nevertheless, several studies indicate that interactions between species within the same plant can affect crop diseases and myco-toxin contamination [3, 4].

Fungi produce low-molecular-weight metabolites that are referred to as secondary metabolites (SMs), because they are not required for growth or development, but instead can provide a selective advantage under certain conditions, such as during patho-genesis or interactions with other microorganisms. Thousands of such metabolites have been described, and they are diverse in structure and biological activity [5]. For example, some fungal SMs are pigments, others are plant hormones, and still others are toxins, including mycotoxins [6, 7]. In fungi, biosynthesis of SMs typically requires the activity of multiple enzymes, and the genes encoding the enzymes are usually located next to one another in a biosynthetic gene cluster. Such clusters can also include genes encoding SM transport proteins and transcription factors that reg-ulate expression of genes in the cluster. Gene clusters responsible for synthesis of many *Fusarium* SMs have been identified, includ-ing those required for synthesis of the pigments aurofusarin, bika-verin, and fusarubins [8–10], the plant hormones gibberellic acids [11], multiple mycotoxins [12–15], and other [16–18].

Fusarium is reported to produce over 50 structurally distinct families of SMs [2, 19]; however, there is marked variation among species in the ability to produce a given SM family [1, 2, 11]. For example, *F. proliferatum* and *F. verticillioides* produce *fumonisins*, whereas *F. subglutinans* and *F. graminearum* do not [2, 20]. In many cases, the ability versus inability to produce a particular SM family results from the presence versus absence of the correspond-ing biosynthetic gene cluster (*see* Fig. 1). For example, Southern blot, PCR, and genome sequence analyses indicated that within the *Fusarium fujikuroi* species complex (FFSC), a lineage of closely related *Fusarium* species that includes *F. proliferatum*, *F. subgluti-nans*, and *F. verticillioides*, some species have the gene clusters responsible for synthesis of fumonisins (*FUM* cluster) and gibberel-lic acids (GA cluster) whereas other species do not have one or both clusters [11, 20–24]. Such variation in the presence and absence of SM clusters can exist even among closely related species. For exam-ple, the *FUM* cluster is present in *F. verticillioides* but absent in its closest known relative *F. musae* [25, 26]. In addition, the genomic location of some SM biosynthetic gene clusters can vary among species, as is the case for the *FUM* cluster among different lineages within the FFSC [21, 23]. Recent analyses of genome sequences of two or more strains of the same species of *Fusarium* indicate that strains within a species can vary in the presence and absence of genes encoding polyketide synthases, a class of enzymes required for the synthesis of numerous SMs [23, 27]. Thus, variation in the genomic location as well as the presence versus absence of SM

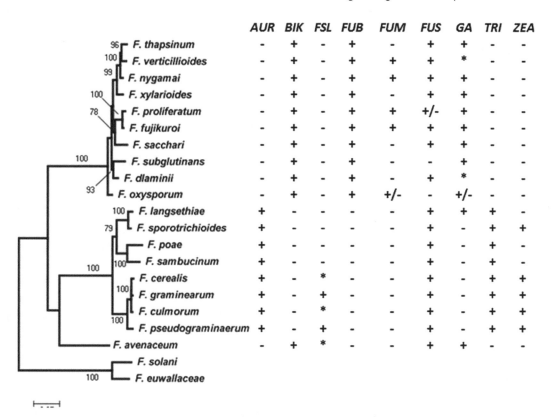

Fig. 1 Variation in the presence and absence of selected SM biosynthetic gene clusters in a subset of *Fusarium* species. SM gene cluster abbreviations are *AUR*, aurofusarin; *BIK*, bikaverin; *FSL*, fusareilin; *FUB*, fusaric acid; *FUM*, fumonisin; *FUS*, fusarin; *GA*, gibberellic acid; *TRI*, trichothecene; *ZEA*, zearalenone. "+" indicates that a gene cluster is present; "–" indicates that a cluster is absent; "+/–" indicates that the cluster is present in some strains of a species but absent in others; and "*" indicates that only part of the cluster is present. The phylogenetic tree to the left was inferred by maximum likelihood analysis from the full-length coding region sequence of *RPB1*, the gene that encodes the largest subunit of RNA polymerase. The alignment length was 5370 bases. Numbers near branches are bootstrap values generated from 500 pseudoreplications. Information on presence/absence SM biosynthetic genes was obtained from published PCR, Southern, and genomic sequence data [21, 23, 29, 30, 32, 43] as well as unpublished genome sequence data

biosynthetic genes represents a source of genetic diversity among, and in some cases within, species of *Fusarium*. Identifying differences in the presence/absence and genomic location of SM biosynthetic gene clusters has been determined empirically by PCR, Southern blot, and genome sequence analyses [11, 20, 21, 23, 28–30]. However, identification of such differences can be done most effectively by comparing whole genome sequences of *Fusarium* species. There are currently genome sequences for 15 described species of *Fusarium* available from the Broad Institute, Joint Genome Institute (JGI), Munich Information Center for Protein Sequences (MIPS), and the National Center for Biotechnology Information (NCBI), and more species are added to these databases each year [23, 29, 31–34].

Studies of the interaction of fungal species *in planta* often require a method that can distinguish between and quantify biomass of each species. Multiple PCR assays have been developed to distinguish between and, in some cases, quantify *Fusarium* species. Such assays have exploited variation in sequences of orthologues of genes involved in primary or secondary metabolism [3, 35–37]. Despite the fact that differences in the presence/absence or genomic locations of SM gene clusters constitute a significant source of interspecies genetic diversity within *Fusarium*, they have been underutilized in PCR-based experiments to detect and quantify different species. Here, we have developed a PCR assay to distinguish between and quantify the fumonisin-producing species *F. proliferatum* and *F. verticillioides* in maize based on variation in the genomic location of the *FUM* cluster in the two species (*see* Fig. 2). The successful development of the method demonstrates that sequence variation associated with SM biosynthetic gene clusters among fungi is a valuable source of genetic diversity among fungi that can be exploited for quantitative analysis of interactions between fungal species.

2   Materials

2.1   DNA Sequences

A previous study indicated that the sequences flanking the *FUM* cluster in *F. proliferatum* and *F. verticillioides* are different (*see* Fig. 2) [21]. Sequence data exhibiting this variation are available from the GenBank database at the NCBI): accessions for *FUM* cluster sequences in *F. proliferatum* (strain ITEM 2287) and *F. verticillioides* (strain FRC M-3125) are KF482467 and AF155773, respectively. We used these sequence data to design PCR primers for an initial qualitative PCR assay to assess sequence variation among multiple strains of both species (*see* Note 1) as well as for the quantitative real-time PCR (qPCR) assay to distinguish between the two *Fusarium* species. The qualitative PCR primers were as follows: primer 2270 (5′-CTSAGCTYCTGGAAKCGAAAGAG-3′) is a forward primer within the *FUM19* coding region and is complementary to sequences in both species; 2271 (5′-CCTGCGCAATGTC TAGAATAATG-3′) is a reverse primer that is complementary to the *FUM19* 3′ flanking region in *F. proliferatum* but not *F. verticillioides*; and 2276 (5′-TAGGCCTGTTCAGAGTCTTATCC-3′) is a reverse primer that is complementary to the *FUM19* 3′ flanking region in *F. verticillioides* but not *F. proliferatum*. The relative locations and other information for the qPCR primers specific to *F. proliferatum* (Fp-F and Fp-R) and *F. verticillioides* (Fv-F and Fv-R) are presented in Fig. 2 and Table 1. In addition to these primers, we also designed primers based on the maize ribosomal protein L17 (NCBI accession NM_001111420).

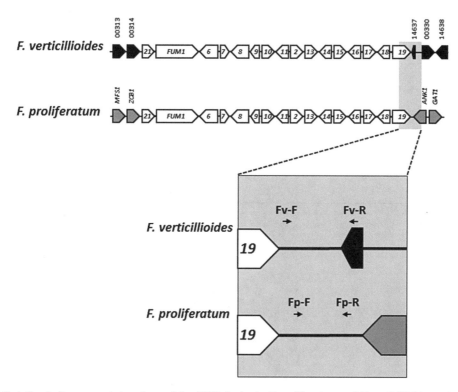

Fig. 2 Variation in the genomic locations of the *FUM* cluster in *F. proliferatum* and *F. verticillioides* as indicated by differences in genes flanking the cluster in each species. The enlarged area (*shaded grey*) depicts the region downstream of cluster gene *FUM19* that was used to design PCR primers that could be used to distinguish between the two *Fusarium* species. Arrows indicate the positions and orientation of genes. White arrows represent *FUM* cluster genes; numbers within the white arrows are *FUM* gene numbers (e.g., *2, 3, 19* correspond to genes *FUM2*, *FUM3*, and *FUM19*, respectively). *Grey and black arrows* indicate *FUM* cluster flanking genes in *F. proliferatum* and *verticillioides*, respectively. The numbers above the black arrows correspond to gene model designations (e.g., 00313 corresponds to FVEG_00313) in the *F. verticillioides* genome sequence databases at the Broad Institute, NCBI, and the Munich Information Center for Protein Sequences (MIPS). The four-character notations above the grey arrows correspond to previously described gene designations [21]

2.2 Additional Materials

While the approach described here may represent a unique combination of methods, the individual methods described are well established. As a result, it is likely that other reagents (e.g., DNA isolation and PCR reagents) and equipment (e.g., growth chamber and thermal cycler) could be substituted for those described below.

1. Maize seeds: Sweet corn variety Silver Queen.

2. V8 juice agar medium: 200 mL Original V8 Juice (Campbell Soup Company), 3 g calcium carbonate, and 15 g agar, water to 1 L. Ingredients are combined and autoclaved for 20 min.

3. Mung bean medium: 40 g Dried mung beans, 1 L of boiling water. Boiling water and mung beans are combined and boiled

Table 1
Quantitative PCR primers for F. proliferatum, F. verticillioides, and maize

Species	Name	Sequence	Location	Fragment (bp)	R^2	Slope	y-intercept
F. proliferatum	Fp-F Fp-R	GGTTCCAAACACAAGTAAG CCTGCGCAATGTCTAGAATAATG	*FUM19* 3′ flank	156	0.99	−3.66	42.3
F. verticillioides	Fv-F Fv-R	GTTGATCATGAAGGTAATC TAGGCCTGTTCAGAGTCTTATCC	*FUM19* 3′ flank	177	0.99	−3.35	35.9
Maize	Zm-F Zm-R	CAAAGTCTCGCCACTCCA CGTCCGTGAGCACGGTA	L17[a]	190	0.99	−3.47	37.2

[a]Primers designed from the coding region of maize ribosomal protein L17 (NCBI Accession number: NM_001111420)

for an additional 10 min; mung beans are removed by filtration, and resulting medium is autoclaved for 20 min.

4. 50 mL Conical propylene tubes with caps for *Fusarium* cultures.

5. Rotary shaker incubator for liquid cultures of *Fusarium*.

6. Substrate: Turface clay substrate (PROFILE Products LLC) (*see* **Note 2**).

7. Planting box: GD GasPak EZ clear plastic container (Becton, Dickinson & Co): dimensions 33 cm × 16 cm × 10 cm, width × height × depth.

8. Plant growth chamber.

9. 2010 Geno/Grinder (SPEX SamplePrep, Metuchen, New Jersey, USA) for plant tissue maceration.

10. DNA isolation: ZYMO Research (ZR) Fungal DNA Kit.

11. Qualitative PCR Reagents: Platinum PCR SuperMix High Fidelity.

12. qPCR Reagents—SsoAdvanced SYBR Green SuperMix.

13. Standard thermal cycler and qPCR machine (e.g., CFX-Connect Real Time System (BioRad).

14. Equipment and reagents for standard agarose gel electrophoresis.

3 Methods

qPCR assay based on variation in the genomic location of the *FUM* cluster to distinguish between and quantify *F. proliferatum* and *F. verticillioides* in maize seedlings.

3.1 Primer Design and Validation

Analysis of *FUM* cluster flanking regions indicated that the regions in *F. proliferatum* and *F. verticillioides* do not consist of homologous sequences and, therefore, could be used to design a qPCR strategy to detect and quantify these species in experiments aimed at studying their interactions in plants and other environments.

1. Suitability of DNA target region—The suitability of the *FUM19* flanking region as a target for qPCR analysis was first evaluated by standard qualitative PCR of ten isolates each of *F. proliferatum* and *F. verticillioides* (*see* **Note 1**). For this PCR we used reagents of the platinum PCR SuperMix and the protocol specified by the manufacturer. DNA of each species was prepared by growing strains of the fungi on V8 juice agar media and then isolating DNA from the resulting growth using the ZR Fungal DNA Kit. The initial qualitative PCR employed a three-primer combination consisting of primers 2270, 2271, and 2276. Based on the published sequence data for the *FUM* cluster

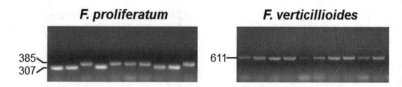

Fig. 3 Image of agarose gel showing the results of qualitative PCR analysis with primers 2270, 2271, and 2276 to assess sequence variation within the 3′ flanking region of *FUM19* orthologues among ten field isolates each of *F. proliferatum* and *F. verticillioides*. Each lanes has the PCR product amplified from genomic DNA of a unique isolate of *F. proliferatum* (*left*) or *F. verticillioides* (*right*). The values to the left of each gel image are base pairs (bp) of amplicons as determined by Sanger sequence analysis, which indicated that the size variation in *F. proliferatum* amplicons resulted from a 78 bp insertion/deletion

flanking regions [21, 38], primers 2270 and 2271 were predicted to amplify a 300 bp fragment from the 3′ flanking region of *FUM19* of *F. proliferatum*; whereas primers 2270 and 2276 were predicted to amplify a 380 bp fragment from the 3′ flanking region of *FUM19* in *F. verticillioides*. The PCR analysis indicated uniformity of the region in *F. verticillioides*, but variability in *F. proliferatum* (*see* Fig. 3). Subsequent Sanger sequencing revealed that the polymorphism was due to a 78 bp insertion/deletion located 36 bp downstream of the *FUM19* stop codon. Given the sequence variability in *F. proliferatum*, qPCR primers for this species were designed from sequences downstream of the insertion/deletion.

2. qPCR conditions—Primer pairs for qPCR were designed using standard criteria [39] so that both primers in the pair were complementary to sequences that are unique to each species (*see* Table 1). Primers specific to maize were also employed so that the level of *Fusarium* DNA could be expressed relative to maize DNA (*see* Table 1). To confirm the specificity of the primers, each primer pair (Fp-F/Fp-R, Fv-F/Fp-R, and Zm-F/Zm-R) was assessed separately in qPCR with genomic DNA of *F. proliferatum*, *F. verticillioides*, or maize and in mixtures of the three genomic DNAs.

   (a) PCR setup: qPCR was performed in a 20 µL volume with 10 µL of 2× SsoAdvanced SYBR Green Supermix, 300 nM of each primer, and 1 µL of template. Sample reactions were performed in triplicate. Genomic DNA samples were diluted tenfold before being used as template in PCR.

   (b) Thermal cycler conditions: initial denaturation at 98 °C for 2 min; 40 cycles each of 98 °C denaturation for 15 s and 60 °C annealing/elongation for 1 min; and a final dissociation curve from 65 to 95 °C.

3. Primer specificity—The ability of the primer pairs to amplify single, specific products for which they were designed and from only the appropriate template was assessed using both a visual

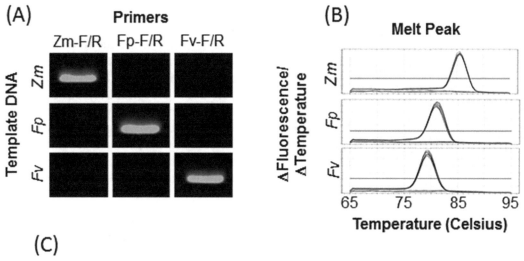

Fig. 4 Assessment of specificity of qPCR primers to distinguish between *F. proliferatum* and *F. verticillioides*. (**a**) Standard gel electrophoresis analysis of PCR products amplified with maize-specific primers Zm-F and Zm-F (=Zm – F/R), *F. proliferatum* -specific primers Fp-F and Fp-R (=Fp - F/R), and *F. verticillioides*-specific primers Fv-F and Fv-R (=Fv - F/R). Amplification products were observed only when PCR primers were combined with the genomic DNA for which they were designed. Template DNA abbreviations Zm, Fp, and Fv correspond to maize, *F. proliferatum*, and *F. verticillioides* genome DNA, respectively. (**b**) Melt curve analysis to assess the specificity of the maize (Zm), *F. proliferatum* (Fp) and *F. verticillioides* (Fv) primers. The presence of a peak at one location along the *x*-axis is indicative of amplification of one PCR product. (**c**) Verification of ability of qPCR primers to estimate known levels of genomic DNA

assessment and melt curve analysis. Both methods indicated that each primer pair exhibited a high level of specificity.

(a) Visual assessment: PCR samples were analyzed by standard agarose gel electrophoresis. In this analysis, PCR products matched expected sizes and were observed only when primers matched the genomic DNA templates for which the primers were designed (e.g., *F. proliferatum* primers and *F. proliferatum* DNA, Fig. 4a).

(b) Melt curve analysis: Following qPCR a melt curve analysis was performed [39, 40]. The result of this analysis indicated that each primer pair amplified a single PCR product from the DNA of the organism for which the primers were designed, but not from DNA of the other organisms tested (*see* Fig. 4b).

4. DNA quantification—The utility of the primer pairs for quantifying genomic DNA was assessed by generating a standard curve from a sixfold dilution series (from 100 to 0.001 ng) of genomic DNA from each organism. The standard curve was generated by plotting quantitation cycle (Cq) values against the log of the DNA concentration (log[DNA]). The R^2, slope, and y-intercept values for each primer pair are shown in Table 1. The standard curve was validated by running samples of known genomic DNA concentration as determined by spectrophotometry. The values obtained from the standard curve were then compared to the known genomic DNA template concentrations added to the sample (*see* Fig. 4c).

The assessment of the qPCR primers indicated that they could effectively distinguish between and quantify *F. proliferatum* and *F. verticillioides* genomic DNA isolated from pure cultures. Given this, we then used the qPCR assay to estimate the biomass of the two species in maize seedlings.

3.2 Maize Seedling Infection Assay

1. *Fusarium* cultures were prepared by adding approximately 30 μL of a frozen glycerol stock of either species to 30 mL of mung bean medium in a 50 mL conical polypropylene tube. Tubes were loosely capped to allow air exchange, and incubated at 28 °C with shaking at 200 rpm.

2. Maize seeds were surface sterilized by soaking in 0.8% sodium hypochlorite solution and then rinsed twice with sterile water for 1 min each time.

3. In a 30 mL conical tube, 30 surface-sterilized seeds were soaked for 48 h without shaking in 30 mL of mung bean medium (control), or 30 mL of a 4-day-old mung bean culture of (a) *F. proliferatum*, (b) *F. verticillioides*, or (c) a 1:1 mixture of *F. verticillioides* and *F. proliferatum* (15 mL aliquots of cultures of each species were mixed immediately before adding maize seeds).

4. Seeds were then sown on the surface of 1 L of Turface clay substrate contained in a GD GasPak EZ box. The lid of the box was fitted loosely onto the base of the box to allow for air exchange.

5. The box was incubated in the growth chamber under the following conditions: 50% relative humidity, a 12-h photoperiod, and a temperature of 25 °C during the light period and 20 °C during the dark period.

3.3 PCR Analysis of Infected Maize Seedlings

1. After a 7-day incubation period in the growth chamber, the resulting seedlings were uprooted (*see* **Note 2**), immediately frozen in liquid nitrogen, and then stored at −80 °C until they were processed for biomass determination.

2. For each replicate, the roots and shoots of two frozen seedlings were pulverized in a 2010 Geno/Grinder. Five replicates were analyzed for each treatment (i.e., control, *F. proliferatum* alone, *F. verticillioides* alone, and *F. proliferatum* and *F. verticillioides* combined).

3. DNA was isolated from the pulverized tissue as specified in the ZR Fungal DNA extraction Kit.

4. The resulting DNA preparations were then used to estimate the biomass of *F. proliferatum* and *F. verticillioides* in the seedlings using the qPCR assay described above in Subheading 3.1.

5. Each DNA sample was subjected to qPCR with the three primer pairs listed in Table 1, and the resulting Cq value from each reaction was used to determine the concentration of genomic DNA for each organism based on the previously determined standard curve. The biomass of *F. proliferatum* and *F. verticillioides* was expressed as a ratio of the amount of *Fusarium* DNA relative to the amount of maize DNA [41, 42]. The results of the analysis indicated that maize seedlings resulting from seed inoculated with a single species were infected only with the species used as inoculum (*see* Fig. 5), providing further evidence that the qPCR assay can distinguish between the two *Fusarium* species.

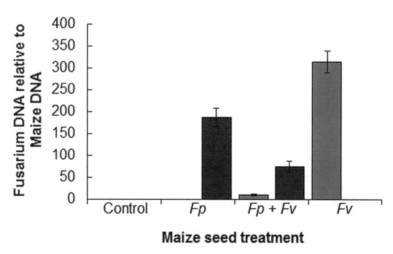

Fig. 5 Results of qPCR analysis to assess biomass of *F. proliferatum* and *F. verticillioides* in maize seedlings. Biomass is expressed as weight (pg) of *Fusarium* DNA per weight (ng) of maize DNA. *Dark grey bars* indicate biomass value obtained with *F. proliferatum*-specific primers Fp-R and Fp-R, and light grey bars indicate biomass values obtained with *F. verticillioides*-specific primers Fv-F and Fv-R. *X*-axis labels indicate data resulting from the following treatments: Control, seeds soaked in mung bean medium; *Fp*, seeds soaked in *F. proliferatum* culture; *Fv*, seeds soaked in *F. verticillioides* culture; and *Fp+Fv*, seeds soaked in a 1:1 mixture of *F. proliferatum* and *F. verticillioides* cultures

4 Notes

1. We have provided a simplified example of a qPCR assay to distinguish between and quantify two species of *Fusarium* in maize seedlings based on interspecies variation in the genomic location of a SM biosynthetic gene cluster. Similar assays could be developed for other species combinations based on differences in the presence and absence of SM biosynthetic gene clusters among species (*see* Fig. 1). When developing qPCR methods based on SM gene clusters, care should be taken when designing primers from published sequences, because such sequences can vary among individuals of the same species. As a result, primers designed based on sequence data for one strain may not yield a PCR product from genomic DNA of a different strain of the same species. Thus, if strains used in an experiment are different from those used to generate the published sequence, a preliminary analysis should be done to determine whether intraspecies variation exists in the target sequence and to take any variation into account when designing qPCR primers.

2. Removal of intact roots from standard peat moss-based soils can be difficult. Use of Turface clay substrate as soil facilitated removal of intact roots. Although we did not compare *Fusarium* biomass in roots versus shoots in the method described above, the ability to remove intact roots from the soil substrate would have facilitated such a comparison.

5 Notice

Mention of trade names or commercial products in this chapter is solely for the purpose of providing specific information and does not imply recommendation or endorsement by the US Department of Agriculture. USDA is an equal opportunity provider and employer. We are grateful for the technical assistance of Stephanie Folmar and Jennifer Teresi.

References

1. Leslie JF, Summerell BA (2006) The *Fusarium* laboratory manual. Blackwell Publishing, Ames
2. Desjardins AE (2006) *Fusarium* mycotoxins chemistry, genetics and biology. APS Press, St. Paul
3. Siou D, Gelisse S, Laval V et al (2015) Interactions between head blight pathogens: consequences for disease development and toxin production in wheat spikes. Appl Environ Microbiol 81:957–965
4. Picot A, Hourcade-Marcolla D, Barreau C et al (2012) Interactions between *Fusarium verticillioides* and *Fusarium graminearum* in maize ears and consequences for fungal development and mycotoxin accumulation. Plant Pathol 61:140–151
5. Cole RJ, Jarvis BB, Schweikert MA (2003) Handbook of secondary fungal metabolites. Academic, San Diego

6. Keller NP (2015) Translating biosynthetic gene clusters into fungal armor and weaponry. Nat Chem Biol 11:671–677

7. Keller NP, Turner G, Bennett JW (2005) Fungal secondary metabolism—from biochemistry to genomics. Nat Rev Microbiol 3:937–947

8. Frandsen RJN, Nielsen NJ, Maolanon N et al (2006) The biosynthetic pathway for aurofusarin in *Fusarium graminearum* reveals a close link between the napthoquinones and naphthopyrones. Mol Microbiol 61:1069–1080

9. Wiemann P, Willmann A, Straeten M et al (2009) Biosynthesis of the red pigment bikaverin in *Fusarium fujikuroi*: genes, their function and regulation. Mol Microbiol 72:931–946

10. Studt L, Wiemann P, Kleigrewe K et al (2012) Biosynthesis of fusarubins accounts for pigmentation of *Fusarium fujikuroi* perithecia. Appl Environ Microbiol 78:4468–4480

11. Bömke C, Tudzynski B (2009) Diversity, regulation, and evolution of the gibberellin biosynthetic pathway in fungi compared to plants and bacteria. Phytochemistry 70:1876–1893

12. Alexander NJ, Proctor RH, McCormick SP (2009) Genes, gene clusters, and biosynthesis of trichothecenes and fumonisins in *Fusarium*. Toxin Rev 28:198–215

13. Brown DW, Lee SH, Kim LH et al (2015) Identification of a 12-gene fusaric acid biosynthetic gene cluster in *Fusarium* species through comparative and functional genomics. Mol Plant Microbe Interact 28:319–332

14. Kim YT, Lee Y-R, Jin J et al (2005) Two different polyketide synthase genes are required for synthesis of zearalenone in *Gibberella zeae*. Mol Microbiol 58:1102–1113

15. Niehaus EM, Kleigrewe K, Wiemann P et al (2013) Genetic manipulation of the *Fusarium fujikuroi* fusarin gene cluster yields insight into the complex regulation and fusarin biosynthetic pathway. Chem Biol 20:1055–1066

16. Sørensen JL, Sondergaard TE, Covarelli L et al (2014) Identification of the biosynthetic gene clusters for the lipopeptides fusaristatin A and W493 B in *Fusarium graminearum* and *F. pseudograminearum*. J Nat Prod 77:2619–2625

17. Sørensen JL, Hansen FT, Sondergaard TE et al (2012) Production of novel fusarielins by ectopic activation of the polyketide synthase 9 cluster in *Fusarium graminearum*. Environ Microbiol 14:1159–1170

18. Kakule TB, Sardar D, Lin Z et al (2013) Two related pyrrolidinedione synthetase loci in *Fusarium heterosporum* ATCC 74349 produce divergent metabolites. ACS Chem Biol 8:1549–1557

19. Vesonder RF, Golinski P (1989) Metabolites of *Fusarium*. *Fusarium Mycotoxins, Taxonomy and Pathogenicity*. In: Chelkowski J (ed) Topics in secondary metabolism, vol 2. Elsevier, Amsterdam, pp 1–39

20. Proctor RH, Plattner RD, Brown DW et al (2004) Discontinuous distribution of fumonisin biosynthetic genes in the *Gibberella fujikuroi* species complex. Mycol Res 108:815–822

21. Proctor RH, Van Hove F, Susca A et al (2013) Birth, death and horizontal transfer of the fumonisin biosynthetic gene cluster during the evolutionary diversification of *Fusarium*. Mol Microbiol 90:290–306

22. Stepien L, Koczyk C, Waskiewicz A (2010) *FUM* cluster divergence in fumonisins-producing *Fusarium* species. Fungal Biol 115:112–123

23. Wiemann P, Sieber CMK, von Bargen KW et al (2013) Deciphering the cryptic genome: genome-wide analyses of the rice pathogen *Fusarium fujikuroi* reveal complex regulation of secondary metabolism and novel metabolites. PLoS Pathog 9, e1003475

24. Malonek S, Bömke C, Bornberg-Bauer E et al (2005) Distribution of gibberellin biosynthetic genes and gibberellin production in the *Gibberella fujikuroi* species complex. Phytochemistry 66:1296–1311

25. Glenn AE, Zitomer NC, Zimeri AM et al (2008) Transformation-mediated complementation of a *FUM* gene cluster deletion in *Fusarium verticillioides* restores both fumonisin production and pathogenicity on maize seedlings. Mol Plant Microbe Interact 21:87–97

26. Van Hove F, Waalwijk C, Logrieco A et al (2011) *Gibberela musae* (*Fusarium musae*) sp. nov.: a new species from banana is sister to *F. verticillioides*. Mycologia 103:570–585

27. Brown DW, Proctor RH (2016) Insights into natural products biosynthesis from analysis of 490 polyketide synthases from *Fusarium*. Fungal Genet Biol 89:37–51

28. Malonek S, Tudzynski B (2003) Evolutionary aspects of gibberellin biosynthesis in the *Gibberella fujikuroi* species complex. Fungal Genet Newslett 50:140

29. Lysoe E, Harris LJ, Walkowiak S et al (2014) The genome of the generalist plant pathogen *Fusarium avenaceum* is enriched with genes involved in redox, signaling and secondary metabolism. PLoS One 9, e112703

30. O'Donnell K, Rooney AP, Proctor RH et al (2013) Phylogenetic analyses of *RPB1* and *RPB2* support a middle Cretaceous origin for a clade comprising all agriculturally and medically important fusaria. Fungal Genet Biol 52:20–31

31. Wingfield BD, Ades PK, Al-Naemi FA et al (2015) IMA genome-F 4: draft genome sequences of *Chrysoporthe austroafricana*, *Diplodia scrobiculata*, *Fusarium nygamai*, *Leptographium lundbergii*, *Limonomyces culmigenus*, *Stagonosporopsis tanaceti*, and *Thielaviopsis punctulata*. IMA Fungus 6:233–248

32. Ma LJ, van der Does HC, Borkovich KA et al (2010) Comparative genomics reveal mobile pathogenicity chromosomes in *Fusarium*. Nature 464:367–373

33. Srivastava SK, Huang X, Brar HK et al (2014) The genome sequence of the fungal pathogen *Fusarium virguliforme* that causes sudden death syndrome in soybean. PLoS One 9, e81832

34. Coleman JJ, Rounsley SD, Rodriguez-Carres M et al (2009) The genome of *Nectria haematococca*: contribution of supernumerary chromosomes to gene expansion. PLoS Genet 5, e1000618

35. Mulè G, Susca A, Stea G et al (2004) A species-specific PCR assay based on the calmodulin partial gene for identification of *Fusarium verticillioides*, *F. proliferatum* and *F. subglutinans*. Eur J Plant Pathol 110:495–502

36. Edwards SG, Pirgozliev SR, Hare MC et al (2001) Quantification of trichothecene-producing *Fusarium* species in harvested grain by competitive PCR to determine efficacies of fungicides against Fusarium head blight of winter wheat. Appl Environ Microbiol 67:1575–1580

37. Fernandez-Ortuno D, Waalwijk C, Van der Lee T et al (2013) Simultaneous real-time PCR detection of *Fusarium asiaticum*, *F. ussurianum* and *F. vorosii*, representing the Asian clade of the *F. graminearum* species complex. Int J Food Microbiol 166:148–154

38. Proctor RH, Brown DW, Plattner RD et al (2003) Co-expression of 15 contiguous genes delineates a fumonisin biosynthetic gene cluster in *Gibberella moniliformis*. Fungal Genet Biol 38:237–249

39. Taylor S, Wakem M, Dijkman G et al (2010) A practical approach to RT-qPCR-Publishing data that conform to the MIQE guidelines. Methods 50:S1–S5

40. Bustin SA, Benes V, Garson JA et al (2009) The MIQE guidelines: minimum information for publication of quantitative real-time PCR experiments. Clin Chem 55:611–622

41. Nicolaisen M, Suproniene S, Nielsen LK et al (2009) Real-time PCR for quantification of eleven individual *Fusarium* species in cereals. J Microbiol Methods 76:234–240

42. Vaughan MM, Huffaker A, Schmelz EA et al (2014) Effects of elevated [CO_2] on maize defence against mycotoxigenic *Fusarium verticillioides*. Plant Cell Environ 37: 2691–2706

43. Proctor RH, McCormick SP, Alexander NJ et al (2009) Evidence that a secondary metabolic biosynthetic gene cluster has grown by gene relocation during evolution of the filamentous fungus *Fusarium*. Mol Microbiol 74:1128–1142

Chapter 14

Targeting Other Mycotoxin Biosynthetic Genes

María J. Andrade, Mar Rodríguez, Juan J. Córdoba, and Alicia Rodríguez

Abstract

Real-time PCR (qPCR) methods are adequate tools for sensitive and rapid detection and quantification of toxigenic molds contaminating food commodities. Methods of qPCR for quantifying zearalenone (ZEA)-, sterigmatocystin (ST)-, cyclopiazonic acid (CPA)-, and patulin (PAT)-producing molds have been designed on the basis of specific target genes involved in the biosynthesis of these mycotoxins. In this chapter reliable qPCR protocols to detect and quantify such toxigenic molds are described. All of these methods are suitable when working with mold pure cultures and mold contaminated foods. For ZEA-producing molds, two qPCR using the SYBR Green fluorochrome and based on two polyketide synthase (PKS) genes are detailed. qPCR protocols relied on the *fluG* and the *idh* genes able to quantify ST- and PAT-producing molds, respectively, which can be performed by both SYBR Green and TaqMan methodologies are described. Regarding CPA-producing molds a TaqManq PCR method including a competitive internal amplification control is detailed. Since DNA extraction is a critical step in the detection and quantification of toxigenic molds by qPCR, a protocol for extracting DNA from mold pure cultures and food is also described.

Key words Molds, Zearalenone, Sterigmatocystin, Cyclopiazonic acid, Patulin, Biosynthesis pathway genes, qPCR

1 Introduction

Accurate methods to detect and quantify toxigenic molds in raw materials as well as in pre-processed foods are necessary to avoid accumulation of mycotoxins in the final products. Polymerase chain reaction (PCR)-based methods are adequate tools for this purpose. To precisely detect toxigenic molds by these methods the correct selection of the target sequence is crucial. One of the best available options consists of using the genes related to the biosynthesis of each mycotoxin given that these genes are only present in molds with the ability to produce the corresponding mycotoxin. Thus the evaluation of the absence or presence of these genes allows detecting potentially toxigenic molds (but not their identification), being of great usefulness for food safety programs. However, it has to be taken into account that the detection of the presence of the mycotoxins biosynthesis pathway genes by PCR

Antonio Moretti and Antonia Susca (eds.), *Mycotoxigenic Fungi: Methods and Protocols*, Methods in Molecular Biology, vol. 1542, DOI 10.1007/978-1-4939-6707-0_14, © Springer Science+Business Media LLC 2017

does not assure the metabolite production since several factors (pH, water activity, temperature, etc.) can inactivate such genes [1] and their expression is necessary for mycotoxin generation.

Furthermore, since PCR-based methods, particularly the quantitative real-time PCR (qPCR) ones, are rapid and sensitive they are of great importance for food routine analysis.

Several qPCR methods to detect and quantify zearalenone (ZEA)-, sterigmatocystin (ST)-, cyclopiazonic acid (CPA)-, and patulin (PAT)-producing molds have been designed on the basis of genes associated with the mycotoxin production. Despite the fact that the biosynthesis pathway of PAT is well known and the sequences of the involved genes are published, there is less available information about genes related to ZEA, ST and CPA biosynthesis. The qPCR methods reported for such toxigenic molds are able to detect and quantify the potentially toxigenic molds in pure cultures and even in food commodities. This allows taking quick and appropriate corrective actions throughout food processing to avoid the growth of toxigenic molds, such as the removal of contaminated batches from the food chain or processing in special conditions of temperature or relative humidity, and consequently the hazard of mycotoxin accumulation in the final products.

Moreover it has to be taken into account that the success of a PCR method depends on the protocol used for mold DNA extraction, especially when working with potentially contaminated foods. A suitable DNA extraction procedure and the subsequent qPCR methods to detect and quantify ZEA-, ST-, CPA-, and PAT-producing molds are detailed in this chapter.

1.1 *DNA Extraction*   DNA extraction is a critical step in the detection and quantification of toxigenic molds by PCR since the lysis of their walls is not easy which may decrease the recovery of the DNA. Besides the presence of inhibitor compounds of DNA polymerase, specially derived from food matrices, can decrease the PCR sensitivity and consequently can provoke false negative results. Thus the availability of suitable procedures for extracting mold DNA is of great importance mainly when working with food commodities.

Mold DNA can be extracted from pure cultures or food using conventional protocols relied on lysis steps such as mechanical disruption (grinding in liquid nitrogen, sonication, bead milling, etc.) and enzymatic digestion (lyticase, proteinase K). Several chemicals (cetyltrimethylammonium bromide (CTAB), β-mercaptoethanol, chloroform, isopropanol, etc.) are also used in such protocols. Alternatively, different commercial kits are available for mold DNA isolation and have been reported as useful for this purpose. A DNA extraction method based on the combination of different reagents and a commercial kit is detailed in this chapter because of its reliability for using in food matrices potentially contaminated with different toxigenic molds [2].

1.2 Zearalenone Biosynthesis Pathway Genes

ZEA is a mycotoxin produced by certain species of *Fusarium/Gibberella*, primarily *F. graminearum* (teleomorph *G. zeae*) and *F. culmorum* [3–5]. PCR detection and quantification of ZEA-producing *Fusarium* have been scarcely described. The reported methods are relied on the genes involved in the biosynthesis pathway of the mycotoxin. Concretely two polyketide synthase (PKS) genes have been described as responsible for ZEA synthesis [3, 6, 7]. Reliable PCR primers have been designed for both genes by using the sequence available in the GenBank database (accession number DQ019316) [8, 9]. Thus the primer sets F1/R1 and ZEA-F/ZEA-R based on the PKS4 [8] and PKS13 [9] genes, respectively, have been used for detecting and quantifying ZEA-producing *Fusarium* species (Table 1). Both methods have been described as suitable for mold cultures as well as for potentially mold contaminated food (specifically maize and maize flour). They are qPCR methods which use the SYBR Green fluorochrome.

Table 1
Primers and probes used for detecting and quantifying zearalenone (ZEA)-, sterigmatocystin (ST)-, cyclopiazonic acid (CPA)-, and patulin (PAT)-producing molds by using the quantitative real-time PCR methods described

Mycotoxin	Gene	Primer/ probe name	Sequence (5'–3')	PCR product size (bp)	References
ZEA	PKS4	F1	CGTCTTCGAGAAGATGACAT	279	[8]
		R1	TGTTCTGCAAGCACTCCGA		
	PKS13	ZEA-F	CTGAGAAATATCGCTACACTACCGAC	192	[9]
		ZEA-R	CCCACTCAGGTTGATTTTCGTC		
ST	*fluG*	FluGF1	GAGTGCCACCGTGATGACC	172	[13]
		FluGR1	TGATGGGTCGGTGGTTGG		
		FluGp	[FAM]-CTCAACATAAACAACAAAC-[TAMRA]		
CPA	*dmaT*	dmaTF	TTCACGCTCGTGGAACTTCT	64	[22]
		dmaTR	GGGTCACAAAGATCGCAAGAT		
		dmaTp	[HEX]-TACTGCCTCCCCCCGAC-[BHQ1]		
	IAC	IACp	[FAM]-CGCCTGCAAGTCCTAAGACGCCA-[TAMRA]	105	
PAT	*idh*	F-idhtrb	GGCATCCATCATCGT	229	
		R-idhtrb	CTGTTCCTCCACCCA		[28]
		IDHprobe	[FAM]-CCGAAGGGCATCCG-[TAMRA]		

1.3 Sterigmatocystin Biosynthesis Pathway Genes

ST is a polyketide secondary metabolite mainly produced by species of *Aspergillus* genus, but it has been reported for species belonging to *Penicillium*, *Emericella*, *Bipolaris*, and *Chaetomium* genera too [10–12]. For the detection and quantification of these toxigenic molds, two qPCR methods have been reported [13]. In these methods specific primers and probe have been designed from the sequences of the *fluG* gene (Table 1) encoding a cytoplasmically localized protein involved in the biosynthesis of ST [14, 15]. The qPCR methods, which can be performed by both SYBR Green and TaqMan methodologies, have been described as sensitive and specific to be used with mold pure cultures as well as with potentially mold contaminated food matrices.

1.4 Cyclopiazonic Acid Biosynthesis Pathway Genes

CPA is a potent mycotoxin classified as nephrotoxin [16] produced by different *Penicillium* species including *P. camemberti*, *P. chrysogenum*, *P. commune*, *P. crustosum*, *P. griseofulvum*, *P. hirsutum*, *P. melanoconidium*, and *P. viridicatum* [17–19]. In addition, some *Aspergillus* species including *A. oryzae*, *A. tamarii*, and *A. versicolor* and the aflatoxigenic mold species *A. flavus* may also produce CPA [20, 21]. PCR detection and quantification of CPA-producing molds have been scantily reported. Therefore only a unique qPCR protocol to detect and quantify those toxigenic molds has been published [22]. This method is relied on specific and effective primers targeted the *dmaT* gene (Table 1) which encodes the enzyme dimethylallyl tryptophan synthase (DMAT) involved in the CPA biosynthesis [23, 24]. This qPCR method uses the TaqMan methodology and includes a competitive internal amplification control (IAC) to avoid false-negative results (Table 1). It is suitable for both mold pure cultures and foods usually contaminated with these toxigenic molds.

1.5 Patulin Biosynthesis Pathway Genes

PAT is a mycotoxin produced by some *Penicillum* species, particularly *P. expansum* and *P. griseofulvum* [25, 26]. Species belonging to *Aspergillus*, *Emericella*, *Paecilomyces*, and *Byssochlamys* genera have been also reported as PAT producers [25–28]. Detection and quantification of PAT-producing molds by qPCR have been scarcely reported. Concretely only two qPCR protocols able to detect and quantify these toxigenic molds have been developed [28]. Such methods are based on reliable primers and probes designed from the sequence of the *idh* gene encoding the isoepoxydon dehydrogenase (IDH) enzyme (Table 1) involved in the PAT biosynthesis [27, 28]. These methods, which can be performed by both SYBR Green and TaqMan methodologies, have been described as sensitive and specific to detect this kind of toxigenic molds in pure cultures and also in potentially contaminated food matrices.

2 Materials

Prepare all reagents and stock solutions using ultrapure water and analytical grade chemicals. Unless indicated otherwise, store at room temperature after the preparation. Consult the safety data sheet provided for each supplier. Follow the official waste disposal regulations.

2.1 DNA Extraction

1. 1 M Tris–HCl buffer, pH 8.0.

2. CTAB lysis buffer: 140 mM D-sorbitol, 34 mM N-lauroylsarcosine, 4.4 mM CTAB, 1.4 M NaCl, 20 mM Na_2EDTA, 2 g PVPP, 0.1 M Tris–HCl, pH 8.0.

3. β-Mercaptoethanol. Store at 4 °C.

4. 10 mg/mL Proteinase K solution. Store at –20 °C.

5. Chloroform (HPLC grade).

6. 10 mg/mL RNAse solution: before use, the enzyme can be heat treated according to the manufacturer's instructions. Store at –20 °C.

7. Cold isopropanol. Store at –20 °C.

8. 70 % (v/v) ethanol.

9. Tris-EDTA (TE) buffer, pH 8.0: 1 M Tris–HCl, pH 8.0, 0.1 M EDTA.

10. EZNA® Fungal DNA Mini Kit (Omega Bio-Tek Inc., Norcross, GA, USA).

11. Liquid nitrogen.

12. Mortar and pestle.

13. Homogenizer stomacher.

14. Microcentrifuge with temperature control.

15. Biophotometer.

2.2 Real-Time Quantitative PCR

1. SYBR Green Master Mix or TaqMan Master Mix. Store SYBR Green reagent at –80 °C for long storage or 4 °C for short storage (less than 6 months). Store TaqMan Master Mix at –20 °C.

2. Oligonucleotide primers and probes (Table 1). Store at –20 °C.

3. Sterile PCR tubes (with optical caps) or PCR plates with optical adhesive films.

4. PCR workstation.

5. qPCR system.

6. Centrifuge for 96-well plates.

7. 96-Well tray for centrifuge.

3 Methods

3.1 DNA Extraction

The DNA extraction method optimized by Rodríguez et al. [2] for moldy food matrices and the slight modifications proposed by Mohale et al. [29] for extracting DNA from mycelium is described below. This method is suitable for extracting DNA from pure culture of toxigenic molds and directly from mold-contaminated food but several steps show some differences depending on such types of analyzed sample (Fig. 1). These differences are specified below.

Some initial considerations have to be taken into account before extracting DNA:

1. Switch on the microcentrifuge, water baths, and block heaters at the adequate working temperatures before starting the DNA extraction procedure.

2. Lysis and elution buffer and enzyme solutions should be kept at pH value 8.0 for stabilizing the DNA for longer time.

3. When working with pure cultures, remove the mycelial biomass from toxigenic molds grown on an adequate solid culture media for 7 days at 25–30 °C (depending on the optimal growth temperature of the tested mold strain). After taking the mycelium, it should be kept at –80 °C until DNA extraction.

4. When working with mold contaminated foods, store them at –80 °C until DNA extraction.

3.1.1 Extraction Procedure

1. When working with pure cultures, grind the mycelium by using mortar and pestle for 2 min after adding liquid nitrogen (*see* **Note 1**). When working with moldy foods, add 5 g of them together with 10 mL of Tris–HCl buffer (pH 8.0) to a stomacher bag with filter and homogenize for 5 min by using a Stomacher machine.

2. Transfer the powder from the ground mycelium or the filtrate obtained from the food to a clean 2 mL safe-lock tube and add 500 µL of CTAB buffer (*see* **Note 2**) containing 5 µL of β-mercaptoethanol (*see* **Note 3**). Shake tube by hand for 15 s.

3. Add 10 µL of proteinase K solution (10 mg/mL) and vortex for 30 s before incubation at 65 °C for 1 h (*see* **Note 4**).

4. Centrifuge the sample at $15,000 \times g$ for 5 min at 4 °C.

5. Transfer the supernatant by pipetting to a clean 2 mL safe-lock tube and add 500 µL of chloroform.

6. Vortex for 30 s and centrifuge the sample at $15,000 \times g$ for 20 min at 4 °C.

7. Transfer the upper layer to a fresh 2 mL safe-lock tube and add 10 µL of RNase solution (10 mg/mL).

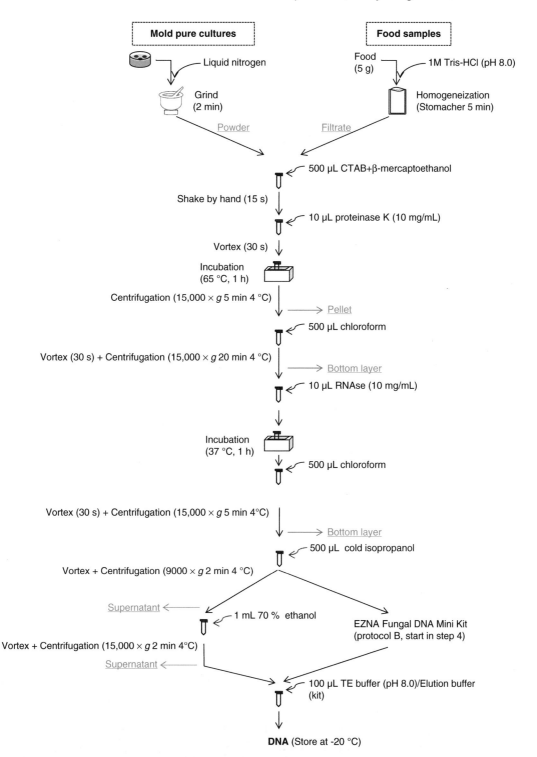

Fig. 1 Flowchart depicting a protocol for DNA extraction from pure cultures of toxigenic molds and from potentially mold-contaminated foods to be used in real-time PCR methods. The differences based on such types of samples are indicated

8. Incubate at 37 °C for 1 h (*see* **Note 4**).

9. Add 500 μL of chloroform, vortex for 30 s and centrifuge at 15,000×*g* for 5 min at 4 °C.

10. Transfer the aqueous phase to a clean 2 mL safe-lock tube and add 500 μL of cold isopropanol. Mix them properly. When working with pure cultures, continue with **steps 11–17** above (*see* **Notes 5** and **6**). When working with moldy foods, such mix (aqueous phase-isopropanol) must be processed according to the instructions of the commercial DNA extraction kit "EZNA Fungal DNA Mini Kit" starting in **step 4**, protocol B [2]. Continue with **steps 16** and **17** above.

11. Centrifuge at 9000×*g* for 2 min at 4 °C (*see* **Note 7**).

12. Remove the supernatant using a pipette (*see* **Note 8**).

13. Wash the pellet with 1 mL of 70% (v/v) ethanol and vortex the sample briefly.

14. Centrifuge at 15,000×*g* for 2 min at 4 °C.

15. Discard the supernatant (*see* **Note 8**).

16. Resuspend the obtained DNA in 100 μL of TE buffer or elution buffer provided by EZNA Fungal DNA Mini Kit (*see* **Note 9**) by passing the solution up and down several times through a pipette tip.

17. Store at –20 °C until use (*see* **Note 10**).

3.1.2 Measurement of DNA Quality and Quantity

1. Switch on the biophotometer at least 10–15 min before measuring quality and quantity of DNA samples.

2. Thaw DNA samples on ice and vortex for 5 s.

3. Pipette an appropriate amount of buffer (*see* **Note 11**) into the cuvette shaft of the biophotometer according to the user guide's instrument and close the cuvette shaft cover.

4. Measure blank.

5. Mix DNA samples by passing the solution up and down several times through a pipette tip.

6. Pipette an appropriate volume of DNA sample into the cuvette shaft of the biophotometer according to the user guide's instrument and then close the cuvette shaft cover.

7. Measure DNA sample. Quality and quantity of DNA are displayed by the biophotometer (*see* **Notes 12–15**).

3.2 Zearalenone Biosynthesis Pathway Genes

The qPCR protocols described by Meng et al. [8] and Atoui et al. [9] based on the PKS4 and PKS13 genes, respectively, for the detection and quantification of ZEA-producing molds are detailed below (*see* **Note 16**).

1. Thaw all the PCR components (SYBR Green Master Mix, primers, and DNA) on ice (*see* **Notes 17–19**). All the reagents have to be prepared according to the manufacturer's instructions.

2. Switch on the qPCR equipment and open the software of qPCR. The run method should be set up according to the user guide's instrument (*see* **Notes 20–23**). Program the thermal cycler with the temperature and time conditions (*see* **Note 24**) detailed in Table 2.

3. Prepare template DNA by diluting the stock solution using sterile ultrapure water (*see* **Note 25**). It should be performed on ice.

4. Prepare the PCR mix (Table 3), including the reagents for all samples excepting template DNA (or non-template samples), in 1.5 mL microcentrifuge tubes to maximize uniformity, minimize the possibility of pipetting errors, and optimize labor when working with multiple tubes (*see* **Notes 26–29**). Before use resuspend all the reagents by vortexing and centrifuge briefly. Each sample is performed by triplicate. Non-template samples should be also prepared by triplicate (*see* **Note 30**). Pipet all the components on ice. Mix the PCR mixture thoroughly and dispense equal aliquots (18 μL for the PKS4 gene method and 23 μL for the PKS13 one) into sterile PCR tubes or PCR plates (*see* **Note 31**).

5. Add the template DNA (or non-template sample; 2 μL) into the corresponding PCR tubes or plate wells (*see* **Notes 32** and **33**).

6. Briefly centrifuge (15 s at maximum speed) the PCR tubes or plate before inserting into the qPCR system (*see* **Note 34**). After centrifuging, check that the mixture is at the bottom of each well or tube. If not, centrifuge again for a longer time. The absence of bubbles should be also checked since they can negatively affect the qPCR results.

7. Load the plate or the tubes on the qPCR instrument (*see* **Notes 35** and **36**) and run the qPCR protocol.

Table 2
Amplification conditions of the SYBR Green qPCR protocols for detecting and quantifying zearalenone-producing molds [8, 9]

Method	Thermal cycling conditions
PKS4 gene	95 °C/6 min *45 cycles*: 94 °C/30 s, 60 °C/30 s, 72 °C/30 s, 80 °C/10 s 72 °C/7 min *Melting curve analysis*: 65–95 °C at the rate of 0.5 °C per s
PKS13 gene	95 °C/4 min *40 cycles*: 94 °C/45 s, 60 °C/45 s, 72 °C/45 s, 80 °C/10 s 72 °C/7 min *Melting curve analysis*: 65–95 °C at the rate of 0.5 °C per s

Table 3
PCR mixture composition of the SYBR Green qPCR protocols for detecting and quantifying zearalenone-producing molds [8, 9]

	Volume (µL) per reaction	
PCR components	PKS4 gene method	PKS13 gene method
SYBR Green Master Mix (2×)	10	12.5
F1 (10 µM)	1	–
R1 (10 µM)	1	–
ZEA-F (10 µM)	–	1
ZEA-R (10 µM)	–	1
DNA	2	2
Sterile ultrapure water	6	8.5
Total volume	20	25

8. Analyze the data by using the software of the qPCR instrument according to the user guide (*see* **Note 37**). The amplification of the DNA is observed at the channel whose fluorescence emission is about 520 nm (fluorophore SYBR). In the qPCR method for the PKS4 gene, the expected amplification product of 279 bp gives only one melting peak at 89 °C in the generated melting curve. In the case of the method for the PKS13 gene, the expected PCR product of 192 bp produces a unique melting peak at 85 °C (*see* **Note 38**). Quantification values (Cq), also called threshold cycle (Ct), are automatically determined by the qPCR instrument.

9. A standard curve by using standards should be also generated in order to quantify an unknown sample. These standards may be prepared using ten-fold dilutions of a template DNA aliquot of a known concentration (*see* **Note 39**). The curve is a plot of the Cq versus log DNA concentration. The curve is used for quantifying an unknown sample. The correlation coefficient (R^2) of the standard curve should be calculated. Ideally this parameter should be 1, although 0.999 is generally the maximum value [30]. The efficiency based on the slope of such curve (efficiency $= 10^{(-1/\text{slope})} - 1$) should be also calculated. The efficiency range of a well-designed qPCR method should range between 90 and 110%, which corresponds to a slope of between −3.58 and −3.10 [30]. The concentration values of the unknown samples should be extrapolated from the standard curve by using the Cq values obtained for such samples.

3.3 Sterigmatocystin Biosynthesis Pathway Genes

The qPCR protocols described by Rodríguez et al. [13] relied on the *fluG* gene for the detection and quantification of ST-producing molds are detailed below.

1. Thaw all the PCR components [SYBR Green Master Mix or TaqMan Master Mix, primers, probe (when TaqMan used) and DNA] on ice (*see* **Notes 17–19, 40** and **41**). All the reagents have to be prepared according to the manufacturer's instructions.

2. Switch on the qPCR equipment and open the software of qPCR. The run method should be set up according to the user guide's instrument (*see* **Notes 20–23**). Program the thermal cycler with the temperature and time conditions (*see* **Note 24**) indicated in Table 4.

3. Prepare template DNA by diluting the stock solution up to about 1 ng using sterile ultrapure water (*see* **Note 25**). It should be performed on ice.

4. Prepare the PCR mix (Table 5), including the reagents for all samples excepting template DNA (or non-template samples), in 1.5 mL microcentrifuge tubes to maximize uniformity, minimize the possibility of pipetting errors, and optimize labor when working with multiple tubes (*see* **Notes 26–29**). Before use resuspend all the reagents by vortexing and centrifuge briefly. Each sample is performed by triplicate. Non-template samples should be also prepared by triplicate (*see* **Note 30**). Pipet all components on ice. Mix the PCR mixture thoroughly and dispense equal aliquots (20 μL) into sterile PCR tubes or PCR plates (*see* **Note 31**).

5. Add template DNA (or non-template sample; 5 μL) into the corresponding PCR tubes or plate wells (*see* **Notes 32** and **33**).

6. Briefly centrifuge (15 s at maximum speed) the PCR tubes or plate before inserting into the qPCR system (*see* **Note 34**). After centrifuging, check that the mixture is at the bottom of each well or tube. If not, centrifuge again for a longer time.

Table 4
Amplification conditions of the SYBR Green and TaqMan qPCR protocols for detecting and quantifying sterigmatocystin-producing molds [13]

Method	Thermal cycling conditions
SYBR Green	95 °C/10 min *40 cycles*: 95 °C/15 s, 60 °C/1 min *Melting curve analysis*: 60–99 °C
TaqMan	95 °C/10 min *40 cycles*: 95 °C/15 s, 59 °C/30 s, 61 °C/30 s

Table 5
PCR mixture composition of the qPCR protocols for detecting and quantifying sterigmatocystin-producing molds [13]

PCR components	Volume (µL) per reaction	
	SYBR Green method	TaqMan method
SYBR Green Master Mix (2×)	12.5	–
TaqMan Master Mix (2×)	–	12.5
FluGF1 (10 µM)	0.75	0.5
FluGR1 (10 µM)	0.75	0.5
FluGp (10 µM)	–	0.5
DNA	5	5
Sterile ultrapure water	6	6
Total volume	25	25

The absence of bubbles should be also checked since they can negatively affect the qPCR results.

7. Load the plate or the tubes on the qPCR instrument (*see* **Notes 35** and **36**) and run the qPCR protocol.

8. Analyze the data by using the software of the qPCR instrument according to the user guide (*see* **Note 37**). When SYBR Green is used, the expected amplification product of 172 bp gives only one peak at 83.9–84.5 °C in the generated melting curve (*see* **Note 38**). Cq values are automatically determined by the software of the qPCR instrument. The amplification of the DNA is observed at the channel whose emission wavelength is about 520 nm (both fluorophores SYBR and FAM).

9. Generate a standard curve as previously described (*see* Subheading 3.2, **step 9**) for obtaining the concentration values of unknown samples.

3.4 Cyclopiazonic Acid Biosynthesis Pathway Genes

The TaqMan-based qPCR protocol including an IAC described by Rodríguez et al. [22] designed on the basis of the *dmaT* gene for the detection and quantification of CPA-producing molds is detailed below (*see* **Note 42**).

1. Thaw all the PCR components (TaqMan Master Mix, primers, probes, IAC, and DNA) on ice (*see* **Notes 18**, **40** and **41**). All the reagents have to be prepared according to the manufacturer's instructions.

2. Switch on the qPCR equipment and open the software of qPCR. The run method should be set up according to the user guide's instrument (*see* **Notes 20–23**). Program the thermal

cycler using the following temperature and time conditions: 50 °C/2 min; 95 °C/10 min; 40 cycles: 95 °C/15 s, 60 °C/1 min.

3. Prepare template DNA by diluting the stock solution up to about 1 ng by using sterile ultrapure water (*see* **Note 25**). It should be performed on ice.

4. Prepare the PCR mix (Table 6), including the reagents for all samples excepting DNA template (or non-template samples), in 1.5 mL microcentrifuge tubes to maximize uniformity, minimize the possibility of pipetting errors, and optimize labor when working with multiple tubes (*see* **Notes 26–29**, and **43**). Before use resuspend all the reagents by vortexing and centrifuge briefly. Each sample is performed by triplicate. Non-template samples should be also prepared by triplicate (*see* **Note 30**). Pipet all components on ice. Mix the PCR mixture thoroughly and dispense equal aliquots (20 μL) into sterile PCR tubes or PCR plates (*see* **Note 31**).

5. Add template DNA (or non-template sample; 5 μL) into corresponding PCR tubes or plate wells (*see* **Notes 32** and **33**).

6. Briefly centrifuge (15 s at maximum speed) the PCR tubes or plate before inserting into the qPCR system (*see* **Note 34**). After centrifuging, check that the mixture is at the bottom of each well or tube. If not, centrifuge again for a longer time. The absence of bubbles should be also checked since they can negatively affect the qPCR results.

7. Load the plate or the tubes on the qPCR instrument (*see* **Notes 35** and **36**) and run the qPCR protocol.

Table 6
PCR mixture composition of the TaqMan qPCR protocol for detecting and quantifying cyclopiazonic acid-producing molds [22]

PCR components	Volume (μL) per reaction
TaqMan Master Mix (2×)	12.5
dmaTF (10 μM)	0.75
dmaTR (10 μM)	0.75
dmaTp (10 μM)	0.5
IAC (10 μM)	0.5
IACp (100 copies)	0.5
DNA	5
Sterile ultrapure water	4.5
Total volume	25

8. Analyze the data by using the software of the qPCR instrument according to the user guide (*see* **Notes 37** and **44**). Cq values are automatically determined by the software of the qPCR instrument. The fluorescence signal for the expected 64 bp PCR product determined at the channel whose emission wavelength is about 556 nm (fluorophore HEX) whilst the signal for the amplification of the competitive IAC (105 bp amplicon) is measured at the channel whose emission wavelength is about 520 nm (fluorophore FAM) (*see* **Note 38**).

9. Generate a standard curve as previously described (*see* Subheading 3.2, **step 9**) for obtaining the concentration values of unknown samples.

3.5 Patulin Biosynthesis Pathway Genes

The qPCR protocols described by Rodríguez et al. [28] relied on the *idh* gene for detecting and quantifying PAT-producing molds are detailed below.

1. Thaw all the PCR components [SYBR Green Master Mix or TaqMan Master Mix, primers, probe (when TaqMan used) and DNA] on ice (*see* **Notes 17–19, 40** and **41**). All the reagents have to be prepared according to the manufacturer's instructions.

2. Switch on the qPCR equipment and open the software of qPCR. The run method should be set up according to the user guide's instrument (*see* **Notes 20–23**). Program the thermal cycler with the following temperature and time conditions (*see* **Note 24**) indicated in Table 7.

3. Prepare template DNA by diluting the stock solution up to about 0.5 ng (for the SYBR Green method) and 1 ng (for the TaqMan method) using sterile ultrapure water (*see* **Note 25**). It should be performed on ice.

4. Prepare the PCR mix (Table 8), including the reagents for all samples excepting template DNA (or non-template samples), in 1.5 mL microcentrifuge tubes to maximize uniformity, minimize the possibility of pipetting errors, and optimize labor when working with multiple tubes (*see* **Notes 26–29**). Before use resuspend all the reagents by vortexing and centrifuge briefly. Each sample is performed by triplicate. Non-template samples should be also prepared by triplicate (*see* **Note 30**). Pipet all components on ice. Mix the PCR mixture thoroughly and dispense equal aliquots (22.5 µL) into sterile PCR tubes or PCR plates (*see* **Note 31**).

5. Add template DNA (or non-template sample; 2.5 µL) into the corresponding PCR tubes or plate wells (*see* **Notes 32** and **33**).

6. Briefly centrifuge (15 s at maximum speed) the PCR tubes or plate before inserting into the qPCR system (*see* **Note 34**).

Table 7
Amplification conditions of the qPCR protocols for detecting and quantifying patulin-producing molds [28]

Method	Thermal cycling conditions
SYBR Green	95 °C/10 min *40 cycles:* 95 °C/15 s, 60 °C/1 min *Melting curve analysis:* 60–95 °C
TaqMan	95 °C/10 min *40 cycles:* 95 °C/15 s, 60 °C/1 min

Table 8
PCR mixture composition of the qPCR protocols for detecting and quantifying patulin-producing molds [28]

PCR components	Volume (μL) per reaction	
	SYBR Green method	TaqMan method
SYBR Green Master Mix (2×)	12.5	–
TaqMan Master Mix (2×)	–	12.5
F-idhtrb (10 μM)	1	1.15
R-idhtrb (10 μM)	1.75	1.15
IDHprobe (10 μM)	–	1.15
DNA	2.5	2.5
Sterile ultrapure water	7.25	6.55
Total volume	25	25

After centrifuging, check that the mixture is at the bottom of each well or tube. If not, centrifuge again for a longer time. The absence of bubbles should be also checked since they can negatively affect the qPCR results.

7. Load the plate or the tubes on the qPCR instrument (*see* **Notes 35** and **36**) and run the qPCR protocol.

8. Analyze the data by using the software of the qPCR instrument according to the user guide (*see* **Note 37**). When SYBR Green is used, the expected amplification product of **229** bp gives only one melting peak ranging from 86.9 to 88.5 °C in the generated melting curve (*see* **Note 38**). Cq values are automatically determined by the software of the qPCR instrument. When both methodologies are used, the fluorescent signal of expected PCR product is observed at the channel whose emission wavelength is about **520** nm (both fluorophores SYBR and FAM).

9. Generate a standard curve as previously described (*see* Subheading 3.2, **step 9**) for obtaining the concentration values of unknown samples.

4 Notes

1. Mortar and pestle have to be pre-frozen at −80 °C in order to avoid the use of high amounts of liquid nitrogen to cool down them before starting. Previously mortar and pestle may be sterilized at 121 °C for 20 min. Between samples mortar and pestle have to be deeply cleaned and disinfected with 70% (v/v) ethanol.

2. Pre-heat the CTAB lysis buffer in a water bath at 65 °C before using.

3. Add β-mercaptoethanol into the tube in a fume cupboard and always wear globes since it is toxic by inhalation, ingestion and skin absorption. Its smell is also quite unpleasant. Consult the safety data sheet provided by the supplier.

4. The incubation could be performed in a water bath or in a block heater.

5. Ethanol or isopropanol:ethanol (1:1) could be also used as DNA precipitation solvents. Isopropanol is normally added at 0.7–1 volumes of sample and ethanol at 2–2.5 volumes of sample.

6. Sample may be processed immediately or after being stored overnight at −20 °C. Sometimes sample storage for few hours helps DNA precipitation.

7. The DNA precipitate is often invisible before centrifugation. After centrifuging, it forms a white or see-through pellet that may be observed on the side or at the bottom of the safe-lock tube.

8. When the pellet is still wet, it is necessary to dry it more thoroughly. For this, put the tube open on a liquid soaking paper in upside down. Wait for few minutes and be very careful to avoid any losing of pellet. If the pellet remains wet, air-dry the DNA pellet for additional 5–10 min.

9. DNA can be also resuspended in other elution buffers provided by commercial extraction kits or sterile ultrapure water.

10. Aliquot the extracted DNA into three to four tubes to avoid DNA denaturalization because of numerous freeze and thaw cycles.

11. The buffer (TE buffer or the elution buffer from the extraction kits) or sterile ultrapure water used for resuspending the obtained DNA is used to measure the blank.

12. Amount of DNA is normally shown in ng/μL.

13. DNA is a molecule which absorbs at 260 nm. Therefore, to assess the purity of the extracted DNA, the ratio of absorbance at 260 and 280 nm (A260/280) is determined. A ratio of about 1.8 is generally accepted as "pure" for DNA. If the ratio is lower, it may indicate the presence of protein, phenol or other contaminants that absorb strongly at or near 280 nm.

14. The ratio of absorbance at 260 and 230 nm (A260/230) may be used as a secondary measurement of DNA purity. A ratio of about 2.0–2.2 is generally accepted as "pure" for DNA. If the ratio is lower, it may indicate the presence of contaminants which absorb at 230 nm.

15. The quantity and quality of DNA samples may be also checked in agarose gel electrophoresis by comparison to DNA standards [31]. To prepare the gel and carry out the electrophoresis check Sambrook et al. [32].

16. Alternatively, the primers F1 and R1 have been also used in a conventional PCR method for detecting potentially ZEA-producing molds [33].

17. The SYBR Green mix (and the later prepared PCR mixture) has to be protected from an excessive exposure to light since the fluorescent dye is light sensitive.

18. Store the primers (and probes when used) in small aliquots to avoid contamination and multiple freeze and thaw cycles.

19. If the SYBR Green Mix does not have passive reference dye (normally ROX), it should be added as additional PCR component. ROX reference dye normalizes signal and ensures data integrity.

20. Make sure that the instrument is calibrated for the dyes (e.g., SYBR Green, FAM, and HEX) used in the protocol.

21. A quantification assay must be selected. Normally, when DNA is used as target, absolute quantification using the standard curve method is performed.

22. In most qPCR software at least three types of samples can be selected: (a) non-template samples consisting of a sample which does not contain template; (b) standard which consists of a sample with known concentration used to construct a standard curve. By running standards of several concentrations, a standard curve is created for extrapolating the quantity of an unknown sample; (c) unknown sample consisting of a sample containing an unknown quantity of template to be quantified.

23. If the used SYBR Green mix (or TaqMan mix when used) contains uracil DNA glycosylase (UDG) for preventing non-template DNA amplification an optional short incubation at 50 °C is performed prior to the PCR cycling [30]. After this

step necessary for the cleaving of the enzyme to the uracil residues from any contaminating DNA, the UDG is inactivated in the ramp up to 95 °C in the initial denaturation step [30].

24. When carrying out a SYBR Green method, a melting curve (also called dissociation curve) has to be included immediately following the thermal cycling protocol [30]. The melting curve is generated for checking the specificity of the amplification products.

25. DNA can be also diluted using the elution buffer provided by commercial extraction kits or TE buffer.

26. The PCR mixture must be prepared in a UV-equipped PCR workstation (also called PCR hood) to prevent cross contamination of samples. The working surface of the PCR workstation should be cleaned with 70% (v/v) ethanol before and after use. Besides UV light of the equipment should be switched on before use for its decontamination.

27. Sterile gloves and filter pipette tips should be used for preventing the introduction of contaminating nucleases or DNA. The gloves must be changed whenever there is a suspicion of being contaminated. Additionally, all sample and reagent tubes should be opened and closed carefully and kept capped as much as possible for avoiding contamination.

28. When preparing the PCR mix including the reagents for all samples excepting DNA, an extra volume has to be considered for the calculations because of losses during transfer of reaction components. Take into account the number of all unknown and non-template samples to calculate the number of the reactions.

29. Total volume per reaction could be decreased up to 12.5 μL. For this, the volume of all PCR components should be decreased proportionally in relation to the final reaction volume.

30. It is recommended to include the following non-template samples (or negative controls): (a) non-template control (water + SYBR Green/TaqMan Master Mix + primers); (b) non-primer/non-template control (water + SYBR Green/TaqMan Master Mix); and (c) non-template DNA control (water + SYBR Green/TaqMan Master Mix + primers + non-template DNA).

31. The PCR mixture can be loaded in different formats: PCR plates (48-, 96-, or 384-well plates depending on the necessity) or PCR tubes (single tubes or eight-strip tubes). When using the plates, they have to be sealed by using adhesive films after filling in the wells with the PCR components.

32. When qPCR plate or PCR tubes are loaded, prevent cross contamination. Thus avoid passing hand above the samples when pipetting template DNA (or non-template DNA) and trying

to keep reactions and components capped as much as possible. Open and close tubes carefully too.

33. Avoid open the PCR tubes containing template DNA (or non-template DNA) once it has been added.

34. If PCR tubes are used, they should be previously inserted into a 96-well tray for centrifuging.

35. If the plate or tubes are not going to be loaded into the qPCR instrument immediately, place them at refrigeration in dark conditions.

36. Avoid touch the optical surface of the cap or sealing film without gloves, as fingerprints may interfere with fluorescence measurements.

37. Make sure that the non-template samples did not amplify in order to ratify that the qPCR run was adequate.

38. When optimizing the qPCR protocol the presence or absence of the expected PCR product should be checked in a 1.5% (w/v) agarose gel stained with ethidium bromide and visualized under UV illumination. Since ethidium bromide is a hazardous compound (consult the safety data sheet provided by the supplier), safer commercially available alternatives to this can be used for staining the agarose gels. The gels stained with them can be visualized and photographed with the same equipment used for those stained with ethidium bromide. For a more precise verification of the adequate amplification, the obtained PCR product can be sequenced and its sequence is then compared with the available sequences in GenBank (http://www.ncbi.nlm.nih.gov/genbank/).

39. The standard curve could be also generated by using serial dilutions of a PCR product of the corresponding gene (larger than the expected qPCR fragment). The concentration of such amplicon has to be previously determined to calculate the number of copies. In this case the standard curve consists of a plot of the Cq versus log gene copy number.

40. The TaqMan probes (and the later prepared PCR mixture) are light sensitive. They have to be stored and processed away from light.

41. If the TaqMan Master Mix does not have passive reference dye (normally ROX), it should be added as additional PCR component. ROX reference dye normalizes signal and ensures data integrity.

42. Other competitive IAC could be used provided that the non-target DNA fragment is flanked by the same primer sites.

43. This procedure could be performed without IAC replacing the volume of IAC and IACp from the PCR mix with sterile ultra-pure water.

44. Check that the IAC is amplified in all the samples to test the validity of the qPCR run. This is necessary to evaluate the absence of false-negative results.

Acknowledgments

This work was supported by AGL2013-45729-P, AGL2010-21623, and Carnisenusa CSD2007-00016—Consolider Ingenio 2010 from the Spanish Comision Interministerial de Ciencia y Tecnología projects and GR10162 of the Junta de Extremadura and FEDER.

References

1. Perrone G, Gallo A, Susca A (2010) *Aspergillus*. In: Liu D (ed) Molecular detection of foodborne pathogens. Taylor and Francis Group, LLC, Boca Raton, pp 529–548

2. Rodríguez A, Rodríguez M, Luque MI et al (2012) A comparative study of DNA extraction methods to be used in real-time PCR based quantification of ochratoxin A-producing molds in food products. Food Control 25:666–672

3. Kim Y-T, Lee Y-R, Jin J et al (2005) Two different polyketide synthase genes are required for synthesis of zearalenone in *Gibberella zeae*. Mol Microbiol 58:1102–1113

4. Marasas WFO, Nelson PE, Toussoun TA (1984) Toxigenic *Fusarium* species; identity and mycotoxicology. The Pennsylvania State University Press, University Park, PA

5. Stob J, Baldwin RS, Tuite J et al (1962) Isolation of an anabolic, utertrophic compound from corn infected with *Gibberella zeae*. Nature 196:1318

6. Gaffoor I, Trail F (2006) Characterization of two polyketide synthase genes involved in zearalenone biosynthesis in *Gibberella zeae*. Appl Environ Microbiol 72:1793–1799

7. Lysøe E, Klemsdal SS, Bone KR et al (2006) The PKS4 gene of *Fusarium graminearum* is essential for zearalenone production. Appl Environ Microbiol 72:3924–3932

8. Meng K, Wang Y, Yang P et al (2010) Rapid detection and quantification of zearalenone-producing *Fusarium* species by targeting the zearalenone synthase gene PKS4. Food Control 21:207–211

9. Atoui A, El Khoury A, Kallassy M et al (2012) Quantification of *Fusarium graminearum* and *Fusarium culmorum* by real-time PCR system and zearalenone assessment in maize. Int J Food Microbiol 154:59–65

10. Klich M, Mendoza C, Mullaney E et al (2001) A new sterigmatocystin-producing *Emericella* variant from agricultural desert soils. Syst Appl Microbiol 24:131–138

11. Frisvad JC, Skouboeb P, Samson RA (2005) Taxonomic comparison of three different groups of aflatoxin producers and a new efficient producer of aflatoxin B1, sterigmatocystin and 3-O-methylsterigmatocystin, *Aspergillus rambellii* sp. nov. Syst Appl Microbiol 28:442–453

12. Rank C, Nielsen KF, Larsen TO et al (2011) Distribution of sterigmatocystin in filamentous fungi. Fungal Biol 115:406–420

13. Rodríguez A, Córdoba JJ, Gordillo R et al (2012) Development of two quantitative real-time PCR methods based on SYBR Green and TaqMan to quantify sterigmatocystin-producing molds in foods. Food Anal Methods 5:1514–1525

14. Seo JA, Guan Y, Yu JH (2003) Suppressor mutations bypass the requirement of fluG for asexual sporulation and sterigmatocystin production in *Aspergillus nidulans*. Genetics 165:1083–1093

15. Shwab EK, Keller NP (2008) Regulation of secondary metabolite production in filamentous ascomycetes. Mycol Res 112:225–230

16. Wannemacher RW, Bunner DL, Neufeld HA (1991) Toxicity of trichothecenes and other related mycotoxins in laboratory animals. In: Smith JE, Henderson RS (eds) Mycotoxins and animal foods. CRC Press, Boca Raton, pp 499–552

17. Frisvad JC, Smedsgaard J, Larsen TO et al (2004) Mycotoxins, drugs and other extrolites produced by species in *Penicillium* subgenus *Penicillium*. Stud Mycol 49:201–241

18. Núñez F, Westphal CD, Bermúdez E et al (2007) Production of secondary metabolites by some terverticillate *Penicillia* on carbohydrate-rich and meat substrates. J Food Prot 70:2829–2836

19. Sabater-Vilar M, Nijmeijer S, Fink-Gremmels J (2003) Genotoxicity assessment of five tremorgenic mycotoxins (fumitremorgen B, paxilline, penitrem A, verruculogen, and verrucosidin) produced by molds isolated from fermented meats. J Food Prot 66:2123–2129

20. Díaz GJ, Thompson W, Martos PA (2010) Stability of cyclopiazonic acid in solution. World Mycotoxin J 3:25–33

21. Pildain MB, Frisvad JC, Vaamonde G et al (2008) Two novel aflatoxin-producing *Aspergillus* species from Argentinean peanuts. Int J Syst Evol Microbiol 58:725–735

22. Rodríguez A, Werning ML, Rodríguez M et al (2012) Quantitative real-time PCR method with internal amplification control to quantify cyclopiazonic acid producing molds in foods. Food Microbiol 32:397–405

23. Chang PK, Ehrlich KC (2011) Cyclopiazonic acid biosynthesis by *Aspergillus flavus*. Toxin Rev 30:79–89

24. Chang PK, Horn BW, Dorner JW (2009) Clustered genes involved in cyclopiazonic acid production are next to the aflatoxin biosynthesis gene cluster in *Aspergillus flavus*. Fungal Genet Biol 46:176–182

25. Niessen L (2007) PCR-based diagnosis and quantification of mycotoxin producing fungi. Int J Food Microbiol 119:38–46

26. Selmanoğlu G (2006) Evaluation of the reproductive toxicity of patulin in growing male rats. Food Chem Toxicol 44:2019–2024

27. Puel O, Galtier P, Oswald IP (2010) Biosynthesis and toxicological effects of patulin. Toxins 2:613–631

28. Rodríguez A, Luque MI, Andrade MJ et al (2011) Development of real-time PCR methods to quantify patulin-producing molds in food products. Food Microbiol 28:1190–1199

29. Mohale S, Medina A, Rodríguez A et al (2013) Mycotoxigenic fungi and mycotoxins associated with stored maize from different regions of Lesotho. Mycotoxin Res 29:209–219

30. Invitrogen (2008) Real-Time PCR: from theory to practice. Invitrogen Corporation, Carlsbad

31. Sánchez B, Rodríguez M, Casado EM et al (2008) Development of an efficient fungal DNA extraction method to be used in random amplified polymorphic DNA-PCR analysis to differentiate cyclopiazonic acid mold producers. J Food Prot 71:2497–2503

32. Sambrook J, Fritsch EF, Maniatis T (1989) Molecular cloning: a laboratory manual, 2nd edn. Cold Spring Harbor Laboratory Press, New York

33. Castañares E, Dinolfo MI, Moreno MV et al (2013) *Fusarium cerealis* associated with barley seeds in Argentina. J Phytopathol 161:586–589

Chapter 15

Evaluating Aflatoxin Gene Expression in Aspergillus Section Flavi

Paula Cristina Azevedo Rodrigues, Jéssica Gil-Serna, and M. Teresa González-Jaén

Abstract

The determination of aflatoxin production ability and differentiation of aflatoxigenic strains can be assessed by monitoring the expression of one or several key genes using reverse transcription polymerase chain reaction (RT-PCR). We herein describe the methods for RNA induction, extraction, and quality determination, and the RT-PCR conditions used to evaluate the ability of a given *Aspergillus* strain to produce aflatoxins.

Key words *Aspergillus flavus*, Mycotoxins, Aflatoxigenic fungi, RNA extraction, RT-PCR, Gel electrophoresis

1 Introduction

Aflatoxins (AF) are the most widely studied of all mycotoxins produced by *Aspergillus* species. Although aflatoxin production ability has been detected in various species, *A. flavus* Link:Fr. and *A. parasiticus* Speare (belonging to *Aspergillus* section *Flavi*) remain the most important and representative aflatoxin producers occurring naturally in food commodities. Molecular techniques have been widely applied in order to discriminate between aflatoxigenic and non-aflatoxigenic strains, through the correlation of presence/absence of genes involved in the aflatoxin biosynthetic pathway with the ability/inability to produce aflatoxins. However, AF biosynthesis is based on a highly complex pathway which requires at least 25 structural and 2 regulatory genes [1], with possible alternative pathways. Additionally, there are reports on genes that are present but not expressed, even under highly aflatoxin-inductive conditions [2]. Furthermore, it is important to highlight that some genes are not exclusive of the aflatoxin biosynthetic pathway, which could create false positives in the case of sterigmatocystin-producing fungi [3] such as *Aspergillus nidulans*. More recently, aflatoxin

Antonio Moretti and Antonia Susca (eds.), *Mycotoxigenic Fungi: Methods and Protocols*, Methods in Molecular Biology, vol. 1542, DOI 10.1007/978-1-4939-6707-0_15, © Springer Science+Business Media LLC 2017

production and aflatoxigenic strain differentiation are being assessed by monitoring the expression of one or several key genes using reverse transcription polymerase chain reaction (RT-PCR). Such systems have been applied to monitor AF production and biosynthetic gene expression based on various regulatory and structural AF pathway genes in *A. parasiticus* and/or *A. flavus* [2, 4–7]. Although with different levels of success, they were found to be rapid, sensitive, and reliable.

2 Materials

Prepare all solutions using ultrapure water and analytical grade reagents. Prepare and store all reagents at room temperature (unless indicated otherwise). All materials and solutions involved in RNA-handling procedures must be RNase-free. Wear gloves during the whole process when working with RNA to protect samples from degradation by RNases.

2.1 Mycotoxin Safety Precautions

All the necessary safety precautions must be taken into account when handling mycotoxin solutions or other potentially contaminated materials since they are highly toxic and potent carcinogenic compounds. Handle contaminated material with protective gear; decontaminate all disposable materials by autoclaving before being disposed; decontaminate reusable materials by immersion in 10% commercial bleach overnight, followed by immersion in 5% acetone for 1 h and washing with distilled water several times.

2.2 Media Preparation

1. Malt extract agar (MEA): Malt extract 20 g/L, glucose 20 g/L, peptone 1 g/L, agar 20 g/L. Mix the components, autoclave (121 °C, 20 min), and plate in 90 cm Petri dishes.

2. Yeast extract sucrose (YES) broth: Yeast extract 20 g/L, sucrose 150 g/L. Mix the components, autoclave (121 °C, 20 min). Distribute 25 mL of YES in 50 mL Falcon tubes.

3. Yeast extract peptone (YEP) broth: Yeast extract 20 g/L, peptone 150 g/L. Mix the components, autoclave (121 °C, 20 min). Distribute 25 mL of YES in 50 mL Falcon tubes.

2.3 RNA Extraction

1. Paper towels: Cover a stack of paper towels by aluminum foil and sterilize in a sterilization oven at 160 °C, overnight.

2. Spatula, mortar, and pestle: Cover by aluminum foil, sterilize in a sterilization oven at 160 °C, overnight, and refrigerate (−20 °C) before use.

3. Eppendorf tubes, PCR tubes, pipette tips: Sterilize by autoclave (121 °C, 1 h). Whether possible, use RNase-free filter pipette tips.

4. Liquid nitrogen.

5. RNeasy Plant Mini Kit (Qiagen) (*see* **Note 1**).

2.4 RNA Analysis by Gel Electrophoresis

1. RNase-free water: Treat ultrapure water with 0.1% diethyl pyrocarbonate (DEPC) (v/v), mix thoroughly, and store overnight. Autoclave at 121 °C for 1 h to eliminate DEPC. Prepare all solutions with DEPC-treated water (*see* **Notes 2** and **3**).

2. EDTA 0.5 M: pH 8.0: Weigh 93.05 g EDTA-Na$_2$ (FW = 372.2). Dissolve in 400 mL RNase-free water and adjust to pH 8.0 with NaOH. Make up to a final volume of 500 mL with water (*see* **Note 4**).

3. Tris-acetate-EDTA (TAE 50×): 2 M Tris-acetate, 0.05 M EDTA. Weigh 242 g Tris base (FW = 121.14) and dissolve in approximately 750 mL of RNase-free water (*see* **Note 5**). Carefully add 57.1 mL glacial acetic acid and 100 mL of 0.5 M EDTA (pH 8.0) previously prepared. Adjust the solution up to a final volume of 1 L. The pH of this buffer does not need to be adjusted and should be about 8.5. Store in the dark at room temperature.

4. Tris-acetate-EDTA (TAE 1×): Dilute the stock solution TAE 50× in RNase-free water. For example, to prepare 1 L of TAE 1×, dilute 20 mL of TAE 50× in 980 mL of water. Final solution contains Tris–HCl (40 mM), glacial acetic acid (40 mM), and EDTA (1 mM).

5. SDS washing solution (SDS 10%): Weight 50 g of sodium dodecyl sulfate (SDS) and dissolve in RNase-free water to a final volume of 500 mL (*see* **Note 6**).

6. DNA/RNA dye (*see* **Notes 7** and **8**).

7. Non-denaturing agarose gel (*see* **Note 9**): Prepare a 1.2% agarose gel in TAE 1× and add the recommended amount of DNA/RNA dye.

8. RNA loading buffer (6×): 30% (v/v) glycerol; 0.25% (w/v) bromophenol blue. Store at 4 °C.

9. RNA molecular weight marker.

10. Horizontal electrophoresis apparatus (*see* **Note 10**).

11. Ultraviolet (UV) transilluminator (preferentially coupled to a gel image analysis software).

2.5 RNA Analysis by Spectrophotometry

1. Tris-EDTA (TE) buffer: 10 mM Tris–HCl, 1 mM EDTA (pH 8.0). Add 1 mL of a 1 M Tris–HCl (pH 8.0) stock solution and 0.2 mL of 0.5 M EDTA (pH 8.0) stock solution to 98.8 mL of RNase-free water. Store at room temperature.

2. Spectrophotometer.

3. Quartz cuvette.

Table 1
Details of the target genes, primer sequences and expected product length in base pairs (bp) for PCR and RT-PCR

Primer pair	Gene	Primer sequence (5′ → 3′)	PCR product size (bp)	RT-PCR product size (bp)	Reference
Tub1-F	*tub1*	GCT TTC TGG CAA ACC ATC TC	1406	1198	[5]
Tub1-R		GGT CGT TCA TGT TGC TCT CA			
Ord1-gF	*aflQ*	TTA AGG CAG CGG AAT ACA AG	719	599	[4]
Ord1-gR		GAC GCC CAA AGC CGA ACA CAA A			

2.6 Analysis of Gene Expression

1. One-Step RT-PCR Pre-Mix (*see* **Note 11**).
2. Primers for *β-tubulin* and *aflQ* genes (Table 1).
3. RNase-free water.
4. RNase-free filter tips.
5. Agarose gel and electrophoresis apparatus (as described for RNA analysis).
6. DNA molecular weight marker (100 pb ladder or similar).

3 Methods

3.1 Biological Material Preparation

1. Grow the isolates under both AF inductive and noninductive conditions (*see* **Note 12**). For that, inoculate a loop full of spores from a 7-day-old culture in MEA into 25 mL of YES (AF inductive) and YEP (noninductive) broths (in 50 mL Falcon tubes).
2. Incubate the cultures horizontally for 4 days at 28 °C, in the dark, with slight agitation (100 rpm).
3. Collect the mycelium with a sterilized spatula, dry the mycelium in sterilized absorbent paper, and rapidly divide it into aliquots of 100 mg.
4. Preserve the mycelium at −80 °C until use or proceed with the RNA extraction protocol immediately (*see* **Note 13**).

3.2 RNA Extraction

1. The Qiagen RNeasy Plant Mini Kit is used for RNA isolation according to the manufacturer's protocol. Grind 100 mg of mycelium to a fine powder with liquid nitrogen (N_2) in a cold mortar and pestle (*see* **Note 14**).
2. Transfer the powder with a residual amount of N_2 into a 2.0 mL Eppendorf tube previously refrigerated by immersion in N_2.

3. Leave the N_2 to evaporate completely and immediately follow the extraction protocol as described by the manufacturer (*see* **Notes 15–17**).

4. Store RNA at –70 °C in 5 µL aliquots, to avoid repeated freeze and thaw that would damage RNA.

3.3 RNA Analysis

*3.3.1 Native (Non-denaturing) AgaroseGel Electrophoresis (See **Note 9**)*

Determine general quality and yield of extracted RNA, as well as contamination with genomic DNA, by native agarose gel electrophoresis.

1. Thaw a 5 µL aliquot of each RNA sample (at all times kept on ice) and add 1 µL of 6× loading buffer. Gently mix by reflux and load into the gel (on native gels, the samples are loaded directly without heating).

2. Make sure to include an RNA marker and/or a positive control RNA (commercial RNA or one of your samples known to be intact) in the gel to rule out unusual results due to gel artifacts and to aid in yield determination.

3. Run the gel in TAE buffer, at constant voltage of 5 V/cm (measured between the electrodes) for approximately 1 h.

4. Observe the gel under UV light. Compare fluorescence intensities between samples and standards, and estimate RNA concentration. Even though you might want a more accurate RNA quantitation (*see* below), the gel is still essential to determine RNA quality in terms of degradation and contamination (with protein or genomic DNA) (Fig. 1).

*3.3.2 Spectrophotometry (See **Note 18**)*

1. Place the sample in a quartz cuvette. Zero the spectrophotometer with the solvent. For accurate readings, dilute the sample with TE (*see* **Note 19**) to obtain absorbance (optical density, OD) values between 0.1 and 1.0.

2. The OD at 260 nm (OD260) equals 1.0 for a 40 µg/mL solution of RNA. For RNA concentration apply the following calculation: RNA concentration = 40 µg/mL × OD260 × dilution factor.

3. For an indication of RNA purity, calculate the OD260/OD280 and OD A260/A230 ratios. For pure RNA, both ratios should be very close to 2.0 in TE (*see* **Note 20**) [8]. Lower or higher ratios could be caused by protein, salts, or ethanol contamination.

4. Dilute some aliquots of RNA in water to obtain a working solution of approximately 1 µg/mL and retain others at the original concentration (stock solution).

5. Store RNA at –70 °C in 5 µL aliquots.

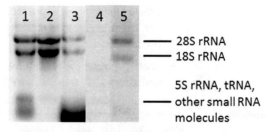

Fig. 1 Native (non-denaturing) agarose gel electrophoresis of RNA samples obtained by different maceration and extraction methods, showing various types of RNA molecules: 1—N$_2$, RNeasy Plant Mini Kit (RLT buffer); 2—N$_2$, RNeasy Plant Mini Kit (RLC buffer); 3—N2, Trizol method (Invitrogen); 4—Glass Beads, RNeasy Plant Mini Kit (RLC buffer); 5—TissueRuptor (Qiagen), RNeasy Plant Mini Kit (RLC buffer)

3.4 Analysis of Key Aflatoxin Gene Expression

1. Perform a Multiplex RT-PCR with the obtained RNA (1 μg/mL) using a One-Step RT-PCR Premix (e.g., iNtRON Biotechnology) (*see* **Note 11**).

2. Prepare the mix as described in Table 2, or adjust to the manufacturer's instructions.

3. Prepare a multiplex reaction by using both primer pairs Ord1-gF/gR and Tub1-F/R (Table 1) in the same tube. Primer pair Ord1-gF/gR will amplify the aflatoxin-related gene *aflQ* (formerly *ord1*) gene (*see* **Note 21**). The pair Tub1-F/R will amplify a part of the housekeeping β-tubulin gene *tub1*, which will be used as internal control of amplification (*see* **Note 22**).

4. Set the amplification program in the thermal cycler as described in Table 2.

5. Check for contamination with genomic DNA. Carry out a PCR as described for the amplification step of RT-PCR (Table 2), using the same primers and 1 μg of total RNA as template (*see* **Note 16**). Use the following PCR mix: Taq buffer 1×, MgCl$_2$ 1.5 mM, dNTPs 0.2 mM, each primer 0.2 μM, Taq 1 U (e.g., GoTaq® Flexi DNA Polymerase, Promega), 1 μg of RNA, make up to 20 μL with ultra pure water.

6. Prepare a 1.2% agarose gel in TAE 1× (not necessary to be cautious such as in the case of gels to run RNA). Confirm that you have amplification for the internal control (Fig. 2). The absence of a band at the internal control position (Fig. 2, lane 4) reflects a failed reaction, potentially due to bad RNA quality or amplification inhibitors (false negative). The presence of a product with the expected RT-PCR size confirms aflatoxin gene expression (Fig. 2, lanes 2 and 3) whereas its absence implies no expression (Fig. 2, lanes 1 and 5). The presence of a band with the PCR expected size confirms genomic DNA contamination, but that will not interfere with your analysis,

Table 2
RT-PCR conditions used for the multiplex amplification of genes *tub1* and *aflQ*

Reaction mix (20 μL)		
One-step RT-PCR pre-mix	8 μL	
Each primer forward	0.2 μM	
Each primer reverse	0.2 μM	
Total RNA	1 μg	
Amplification program		
Reverse transcription	45 °C, 30 min	
Initial denaturation	94 °C, 4 min	
Denaturation	94 °C, 1 min	5×
Annealing	60 °C, 1 min	
Extension	72 °C, 1 min	
Denaturation	94 °C, 1 min	30×
Annealing	55 °C, 1 min	
Extension	72 °C, 1 min	
Final extension	72 °C, 6 min	

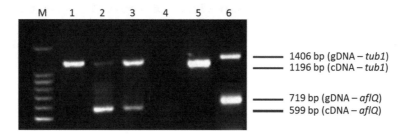

Fig. 2 Electrophoretic pattern of RT-PCR products for *Aspergillus flavus* and *Aspergillus parasiticus* isolates. Lanes: *M*—100 bp DNA ladder (Promega); *1* and *5*—*A. flavus* AF non-producing strain; *2* and *3*—*A. parasiticus* AF producing-strain; *4*—false negative result for *A. parasiticus*; *6*—*A. parasiticus* DNA-PCR control

because primers have been constructed in such a way that genomic DNA and cDNA amplification products will have different sizes (*see* **Note 16**).

4   Notes

1. The *RNeasy Plant Mini Kit* (Qiagen) is one of the most cited methods for fungal RNA extraction and it seems to show the best results for RNA extraction from *Aspergillus* mycelium and conidia. It is, though, more expensive than other routine

protocols. An alternative protocol using the *TRIzol* reagent (Invitrogen) is also available (Fig. 1) [9].

2. Diethylpyrocarbonate (DEPC) treatment is the most commonly used method for eliminating RNase contamination from water, buffers, and other solutions, as it destroys enzymatic activity by modifying –NH, –SH, and –OH groups in RNases. Solutions containing Tris and EDTA cannot be DEPC-treated. Solutions that cannot withstand autoclaving also cannot be DEPC-treated since autoclaving is essential for inactivating DEPC. It is thus preferable to prepare all solutions with DEPC-treated water instead of treating the solutions themselves. DEPC will dissolve some plastics; glass should be used whether possible.

3. DEPC is highly toxic (oral, dermal, and inhalation), so take special care while handling the reagent. DEPC must be always handled at the fume hood wearing high-protection gloves. When mixing DEPC with water, take special care to avoid spilling; make sure that the flask is tightly closed and even cover the flask with absorbent paper. After the treatment, autoclave the gloves and any other material that has been in contact with DEPC. After being autoclaved, DEPC-treated water is safe to be handled without special care.

4. EDTA solutions must be prepared ahead of time since EDTA dissolution only takes place when the pH is about 8.

5. Tris will dissolve better if you already have *ca.* 100 mL of water and a magnetic stirrer in the cylinder before you add the salt.

6. Wear face mask or use the fume hood when preparing SDS solutions to avoid inhalation of SDS dust. SDS is synonymous to sodium lauryl sulfate. 10% SDS solution will precipitate at room temperature and this solution has to be kept at 37 °C.

7. There are numerous new generation fluorescent DNA and RNA dyes designed to replace the highly toxic ethidium bromide (EtBr) such as SYBR Green I, Gel Red, or Green Safe. The amount of dye recommended by the manufacturer is usually excessive, and you can try to reduce it by one-half or one-third. However, depending on the sample a loss of sensibility might occur using these dyes.

8. If you are not able to avoid EtBr, it is preferable to add it directly to the gel (0.5 µg/mL) to avoid the additional step of gel staining (potentially RNase-prone). EtBr is highly toxic and potentially carcinogenic; make sure that you wear protective gear (highly protective gloves and goggles) when handling it and that you use it in a confined and appropriately identified area.

9. A denaturing gel system (which involves the use of acrylamide, TEMED and formamide) is sometimes suggested because

RNA might form secondary structures. Denaturing conditions prevent RNA from migrating strictly according to its size. Native agarose gel electrophoresis is sufficient to judge the integrity and overall quality of a total RNA preparation by inspection of the 28S and 18S rRNA bands (Fig. 1). Bands are generally not as sharp as in denaturing gels, but native gels are safer and easier to prepare.

10. Use electrophoresis equipment (tank, trays, and combs) exclusively for RNA analysis, and wash it regularly with 10% SDS and RNase-free water.

11. You may choose between one-step RT-PCR or two-step RT-PCR procedures. Both have pros and cons. Using sequence-specific primers, it might be better the former since it allows easier processing of large numbers of samples and helps minimize carryover contamination (all steps happen in the same tube). However, in some situations two-step procedures are the best option. Independent PCR reactions need to be performed if: (i) you want to test in the same cDNA sample the expression of several genes which require different amplification conditions; (ii) the amplification products are similar in size or; (iii) some interference or cross reaction might be suspected.

12. In order to confirm that AF genes are only expressed under inductive conditions, some isolates should also be tested on YEP (non-AF inductive) broth. It is important to perform this analysis in a wide range of isolates. While testing your method, the presence or absence of AF must be checked by HPLC in both YEP and YES broths used for fungal growth to confirm the correlation between expression of the test gene and AF production. Because AF production is extremely dependent on growth conditions, it is important to determine aflatoxigenic ability under the same test conditions as gene expression. The description of the HPLC method for AF analysis is not within the scope of this text.

13. RNA extraction should be performed on freshly produced material immediately after harvest to avoid RNA degradation. If you are not able to carry out RNA isolation immediately, you must store the harvested mycelium either at –70 °C or at 4 °C after immersion in an RNA-stabilizing solution (e.g., *RNAlatter*, Ambion) until use.

14. The maceration of biological material for RNA extraction is probably the most important and critical step of the procedure. Maceration with liquid nitrogen will result in higher RNA yield and quality (Fig. 1), but requires extra care and skills to avoid RNA contamination and degradation.

15. Using the Qiagen protocol, we found RLC extraction buffer to work slightly better than RLT buffer for *Aspergillus* mycelium and conidia (Fig. 1).

16. A DNase treatment is recommended to avoid contamination with genomic DNA but it is not mandatory if you choose primers that differentially amplify genomic DNA (gDNA) and complementary DNA (cDNA). Make sure to select primers that span a part of the gene containing at least one intron. That way, you can easily differentiate gDNA from cDNA on the basis of the amplification product size (Fig. 2; Table 1).

17. At the final step of the procedure, elute the RNA in water instead of Elution Buffer to avoid buffer interferences in subsequent reactions.

18. RNA analysis can be performed using a NanoDrop spectrophotometer (ThermoScientific), which is simpler to use and requires smaller amounts of sample than classic spectrophotometers.

19. OD ratios will vary depending on the solvent. While RNA concentration is independent of the solvent you use, OD260/230 and OD260/280 ratios are more reliable if TE is used as solvent (turning pH-dependent variations in the OD230 and OD280 readings null). RNA samples are eluted with water for optimal subsequent reactions, but for spectrophotometer analysis it is better to dilute the samples in TE.

20. In water, OD260/230 and OD260/280 ratios are expected to be 1.8–2 [8].

21. Besides *aflQ*, other key genes in the aflatoxin pathway have been used, e.g., *aflD*, *aflO*, *aflP*, and *aflR* [2, 4–7]. We recommend *aflQ*, since it is the last known gene in the pathway necessary for aflatoxin production and the only one specific for aflatoxin producers.

22. When testing isolates for presence/absence of specific genes, you must include an internal control, which consists of a housekeeping gene, universally expressed in all isolates tested regardless of aflatoxin production ability and culture conditions. Make sure that you choose control primers that work exactly under the same conditions as your test primers. The detection of expression of this internal control is mandatory to rule out false negatives.

References

1. Yu J, Bhatnagar D, Cleveland TE (2004) Completed sequence of aflatoxin pathway gene cluster in *Aspergillus parasiticus*. FEBS Lett 564:126–130

2. Rodrigues P, Venâncio A, Kozakiewicz Z, Lima N (2009) A polyphasic approach to the identification of aflatoxigenic and non-aflatoxigenic strains of *Aspergillus* Section *Flavi* isolated from

Portuguese almonds. Int J Food Microbiol 129:187–193

3. Paterson RPM (2006) Identification and quantification of mycotoxigenic fungi by PCR. Process Biochem 41:1467–1474

4. Sweeney MJ, Pàmies P, Dobson ADW (2000) The use of reverse transcription-polymerase chain reaction (RT-PCR) for monitoring aflatoxin production in *Aspergillus parasiticus* 439. Int J Food Microbiol 56:97–103

5. Scherm B, Palomba M, Serra D et al (2005) Detection of transcripts of the aflatoxin genes *aflD*, *aflO*, and *aflP* by reverse-transcription-polymerase chain reaction allows differentiation of aflatoxin-producing isolates of *Aspergillus flavus* and *Aspergillus parasiticus*. Int J Food Microbiol 98:201–210

6. Degola F, Berni E, Dall'Asta C et al (2007) A multiplex RT-PCR approach to detect aflatoxigenic strains of *Aspergillus flavus*. J Appl Microbiol 103:409–417

7. Jamali M, Karimipour M, Shams-Ghahfarokhi M et al (2013) Expression of aflatoxin genes *aflO* (*omtB*) and *aflQ* (*ordA*) differentiates levels of aflatoxin production by *Aspergillus flavus* strains from soils of pistachio orchards. Res Microbiol 164:293–299

8. Heptinstall J, Rapley R (2000) Spectrophotometric analysis of nucleic acids. In: Rapley R (ed) The nucleic acid protocols handbook. Humana Press, USA

9. Rio DC, Ares M Jr, Hannon GJ, Nilsen TW (2010) Purification of RNA using TRIzol (TRI reagent). Cold Spring Harb Protoc (6):pdb.prot5439

Chapter 16

Evaluating Fumonisin Gene Expression in *Fusarium verticillioides*

Valeria Scala, Ivan Visentin, and Francesca Cardinale

Abstract

Transcript levels of key genes in a biosynthetic pathway are often taken as a proxy for metabolite production. This is the case of *FUM1*, encoding the first dedicated enzyme in the metabolic pathway leading to the production of the mycotoxins Fumonisins by fungal species belonging to the genus *Fusarium*. *FUM1* expression can be quantified by different methods; here, we detail a protocol based on quantitative reverse transcriptase polymerase chain reaction (RT-qPCR), by which relative or absolute transcript abundance can be estimated in *Fusaria* grown in vitro or *in planta*. As very seldom commercial kits for RNA extraction and cDNA synthesis are optimized for fungal samples, we developed a protocol tailored for these organisms, which stands alone but can be also easily integrated with specific reagents and kits commercially available.

Key words *F. verticillioides*, *FUM* pathway, *FUM1*, mRNA, RT-qPCR

1 Introduction

Several fungal species belonging to the *Fusarium* genus are known to synthesize health-hazardous Fumonisins, the most common producer being *Fusarium verticillioides* in infected maize kernels [1]. All Fumonisin producers will coordinately express several Fumonisin-related (*FUM*) genes to generate toxin molecules and extrude them in the environment. The first dedicated biosynthetic gene in the pathway is *FUM1*, encoding a crucial polyketide synthase involved at an early step in the assembly of the Fumonisin backbone. Its transcript levels in the cell are often taken as an indicator of activity in the whole *FUM* pathway and of Fumonisin accumulation in fungal biomass or in different matrices [2]. While this is generally true, conditions do exist that are known to disrupt coordinated transcription of the *FUM* cluster as a whole [3]; because of this, caution should be exerted in inferring metabolite production from transcript quantification data.

Antonio Moretti and Antonia Susca (eds.), *Mycotoxigenic Fungi: Methods and Protocols*, Methods in Molecular Biology, vol. 1542, DOI 10.1007/978-1-4939-6707-0_16, © Springer Science+Business Media LLC 2017

Real-time polymerase chain reaction (PCR), also called quantitative PCR (qPCR), and reverse transcriptase qPCR (RT-qPCR) are PCR-based techniques used to amplify and simultaneously detect or quantify target DNA or cDNA molecules, respectively. The detection range spans six orders of magnitude, which confers both sensitivity and a wide dynamic range upon these techniques. qPCR allows quantification of the desired product at any point in the amplification process by measuring fluorescence in the reaction sample. The alternatives for amplicon detection/quantification in qPCR are (1) nonspecific fluorescent dyes that intercalate into any double-stranded DNA, or (2) sequence-specific DNA probes labelled with a fluorescent reporter. No probes of the latter type have been reported yet for genes in the *FUM* cluster at present, while several primer pairs have been designed to work in conventional PCR [4] and in qPCR reactions followed by detection with intercalating dyes such as SYBR Green. The protocols described here will allow rather straightforward quantification of *FUM1* transcript abundance in *F. verticillioides* grown in different matrices.

2   Materials

2.1   Media and Solutions

1. Czapek medium amended with yeast extract (CDY; not inductive for Fumonisin production, useful as a negative control): $ZnSO_4$ 5 mg/L; $NaMoO_4$ 1 mg/L; yeast extract 0.5 % w/v; $NaNO_3$ 2 g/L; KH_2PO_4 1 g/L; $MgSO_4$ 0.4 g/L; KCl 0.5 g/L; $FeSO_4$ 0.016 g/L; sucrose 3 g/L.

2. Czapek medium amended with yeast extract and 2% w/v cracked maize (CDYM) [5] (*see* **Note 1**): $ZnSO_4$ 5 mg/L; $NaMoO_4$ 1 mg/L; yeast extract 0.5 % w/v; $NaNO_3$ 2 g/L; KH_2PO_4 1 g/L; $MgSO_4$ 0.4 g/L; KCl 0.5 g/L; $FeSO_4$ 0.016 g/L; sucrose 3 g/L. Strongly inducing Fumonisin synthesis. Cracked maize is prepared from sterilized (autoclaved) and lyophilized maize kernels subsequently homogenized and powdered in a blender at maximum speed for 2 min.

3. Hoagland solution: NH_4HSO_4 463.0 mg/L, H_3BO_3 1.6 mg/L, $CaCl_2$ 125.33 mg/L, Na_2-EDTA 37.25 mg/L, $FeSO_4 \times 7H_2O$ 27.85 mg/L, $MgSO_4$ 90.37 mg/L, $MgSO_4 \times H_2O$ 3.33 mg/L, KI 0.8 mg/L, KNO_3 2830.0 mg/L, KH_2PO_4 400 mg/L, $ZnSO_4 \times 7H_2O$ 1.5 mg/L.

2.2   RNA Extraction and Quantification

Prepare all solutions using ultrapure water, made DNase and RNase free by DEPC treatment (see **item 1** below). Several solutions and buffers can be purchased as ready to use (see below). All the devices and consumables used for RNA, mRNA, and cDNA manipulation must be purchased DNase and RNase free, and further sterilized in aliquots for more convenient use.

1. Diethylpyrocarbonate water (DEPC water): 0.1 % v/v. Prepare a solution of DEPC in ultrapure water, mix thoroughly and let sit at room temperature overnight. Then autoclave.

2. Extraction buffer: 2 % w/v Hexadecyltrimethylammonium bromide (CTAB), 2 % w/v polyvinylpyrrolidone K 30 (PVP), 100 mM Tris–HCl pH 8.0, 25 mM EDTA, 2.0 M NaCl, 0.5 g/L spermidine. Mix well (may need overnight stirring) and autoclave. Add 2 % v/v β-mercaptoethanol just before use.

3. SSTE buffer: 1 M NaCl, 0.5 % w/v SDS, 10 mM Tris–HCl (pH 8.0), 1 mM EDTA (pH 8.0).

4. Ethanol 70 %: Mix 70 mL of ethanol (≥99.8 % pure) and 30 mL of DEPC water.

5. Dithiothreitol (DTT): Prepare a 100 mM solution in DEPC-treated water.

6. DNase enzyme with related reaction buffer and stop solution. Can be conveniently purchased from any commercial source.

7. dNTP solutions: Dilute the dATP, dCTP, dGTP, and dTTP stocks from 100 mM (the concentration at which they are normally sold) to 10 µM with DEPC water. To do so, add 5 µL of each dNTP to 30 µL of DEPC water.

8. RNase inhibitor, reverse transcriptase and its reaction buffer can be purchased from any commercial source.

9. The primers for the housekeeping gene (*β-TUB*, coding for β-tubulin) are designed on the FVEG_05512.3 accession in NCBI: βtub_F (ACATTCGTCGGAAACTCCAC) and βtub_R (CAGCATCCTGGTACTGCTGA). Efficient primers for *FUM1* amplification in qPCR are Fum1_for (GAGCCGA GTCAGCAAGGATT) and Fum1_rev (AGGGTTCGTGAGCC AAGGA) [3, 6]. For RNA extraction from *F. verticillioides*-infected maize cobs an additional primer pair must be used to normalize samples onto total RNA including plant-derived RNA. Several options are available, but we find the primer pair Zm_actin_for (TCCTGACACTGAAGTCCCGATTG) and Zm_actin_rev (CGTTGTAGAAGGTGTGATGCCAGTT) designed on *α−ACTIN* from *Zea mays* (accession DQ492681.1) to work well [7].

10. Polysaccharide precipitation solution: Prepare a 4 M solution of ammonium acetate in ultrapure water, and then autoclave.

11. SYBR Green dye: It is available as a pure dye or as a ready-to-use PCR mix from several commercial sources.

3   Methods

3.1   F. verticillioides Culture

1. Grow the *F. verticillioides* strains for spore production in potato dextrose agar at 25 °C. Wash spores off the plate with sterile water and under sterile conditions to prepare a stock conidia suspension of 1×10^7 conidia/mL of *F. verticillioides* (*see* **Notes 2** and **3**).

2. With a cut tip, inoculate CDY or CDYM medium (usually, 50 mL in 100 mL Erlenmeyer flasks work fine) and grow in the dark and at 25 °C under shaken conditions (150 rpm). An initial inoculum of 10^6 spores in 50 mL should allow appropriate amounts of mycelia and spores being produced within 10 days by most *F. verticillioides* strains.

3. Harvest at different time points after inoculation, depending on the hypothesis being investigated and your experimental setup. Mycelia is easily harvested as a more or less compact disk floating over liquid medium; it should be washed thoroughly on filter paper with distilled water, then excess water should be dried away with clean absorbing paper.

4. Quantify fungal growth by weighing freeze-dried mycelium (lyophilized at 0.1 hPa/−100 °C for 24 h). Alternatively, samples can be deep-frozen with liquid nitrogen and kept at −80 °C until use; in this case, be careful to dry away most water quickly and without damaging the cells.

   To inoculate maize cobs, proceed as follows:

5. Produce and quantify an inoculum as at Subheading 3.1, **step 1**. Adjust concentration to 10^5 spores/mL.

6. Collect maize cobs at dough stage (i.e., R4: when kernels are about 50% their final size, usually around 24–28 days after silking).

7. Infect them with the pin bar technique. To do so, wound with a pin bar all around the middle area of the cob, without leaf removal, obtaining three portions with visible holes. Place 100 μL of the spore suspension in each hole.

8. Place cobs in plastic bottles containing 50 mL of Hoagland's solution at the bottom as nutritional source and incubate at room temperature (around 25 °C). Include non-inoculated cobs. Harvest at the desired time after inoculation (usually, symptoms are visible within a few days) and store at −20 °C until analysis.

3.2   RNA Extraction

3.2.1   Starting from In Vitro-Grown F. verticillioides Pure Culture

1. Grind 20 mg of freeze-dried mycelium with liquid nitrogen in clean and sterilized mortar and pestle.

2. Warm 5 mL of extraction buffer to 65 °C in a water bath, then add the ground mycelium, and mix by inverting the tube and vortexing (*see* **Note 4**).

3. Add an equal volume of chloroform:isoamyl alcohol (24:1, v/v), vortex, and centrifuge at $13,680 \times g$ and 4 °C for 10 min (extend centrifugation time if phases are not well separated).

4. Recover the aqueous phase, being careful not to pick the interphase.

5. Repeat the extraction step with chloroform:isoamyl alcohol.

6. Add 1/4 volume 10 M LiCl to the supernatant, mix gently, and store samples at 4 °C overnight (*see* **Note 5**). This step allows precipitation of RNA, so most genomic DNA will be eliminated in the following steps.

7. Centrifuge at $18,620 \times g$ for 20 min and 4 °C. Discard the supernatant.

8. Dissolve the pellet in 1 mL of SSTE pre-warmed to 65 °C.

9. Add an equal volume of chloroform:isoamyl alcohol, vortex, and centrifuge for 10 min at $13,680 \times g$ and 4 °C.

10. Recover the aqueous phase, being careful not to pick the interphase.

11. Add 0.7 volumes of isopropanol and 0.1 volumes of 3 M Na acetate (pH 5.2).

12. Centrifuge for 20 min at $18,620 \times g$ and 4 °C and discard the supernatant.

13. Add 1 mL of 70% ethanol (DEPC) and centrifuge for 10 min at $18,620 \times g$ and 4 °C.

14. Discard the supernatant and dissolve the pellet in 20–50 μL of DEPC water.

3.2.2 Starting from F. verticillioides-Infected Maize Kernels

1. Lyophilize (at 0.1 hPa and –100 °C for 24 h) 100 g of kernels from *F. verticillioides*-infected maize cobs, and grind it with liquid nitrogen in clean and sterilized mortar and pestle.

2. Take an aliquot (30 mg) of the homogenized kernels and transfer into a DEPC-treated, sterile 2 mL tube.

3. Warm 5 mL of extraction buffer (*see* **step 2**) to 65 °C in a water bath, then add the ground sample, and mix by inverting the tube and vortexing.

4. Put the sample in ice for 10 min and then centrifuge at $13,680 \times g$ or 15 min and 4 °C to allow polysaccharide precipitation. Recover the aqueous phase and transfer into a DEPC-treated, sterile 2 mL tube.

5. Add 0.3 volumes of ammonium acetate solution and put the tube in ice for 30 min. Then centrifuge at $13,680 \times g$ for 15 min and 4 °C. Recover the aqueous phase and then put it in DEPC-treated, sterile 2 mL tubes. From this point on, the protocol follows **steps 3–14** above (*see* **steps 3–14**).

3.3 Total RNA Quantification

The extracted RNA is quantified spectrophotometrically. For RNA quantification, readings should be taken at wavelengths of 260 and 280 nm (*see* **Note 6**). For most samples, 1000-fold dilutions will be in the right absorbance range (dilute 1 μL of total RNA in 1 mL of DEPC water, mix by shaking without forming bubbles and measure). Please be aware that poor RNA integrity might cause serious problems to reverse transcription and amplification steps downstream (*see* **Note 7**).

3.4 cDNA Synthesis

Reverse transcriptase reactions are usually performed on 1 μg of total RNA per sample.

1. Add 1 μg of total RNA to 1 μL of 10× buffer, 1 μL of DNase (1 enzymatic unit/μL) and DEPC-treated water to the final volume of 8 μL. Run the reaction for 25 min at 25 °C, and then stop it by heating at 70 °C for 10 min (*see* **Note 8**).

2. Add 1 μL of oligo-dT solution (5′-TTTTTTTTTTTTTTTTT-3′, 10 μM in DEPC water) and 1 μL of each individual dNTP solution (all 10 mM) to the samples; incubate at 65 °C for 5 min.

3. Incubate the samples at 4 °C for 5 min. During this incubation add to each sample 8 μL of the following mix: 4 μL of buffer 5×, 2 μL of 100 mM DTT, 0.5 μL of reverse transcriptase enzyme, 0.5 μL of RNase inhibitor, 1 μL of DEPC water.

4. Incubate at 42 °C for 30 min, and then at 70 °C for 15 min. Store at –20 °C.

3.5 Real-Time Amplification of FUM1 Transcript in F. Verticillioides

FUM1 transcript is quantified based on a relative standard curve (*see* **Note 9**) on a real-time PCR system with SYBR Green dye. This method allows analyzing the transcript level of the target gene (*FUM1*) by normalizing its transcript onto the transcript of a housekeeping gene (*β-TUB*) and onto its value in another sample named the "calibrator" (*see* **Note 10**). A calibrator is a sample used as the basis for comparing results and all other quantities are expressed as an n-fold ratio relative to the calibrator; for instance in a time course experiment, the time of inoculation is typically taken as the calibrator [7, 8].

1. Use fivefold dilutions of a reverse transcription product (cDNA) or plasmid to prepare the standard curve; the units used to express the dilutions are not important (*see* **Note 11**).

2. RT-qPCR reactions: Set up in 10 μL (final volume) of SYBR Green PCR Master Mix, with primers (500 nM each) and cDNA (10 ng). Transcripts from the target and reference gene are quantified in each of three independent biological replicates per experimental condition, in analytical triplicates.

3. PCR cycling conditions consist of 10 min at 95 °C (1 cycle) and 15 s at 95 °C followed by 1 min at 60 °C (40 cycles).

Transcript abundance values for *FUM1* and *β-TUB* are the means from three biological replicates and three analytical repetitions (*see* **Note 9**).

4. Analyze the results by the software installed in the thermocycler, which provides the values of the relative or absolute expression of the gene of interest (*see* **Note 11**). These values can then be expressed as referred to the dry or fresh weight of the mycelium corresponding to the cDNA loaded in the reaction tube. For *in planta* estimation of *FUM1* expression, normalize over a plant housekeeping gene as well, and/or to the fresh weight of extracted tissue (the latter however is more error-prone).

4   Notes

1. Other synthetic media are known to be inductive for Fumonisin production and can be used instead of CDYM (*see* ref. [3]).

2. *F. verticillioides* strains used for the production of conidia inoculum can be grown also in potato dextrose liquid broth (26.5 g of the commercial powder—e.g., Sigma catalogue number P6685 in 1 L of distilled water) for 7 days at 25 °C. Then the liquid culture is filtered on 0.45 μm Millipore filters (Millipore, Billerica, USA) and the conidia are resuspended in water.

3. To adjust your stock suspension to 1×10^7 conidia/mL, conidia must be counted under an optical microscope and in a Burker chamber. Initially the conidia suspension is serially diluted at least thrice 1:1000, and then 100 μL of the least concentrated aliquot are counted. Sterile water is then added to reach the desired concentration.

4. A commercial alternative to the CTAB method for RNA isolation is the NucleoSpin RNA Plant kit (Macherey–Nagel, www.mn-net.com) or the extraction by Trizol (Sigma-Aldrich). For the latter method swap **steps 2–12** of Subheading 3.2 with the following:

   – Add 1 mL of Tri-Reagent (Sigma-Aldrich) to the powdered mycelia in a 2 mL Eppendorf tube and chill on ice.

   – Vortex and incubate the samples for 10 min at room temperature.

   – Centrifuge the samples for 10 min at 4 °C and $13,680 \times g$.

   – Add 200 μL of chloroform to the supernatant, shake samples gently for 5 min and incubate for another 5 min at room temperature.

   – Centrifuge the samples for 15 min at 4 °C and $13,680 \times g$.

   – Recover the aqueous phase, being careful to not pick the interphase, and transfer to a 2 mL tube.

– Add an equal volume of chloroform:isoamyl alcohol to the recovered aqueous phase and shake gently.

– Centrifuge the samples for 15 min at 4 °C and $13,680 \times g$.

– Recover the aqueous phase in a 2 mL tube and add an equal volume of chloroform:isoamyl alcohol; shake gently.

– Centrifuge the samples for 15 min at 4 °C and $13,680 \times g$.

– Recover the aqueous phase in a 1.5 mL tube and add 500 mL of isopropanol; mix for 5 min by inverting the tubes.

– Centrifuge for 10 min at 4 °C and $13,680 \times g$. Then continue from **step 13**.

5. Shorter precipitation times may also be used, but with lower yield: 1 h leads to ~30 % yield of total input.

6. The 260 nm reading allows calculation of the concentration of nucleic acid in a given sample. Absorbance at 280 nm is affected by the amount of protein in the sample, and will influence the final OD_{260}/OD_{280} ratio. Pure preparations of RNA and DNA will have OD_{260}/OD_{280} values of 1.8–2.0, respectively. If the nucleic acid is contaminated by protein or phenol, this ratio will be significantly lower than the above values, and accurate quantitation of nucleic acids will not be possible. In this case, further purification is needed, for example by repeating the phenol:chlorophorm extraction and washes, followed by precipitation. Yields will be lower but clean samples are needed for reliable quantification and efficient amplification.

7. Quality of extracted RNA can be visually checked by agarose gel electrophoresis. For eukaryotic samples, intact total RNA run on a denaturing gel will give sharp 28S and 18S rRNA bands. A 2:1 (28S:18S) band intensity ratio is a good indication that the RNA is intact. The area between 1.5 and 2 kb might show a smear due to the presence of abundant mRNAs, depending on the sample. Partially degraded RNA will appear smeared in a wider range of sizes, will lack the sharp rRNA bands, or will not exhibit the 2:1 ratio of high-quality RNA. Completely degraded RNA will appear as a very low molecular weight smear.

8. Add 1 μL of stop solution (provided together with DNase solution by most producers) after the samples have been at 25 °C for at least 20 min.

9. Relative standard curves are needed in case of different amplification efficiencies of the target and endogenous control (which should anyway be comprised between 90 and 110 %). The comparative C_T method ($-2^{\Delta\Delta CT}$) is a way of quantifying the fold-change of a gene compared to a calibrator, and standard curves are not strictly required, although comparisons are meaningful only if reactions are performed under the exact same conditions (operationally, in the same run).

10. Transcript levels of target gene should fall within the limits of standard curve. If the sample C_T value is outside of the standard curve, dilute the sample so that the C_T value falls between the most and the least diluted points of your standard curve.

11. For relative quantitation, any stock cDNA or DNA containing the appropriate target sequence can be used to prepare standards; for example, plasmids in which the target sequence was cloned. In this latter case, amplification could become absolute and lead to an estimate of the number of target sequences in your samples. To do so, pure plasmid DNA preps should be quantified spectrophotometrically to obtain the concentration in ng/mL. Purity will be easily obtained by column chromatography; commercial kits are available to this purpose from several producers. Knowing the molecular mass of the whole plasmid, the number of molecules and therefore of target sequences can be accurately estimated in your plasmid solution. Thereby, the calibration curve can be prepared starting from that stock solution, and its measure unit converted to number of transcript molecules/volume unit (usually, mL).

Acknowledgement

This work was supported by the Ministry of Research and Education through the project FIRB2008 "Futuro in Ricerca," grant N. FIRB-RBFR08JKHI to V.S.

References

1. Munkvold GP (2003) Cultural and genetic approaches to managing mycotoxins in maize. Annu Rev Phytopathol 41:99–116

2. Sanchez-Rangel D, SanJuan-Badillo A, Plasencia J (2005) Fumonisin production by *Fusarium verticillioides* strains isolated from maize in Mexico and development of a polymerase chain reaction to detect potential toxigenic strains in grains. J Agric Food Chem 53(22):8565–8571

3. Visentin I, Montis V, Doll K et al (2012) Transcription of genes in the biosynthetic pathway for fumonisin mycotoxins is epigenetically and differentially regulated in the fungal maize pathogen *Fusarium verticillioides*. Eukaryot Cell 11(3):252–259

4. Visentin I, Valentino D, Cardinale F, Tamietti G (2010) DNA-based tools for the detection of *Fusarium* spp. pathogenic on maize. In: Gherbawy Y, Voigt K (eds) Molecular identification of fungi. Springer-Verlag, Berlin Heidelberg

5. Scala V, Camera E, Ludovici M et al (2013) *Fusarium verticillioides* and maize interaction *in vitro*: relation between oxylipin cross-talk and fumonisin synthesis. World Mycotoxin J 6:343–351

6. Flaherty JE, Pirttila AM, Bluhm BH, Woloshuk CP (2003) *PAC1*, a pH-regulatory gene from *Fusarium verticillioides*. Appl Environ Microbiol 69:5222–5227

7. Scala V, Giorni P, Cirlini M et al (2014) LDS1-produced oxylipins are negative regulators of growth, conidiation and fumonisin synthesis in the fungal maize pathogen *Fusarium verticillioides*. Front Microbiol 5:669

8. Schefe JH, Lehmann KE, Buschmann IR et al (2006) Quantitative real-time RT-PCR data analysis: current concepts and the novel "gene expression's C_T difference" formula. J Mol Med 84(11):901–910

Part III

Polymerase Chain Reaction (PCR)-Based Methods for Multiplex Detection of Mycotoxigenic Fungi

Chapter 17

Multiplex Detection of *Aspergillus* Species

Pedro Martínez-Culebras, María Victoria Selma, and Rosa Aznar

Abstract

Multiplex real-time polymerase chain reaction (PCR) provides a fast and accurate DNA-based tool for the simultaneous amplification of more than one target sequence in a single reaction. Here a duplex real-time PCR assay is described for the simultaneous detection of *Aspergillus carbonarius* and members of the *Aspergillus niger* aggregate, which are the main responsible species for ochratoxin A (OTA) contamination in grapes. This single tube reaction targets the beta-ketosynthase and the acyl transferase domains of the polyketide synthase of *A. carbonarius* and the *A. niger* aggregate, respectively.

Besides, a rapid and efficient fungi DNA extraction procedure is described suitable to be applied in wine grapes. It includes a pulsifier equipment to remove conidia from grapes which prevents releasing of PCR inhibitors.

Key words Multiplex PCR, *Aspergillus carbonarius*, *Aspergillus niger* aggregate, Grapes, Mycotoxin, Ochratoxin A, Polyketide synthase, Real-time PCR, Wine

1 Introduction

Several studies highlighted *Aspergillus* section *Nigri* (black aspergilli) as the fungal species responsible for ochratoxin A (OTA) contamination of grapes, must and wine [1–3]. Among black *Aspergillus* species, *Aspergillus carbonarius* and species belonging to the *Aspergillus niger* aggregate are reported as the main ochratoxigenic species [1, 4–7]. *A. carbonarius* is considered the mayor responsible for OTA contamination in grapes because the reported percentages of OTA-producing strains in this species are higher than those reported for members of the *A. niger* aggregate. Nevertheless, the black aspergilli most frequently isolated from grapes are species belonging to the *A. niger* aggregate, mainly *A. niger*, *Aspergillus tubingensis* and *A. tubingensis*-like [4, 8–11]. These data indicates an important contribution to OTA contamination in grapes by species belonging to the *A. niger* aggregate. Therefore, quick assays capable of detecting and quantifying the presence of both *A. carbonarius* and the *A. niger* aggregate in grapes would improve OTA risk assessment.

Antonio Moretti and Antonia Susca (eds.), *Mycotoxigenic Fungi: Methods and Protocols*, Methods in Molecular Biology, vol. 1542, DOI 10.1007/978-1-4939-6707-0_17, © Springer Science+Business Media LLC 2017

Real-time PCR (qPCR) assay methods are a good alternative to traditional culturing techniques, and conventional PCR, since they are rapid, automated, high throughput, and sensitive and allow accurate identification and quantification of fungal species. In addition, multiplex qPCR reactions can be arranged for the simultaneous amplification of several target sequences in a single reaction. The method here outlined allows the specific detection of black ochratoxigenic *Aspergillus* species, *A. carbonarius* and species belonging to the *A. niger* aggregate, in combination with a rapid and efficient DNA extraction method [12, 13] that is suitable for wine grapes monitoring. Fungal DNA extraction from grapes is approached combining the use of a Pulsifier equipment for sample homogenization and a commercial kit for DNA purification. Instead of crushing the food sample, we introduced the use of the Pulsifier equipment, which beats the sample bag very rapidly using an oscillating metal ring. In this way, conidia are removed from samples but there is minimal breakdown of the grapes, which prevents releasing of PCR inhibitors. Following the procedure here outlined, fungal DNA extraction from grapes is accomplished in 30 min, and the complete qPCR analysis in less than 2 h.

The duplex qPCR assay for the simultaneous detection of ochratoxigenic black *Aspergilli* targets the β-ketosynthase and the acyl transferase domains of the polyketide synthase of *A. carbonarius* and the *A. niger* aggregate, respectively. Results indicated no differences in sensitivity when using either the two sets of primers and probes in separate or in the same reaction. This qPCR procedure provides a fast, automated, and accurate tool to monitor in a single reaction the presence of OTA-producing species in grapes which, to some extent, will facilitate OTA contamination surveys to guarantee food safety in the wine industry.

2   Materials

Prepare and store all reagents at room temperature (unless indicated otherwise).

2.1 Fungal Strains and Culture Media

1. Fungal strains: *Aspergillus carbonarius* CECT 2088 and *Aspergillus niger* CECT 11380 are recommended as qPCR amplification controls (CECT, Colección Española de CultivosTipo, Spanish Type Culture Collection).

2. Malt extract agar (MEA) plates: 2% (w/v) Maltose, 2% dextrose, 0.1% peptone, 1.5% agar in 1 L distilled water. Autoclave at 121 °C, 15 min (*see* **Note 1**).

3. Dichloran rose-bengal chloramphenicol (DRBC) medium (oxoid) (*see* **Note 1**): 0.5% (w/v) peptone, 1% glucose, 0.1% potassium dihydrogen phosphate, 0.05% magnesium sulfate,

0.0002 dichloran, 0.0025 rose bengal, 1.5% agar, 0.01 chloramphenicol, in 1 L distilled water.

4. Digralsky spreader, l-shaped stick with rounded corners to avoid damaging solid medium surface while spreading.

2.2 Fungal DNA Extraction Reagents

1. Buffered peptone water solution 0.1% (BPW 0.1%): 0.1% (w/v) peptone, 0.05% sodium chloride, 0.035% disodium phosphate, 0.015% potassium dihydrogen phosphate.

2. TE buffer solution: 10 mmol L^{-1} Tris–HCl, 1 mmol L^{-1} EDTA, pH 8.

3. Isopropanol.

4. Ethanol 100%.

5. EZNA Fungal DNA kit (Omega bio-teck, Doraville, USA): FG1 Buffer, FG2 Buffer, FG3 Buffer, Equilibration Buffer, DNA Wash Buffer, Elution Buffer, RNase A, HiBind DNA Mini Column, 2 mL Collection Tubes, Homogenizer Spin Column.

6. Sterile nanopore water.

7. Heating block.

8. Homogenizer equipment (preferably, Pulsifier (Microgen bio-products, Surrey, UK)).

2.3 Multiplex Real-Time PCR Reaction Components (qPCR)

1. Primer and TaqMan probe for *Aspergillus Nigri* detection targeted to a conserved region in the acyl transferase domain of polyketide synthase gene [13]: AtNig1 (5′-GAC TGA GCC CAG ATG ACC TAC A-3′) AtNig2 (5′-CGC TGT CGC CGG ATA CTG-3′) (stock and final concentration of each primer 7.5 μM and 0.2 μM, respectively), AN probe (5′-VIC-TTG ACT ATT GCA TGT TTT AAT AGC CCR AAG AAC C-MGB-3′), Taqman (MGB) probe labeled at the 5′end with VIC® and at the 3′end with a dark quencher (stock and final concentration of probe 5 μM and 0.2 μM respectively).

2. Primer and TaqMan probe for *Aspergillus carbonarius* detection targeted to a conserved region in the β-ketosynthase domain of a polyketide synthase gene [12, 13]: AcKS10R (5′-CCC TGA TCC TCG TAT GAT AGC G-3′), AcKS10L (5′-CCG GCC TTA GAT TTC TCT CAC C-3′) (stock and final concentration of each primer 7.5 μM and 0.1 μM respectively), AC probe (5′-FAM-AGA ACG CTG ATG GGT ATG CGC GG-TAMRA-3′), Taqman probe labeled at the 5′end with 6-carboxy-fluorescein (FAM) and at the 3′end with 6-carboxy-tetramethyl-rhodamine (TAMRA) (stock and final concentration of probe 5 μM and 0.15 μM, respectively).

3. Master mix: TaqMan Core Reagents from Applied Biosystems containing: 1× TaqMan Buffer A, 200 μM each dATP, dCTP, dGTP and 400 μM dUTP; Amperase uracil *N*-glycosidase (*see* **Note 2**); MgCl$_2$; AmpliTaq Gold DNA polymerase.

4. Real-time PCR equipment with color dye flexibility (i.e., FAM, VIC®), e.g., ABI Prism 7000 Sequence Detection System (Applied Biosystems).

2.4 Standard Curves for Quantification Purposes

Use purified DNA from *Aspergillus carbonarius* CECT 2088 and *Aspergillus niger* CECT 11380 from

1. Tenfold serial dilutions of purified DNA in sterile nanopure water covering the range from 10 to 1×10^{-6} ng per reaction. DNA concentration determined fluorometrically using the Fluorescent DNA quantitative kit (Bio-Rad, California, USA) and a VersaFluor® Spectrofluorimeter (Bio-Rad, London, England).
   or
2. DNA extracted from tenfold serial dilutions of conidial suspensions in sterile saline (0.8% NaCl), covering the range from 1 to 1×10^7 conidia mL^{-1}.

3 Methods

3.1 Fungal Strain Culture Conditions and DNA Extraction

1. Grow black *Aspergillus* strains, *A. carbonarius* CECT 2088 and *A. niger* CECT 11380 in MEA plates at 25 ± 1 °C for 5–7 days [14] (*see* **Note 3**).
2. Collect mycelia from the surface of the agar with the aid of a scalpel and put in an Eppendorf tube.
3. Place the Eppendorf tube filled with mycelia in a metallic container and cover with liquid nitrogen for 5 min to frozen.
4. Place the frozen mycelia in a mortar and ground them to a fine powder.
5. Take 100 mg of powdered mycelia for DNA extraction.
6. Use the commercial EZNA Fungal DNA kit according to the manufacturer's instructions.

3.2 Food Sample Processing for Fungi Detection

1. Place grapes (approx. 25 g) in a plastic bag containing 100 mL of sterile BPW and disaggregate for 1 min with the aid of a Pulsifier equipment. Continue as indicated in Subheading 3.4 (*see* **Note 4**).

3.3 Conidial Suspensions for Calibration Curve

1. Prepare conidial suspensions (*see* **Note 5**) of *A. carbonarius* CECT 2088 and *A. niger* CECT 11380 in sterile nanopure water at different concentrations ($10–10^6$ conidia mL^{-1}) (*see* **Note 6**). Continue as indicated in Subheading 3.4.

3.4 Fungal DNA Extraction from Conidial Suspensions or Plant Material (i.e., Wine Grapes)

1. Centrifuge 1 mL aliquots of homogenized grape samples (*see* **Note 7**) or conidial suspensions for 3 min at $13,000 \times g$.
2. Wash pellets in 0.5 mL of TE buffer solution, centrifuge at $13,000 \times g$ for 3 min, and resuspend in 50 μL of sterile nanopure water.

3. Boil (95 °C for 10 min) samples to break down conidia for DNA release and cool on ice for 10 min.

4. Use the commercial EZNA Fungal DNA kit (Omega bio-teck, Doraville, USA) according to the manufacturer's instructions for fresh or frozen specimens adding 600 μL of buffer FG1 to 50 μL of pellets re-suspended in sterile nanopure water. In the final step, elute DNA in 100 μL of sterile nanopure water and kept at −20 °C until use as template for PCR amplification (*see* **Note 8**).

3.5 Real-Time PCR Reaction Assembly (qPCR)

1. Amplification mixtures for qPCR reactions with TaqMan probe (*see* **Note 9**): prepare $3 \times n + 1$ (n = number of samples plus positive and negative controls (*see* **Note 10**) volume of master mix containing in a final volume of 20 μL/sample, 1× TaqMan Buffer A, 200 μM each dATP, dCTP, dGTP and 400 μM dUTP; 1 U of Amperase uracil *N*-glycosidase; 3.5 mM MgCl$_2$; 0.9 U of AmpliTaq Gold DNA polymerase; 200 nM each AtNig1 and AtNig2 primers; 100 nM each AcKS10R and AcKS10L primers; 200 nM AN probe targeting *A. niger* aggregate and 150 nM AC probe targeting *A. carbonarius* (*see* **Note 11**).

2. Distribute the master mix in 15 μL aliquots and add 5 μL of template DNA, in triplicate, for each sample.

3. Standard curve: For quantitative purposes, include a set of tubes, in triplicate, containing as template from 10 to 1×10^{-6} ng of purified DNA or DNA corresponding to 1 to 1×10^{7} conidia mL^{-1}, per reaction.

3.6 Real-Time PCR Amplification (qPCR)

Parameters correspond to an ABI Prism 7000 Sequence Detection System. For multiplex real-time PCR reactions using TaqMan probes it is programmed to hold at 50 °C for 2 min (*see* **Note 2**), to hold at 95 °C for 10 min, and to complete 40 cycles of 95 °C for 15 s and 60 °C for 1 min [13].

3.7 Data Analysis for Detection and/or Quantification

PCR results are given as the increase in the fluorescence signal of the reporter dye detected and visualized by the software provided with the real-time PCR equipment. Threshold cycle (C_T) values represent the PCR cycle in which an increase in fluorescence, over a defined threshold, first occurred, for each amplification plot. $C_T < 40$ corresponds to positive detection. $C_T \geq 40$ indicates negative amplification (*see* **Note 12**).

Standard curves are generated by plotting the genomic DNA and conidia suspensions from *A. carbonarius* and *A. nigri* against the C_T values (*see* **Note 13**) exported from the sequence detection system for each plate (*see* **Note 14**). The C_T values for unknown samples are extrapolated from standard curves.

4   Notes

1. DRBC agar is a selective medium for molds associated with food spoilage. It has been optimized to inhibit bacterial growth and spreading molds, such as *Rhizopus* and *Mucor*. After fungal isolation, each fungal species has to be cultivated on the appropriate medium to achieve typical growth and sporulation. Malt extract agar (MEA), Czapek yeast agar (CYA), and potato dextrose agar (PDA) are suitable media for most black *Aspergillus* species. In general, fungal sporulated cultures can be achieved after 4–8 days of dark incubation at 25 °C without the need to stimulate sporulation. In the case of *A. carbonarius*, it is possible to observe fungal sporulation after 3 days.

2. Amperase uracil *N*-glycosidase (UDG) is used in combination with dUTP instead of dTTP in order to prevent PCR cross contamination arising from PCR products. Amplicons will contain dU instead of dT. Cleavage of contaminant PCR products is conducted by the UDG during the incubation step at 50 °C for 2 min programmed in the PCR run.

3. The lyophilized fungal cultures must be open under a sterile environment. Add 1 mL of sterile water or culture media and let stand for 30 min (rehydration). After that, take a small amount of the suspension and put on the Petri dishes with the culture media. It is desirable to grow fungal cultures from the same original stock. Thus, fungal cultures must be kept in glycerol conidia suspensions to a final concentration of 40 % at −80 °C. Conidia obtained from stock tubes can be used to inoculate agar plates. They must be homogeneously spread on Petri dishes containing CYA or MEA medium and sub-cultured in the dark at 25 °C.

4. Qualitative approach to improve sensitivity of ochratoxigenic fungi detection in environmental samples: prepare tenfold serially dilutions of the homogenized sample and plated 100 μL onto DRBC agar plates. Incubate for 4–5 days at 25 °C to allow black aspergilli to grow. Then recover the mycelia from plates and proceed as indicated in 3.1 (**step 2**).

5. To prepare conidia suspensions, due to the hydrophobicity of fungal conidia, it is advisable to use a sterile solution 0.005 % v/v Tween 80 or other surfactants, instead of distilled water. After 4–8 days of incubation, conidia can be harvested from the plates by adding about 10 mL of sterile solution 0.005 % v/v Tween 80, and swirl handily and gently the surface of the agar with a sterile inoculating loop to favor detachment of conidia.

6. To prepare calibrated suspensions, transfer with a sterile pipette the conidia suspension obtained to a sterile vial or Eppendorf tube with sterile distilled water. Take an aliquot of this suspension

and count conidia with a hemocytometer under the microscope to estimate the conidia concentration. Adjust conidia suspensions from different strains to the same final concentration.

7. Higher volumes, i.e., 10 mL can be centrifuged to improve sensitivity in detection from environmental samples. The resulting pellet will be washed, resuspended and used for DNA extraction as indicated.

8. The DNA extraction method described shall be appropriate to obtain the quality and quantity of nucleic acid required for the subsequent PCR analysis. In case of low DNA recovery we recommend the use of the method described by [15] for rapid DNA extraction as modified by [16].

9. In order to prevent cross-contamination in qPCR, the working area for reaction assembly and materials should be separated from other laboratory uses and they should be exposed to UV light before handling reagents and DNA samples; positive displacement pipettes or specialized barrier materials are recommended.

10. Amplification Positive control is prepared using DNA from reference strains as target DNA. Amplification negative control contains all reagents used in the master mix and nanopure sterile water instead of DNA template.

11. Additionally, PCR enhancers such as BSA (0.1–0.5 μg/μL final concentration) can be added in case of low-quality DNA to facilitate PCR amplification.

12. Negative results might be due to the presence of PCR inhibitors. It can be overcome by diluting the DNA solution to get enough DNA for amplification, but low inhibitors concentration. To check for the presence of PCR inhibitors in the sample, a small aliquot of the positive control (1–2 μL) can be added to the PCR mixture as an internal amplification control. No amplification will be obtained in case of PCR inhibitors. Otherwise it would indicate DNA recovery was too low.

13. Discard C_T values with SD higher than 1, from triplicates.

14. Slope values should be around the theoretical optimum of −3.32 [17] and R^2 values close to 1 for quantification purposes. Preferential amplification of one of the targets in this multiplex PCR can be obtained when they are present in very different concentrations [13]. Therefore, the multiplex qPCR reaction here described is suitable for the simultaneous detection of *A. carbonarius* and the *A. niger* aggregate but for an accurate quantification of each, *A. carbonarius* or *A. niger* aggregate species, the corresponding single qPCR reaction should be used.

References

1. Battilani P, Magan N, Logrieco A (2006) European research on ochratoxinA in grapes and wine. Int J Food Microbiol 111:S2–S4

2. Perrone G, Susca A, Cozzi K et al (2007) Biodiversity of *Aspergillus* species in some important agricultural products. Stud Mycol 59:53–66

3. Martínez-Culebras PV, Crespo Sempere A, Sánchez Hervás M et al (2009) Molecular characterization of the black *Aspergillus* isolates responsible for ochratoxin A contamination in grapes and wine in relation to taxonomy of *Aspergillus* section *Nigri*. Int J Food Microbiol 132:33–34

4. Abarca ML, Accensi F, Bragulat MR et al (2003) *Aspergillus carbonarius* as the main source of ochratoxin A contamination in dried vine fruits from the Spanish market. J Food Prot 66:504–506

5. Samson RA, Houbraken JAMP, Kuijpers AFA et al (2004) New ochratoxin A or sclerotium producing species in *Aspergillus* section Nigri. Stud Mycol 50:45–61

6. Bau M, Abarca ML, Bragulat MR et al (2005) Ochratoxigenic species from Spanish wine grapes. Int J Food Microbiol 98:125–130

7. Sage L, Krivobok S, Delbos E et al (2002) Fungal flora and ochratoxin A production in grapes and musts from France. J Agric Food Chem 50:1306–1311

8. Cabañes FJ, Accensi F, Bragulat MR et al (2002) What is the source of ochratoxin A in wine? Int J Food Microbiol 79:213–215

9. Battilani P, Pietri A (2002) Ochratoxin A in grapes and wine. Eur J Plant Pathol 108:639–643

10. Bellí N, Pardo E, Marín S et al (2004) Occurrence of ochratoxin A and toxigenic potential of fungal isolates from Spanish grapes. J Sci Food Agric 84:541–546

11. Martínez-Culebras PV, Ramón D (2007) An ITS-RFLP method to identify black *Aspergillus* isolates responsible for OTA contamination in grapes and wine. Int J Food Microbiol 113:147–153

12. Selma MV, Martínez-Culebras PV, Aznar R (2008) Real-time PCR based procedures for detection and quantification of *Aspergillus carbonarius* in wine grapes. Int J Food Microbiol 122:126–134

13. Selma MV, Martínez-Culebras PV, Elizaquível P et al (2009) Simultaneous detection of the main black aspergilli responsible for ochratoxin A (OTA) contamination in grapes by multiplex real-time polymerase chain reaction. Food Addit Contam 26:180–188

14. Pitt JI, Hocking AD (1997) Fungi and food spoilage. Blackie Academic and Professional, London

15. Cenis JL (1992) Rapid extraction of fungal DNA for PCR amplification. Nucleic Acids Res 20:2380

16. Crespo-Sempere A, Martínez-Culebras PV, González-Candelas L (2014) The loss of the inducible *Aspergillus carbonarius* MFS transporter MfsA leads to ochratoxin A overproduction. Int J Food Microbiol 181:1–9

17. Higuchi R, Fockler C, Dollinger G et al (1993) Kinetic PCR analysis: real-time monitoring of DNA amplification reactions. Biotechnology 11:1026–1030

Chapter 18

Multiplex Detection of *Fusarium* Species

Tapani Yli-Mattila, Siddaiah Chandra Nayaka, Mudili Venkataramana, and Emre Yörük

Abstract

Multiplex PCR is a powerful method to detect, identify, and quantify the mycotoxigenic fungus by targeting the amplification of genes associated with mycotoxin production and detection, identification, and quantification of *Fusarium* species. As compared with uniplex PCR, it has several advantages such as low cost, shortened time, and simultaneous amplification of more than two genes (in only one reaction tube). Here, we describe multiplex PCR-based detection and identification of trichothecene-, zearalenone-, fumonisin-, and enniatin-producing *Fusarium* species, the use of multiplex PCR in multiplex genotype assay and the use of multiplex TaqMan real-time qPCR.

Key words *Fusarium graminearum*, *F. culmorum*, *F. avenaceum*, *F. tricinctum*, *F. verticillioides*, Trichothecenes, Zearalenone, Fumonisins, Moniliformin, Enniatins

1 Introduction

PCR-based identification, detection, and quantification of mycotoxigenic fungal species play a key role in development/improvement of molecular plant pathology. Specific nucleic acid regions are targeted and then amplified in order to characterize *Fusarium* species by PCR [1–5]. Multiplex PCR has been introduced into several DNA-based research areas from the year 1988, its first recovery [6]. The method provides simultaneously amplification of more than two genomic DNA regions and it facilities the researches including two or more aims/strategies and high number of samples [7, 8]. Moreover, multiplex PCR is fast, reproducible, and cost-effective method. Multiplex PCR presents qualitative and quantitative results. Standard PCR-based multiplex PCR approach is involved in qualitative and also semi-quantitative analysis, while real-time PCR-based multiplex PCR can be used for quantitative analysis.

Trichothecenes and Fumonisins are a group of mycotoxins produced by genus *Fusarium* [9]. Trichothecenes are group of sesquiterpenoid mycotoxins produced by *Fusarium* species, which share

Antonio Moretti and Antonia Susca (eds.), *Mycotoxigenic Fungi: Methods and Protocols*, Methods in Molecular Biology, vol. 1542, DOI 10.1007/978-1-4939-6707-0_18, © Springer Science+Business Media LLC 2017

the 12, 13-epoxytrichothecene skeleton as the common structural feature. Presence or absence of an 8-keto moiety leads to differentiation of group B and group A trichothecenes, respectively [10]. The genetics and regulation of trichothecene biosynthesis have been elucidated in much detail in *F. sporotrichioides*[11] and *F. graminearum* [12]. Sequencing of parts of the trichothecene gene cluster was done for species in the *F. graminearum group*, i.e., *F. crookwellense*, *F. culmorum*, *F. lunulosporum*, and *F. pseudograminearum*.

The *tri5* gene, which codes for trichodiene synthase catalyzing the first specific step in the biosynthesis of all trichothecene producing fungi, was particularly well characterized in *Fusarium* spp. A set of primer pairs developed by Niessen and Vogel [13] targeting *tri5* gene were used by many researchers to detect trichothecene producing fungi from various food matrices [14, 15]. Besides *tri5*, other genes from the trichothecene biosynthesis cluster were used to design species- and group-specific PCR primers. A group-specific PCR assay for the detection of trichothecene producing *Fusarium* spp. involving primers binding to the *tri6* gene (transcription factor), was set up by Bluhm et al. [16] and also used the system together with primers for sensitive detection of fumonisin producers. In order to differentiate DON and NIV producing chemotypes in *Fusarium*, Lee et al. [17] designed primers hybridizing adjacent to an inserted region present in the *tri7* gene. A genus specific primer pair hybridizing to sequences within the *tri13* gene was published by Demeke et al. [15] and Ramana et al. [10].

*F. verticillioide*s and *F. proliferatum* are the major fumonisin producers among the genus *Fusarium*. Waalwijk et al. [18] studied the fumonisin biosyntheses of *F. proliferatum* and identified 19 genes for its synthesis and regulation. It was speculated that the principle ability of a *F. verticillioides* isolate to produce fumonisin will depend on the presence or absence of the *fum1* gene, but additional factors, which regulate the concentrations of fumonisin finally produced, may be necessary. Similar assay for detection of fumonisin producers was published by Bluhm et al. [16] and Ramana et al. [10]. These systems were based on the *fum1* and *fum13* gene sequences of *F. proliferatum* and *F. verticillioides* and were applied for the detection of these fungi by multiplex PCR assay.

Multiplex PCR's both strategies (standard and real-time PCR) had been widely used in simultaneous detection and identification of mycotoxigenic *Fusarium* species [18–25]. By this way, species-specific identification, mycotoxin profile characterization, and mating-type detection in *Fusarium* species have been adopted in plant pathology. SCAR markers specific to several *Fusarium* species including *F. graminearum*, *F. culmorum*, *F. pseudograminearum*, *F. poae*, *F. sporotrichioides*, *F. langsethiae*, and *F. avenaceum*, genes including *tri5*, *tri7*, *tri13*, *tri3*, and *tri12* located in *tri5* gene cluster responsible for trichothecene production and *MAT-1/MAT-2* loci, are amplified in those studies.

1.1 Multiplex PCR Detection and Identification of Trichothecene-Producing Fusarium Species

Here, we describe standard PCR-based multiplex PCR approaches for characterization of phytopathogenic trichothecene-producing *Fusarium* spp. In the first approach genomic DNA molecules were isolated from diseased barley samples, non-diseased samples and single-spore isolates of *F. graminearum*, *F. culmorum*, *F. poae*, and *F. pseudograminearum* species. Before amplification of species-specific SCAR markers and *tri5* gene essential in trichothecene production, primer molecules were tested in terms of controlling the self-dimer, hetero-dimer, and hairpin formation via nucleotide sequence analysis of primers. ΔG values were in the range of "+2 and −6" presenting that no possible secondary structure formation that could inhibit the efficiency of PCR strategies (*see* http://eu.idtdna/calc/analysis/). Both uniplex and multiplex PCR were performed in this studies. *F. graminearum*-, *F. culmorum*-, *F. poae*-, and *F. pseudograminearum*-specific SCAR markers and partial region of *tri5* gene which is targeted for detection of trichothecene producing *Fusarium* spp. were amplified from genomic DNA of diseased barley samples and monosporic isolates [2, 18, 26–28]. PCR bands are analyzed by agarose gel electrophoresis like in other multiplex PCR approaches.

The monosporic isolates could be individually confirmed at species level by amplification of 779, 472, 332, and 220 bp fragments with *F. pseudograminearum*-, *F. culmorum*-, *F. graminearum*-, and *F. poae*-specific primers (Table 1). Moreover, the isolates could be characterized as trichothecene producers via amplification of *tri5*. In multiplex PCR analysis, these five bands were amplified from monosporic isolates' genomes. Consequently, qualitative method comprising of multiplex PCR provided detection of toxigenic *Fusarium* spp. simultaneously from monosporic and barley samples (Fig. 1).

1.2 Multiplex PCR Detection and Identification of Trichothecene-, Fumonisin-, and Zearalenone-Producing Fusarium spp. with Internal Control

In the second approach we describe PCR-based multiplex PCR assays (Fig. 2) for identification of both trichothecene- or zearalenone-producing *Fusarium* isolates and fumonisin-producing *F. verticillioides* and *F. proliferatum* isolates, which are the major fumonisin producers among the genus *Fusarium*. DNA sequences were analysed and aligned by Clustal method [18]. Primers were designed using the aligned GenBank database sequences of *tri6, tri7, tri13,* and rDNA genes for the specific detection of nivalenol- and deoxynivalenol-producing *F. culmorum*. Total of four primer pairs were designed using Gene runner software (http://www.generunner.com). Primer sequences are listed in Table 1. Before standardizing mPCR protocol, all designed primers were evaluated on to array of fungal species to check the specificity and sensitivity.

Table 1
Primers used in multiplex PCR

Primer set	Nucleotide sequence 5′–3′	Target region	Tm (°C)	Reference	Aim	ΔG for hairpin structure
OPT18F	GATGCCAGACCAAGACGAAAG	SCAR marker	57	[26]	*F. culmorum* identification	1.18
OPT18R	GATGCCAGACGCACTAAGAT					−1.68
UBC85F	GCAGGGTTTGAATCCGAGAC	SCAR marker	59	[26]	*F. graminearum* identification	−0.68
UBC85R	AGAATGGAGCTACCAACGGC					−1.93
FPGF	GTCGCCGTCACTATC	SCAR marker	58	[2]	*F. pseudograminearum* identification	1.18
FPGR	CACTTTTATCTCTGGTTGCAG					−0.09
FP82F	CAAGCAAACAGGCTCTTCACC	SCAR marker	58	[2]	*F. poae* identification	−0.38
FP82R	TGTTCCACCTCAGTGACAGGTT					−0.94
Tox5-1	GCTGCTCATCACTTTGCTCAG	*Tri5* gene	65	[27]	Toxigenic *Fusarium* spp. identification	−0.62
Tox5-2	CTGATCTGGTCAGCGTCATC					−0.48
Fa5f	GGGGTCTTGCCACTCAGCTTGT	SCAR marker		[24]	*F. avenaceum/F. arthrosporioides/F. tricinctum* identification	−0.99
Fa5r	GGGGTCTTGCGGATCATGTGCT	SCAR marker		[24]	*F. avenaceum/F. arthrosporioides/F. tricinctum* identification	−0.99
Fa8f	GTGACGTAGGGAAACTGCCTGG	SCAR marker		[24]	*F. avenaceum/F. arthrosporioides/F. tricinctum* identification	−1.02
Fa8r	GTGACGTAGGACCAGAGATGTA	SCAR marker		[24]	*F. avenaceum/F. arthrosporioides/F. tricinctum* identification	−0.66

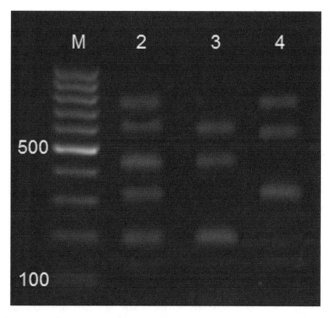

Fig. 1 From *left* to the *right*: First line (*M*): 100 bp DNA size marker; second line (*2*): multiplex PCR assay of monosporic fungal isolates resulting with five bands of 220, 332, 472, 620, and 729 bp specific to *F. poae* (24 F, Cankiri, Turkey), *F. graminearum* (15 F, Kastamonu, Turkey), *F. culmorum* (Ankara, Turkey), *tri5* gene, and *F. pseudograminearum* (Corum, Turkey); third line (*3*): bands of 620, 220, and 472; and fourth line (*4*): 332, 620, and 779 bp amplified from genomic DNA of diseased barley samples

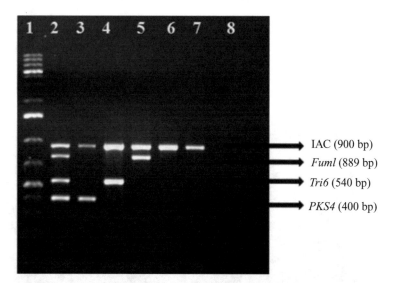

Fig. 2 Agarose gel representing a multiplex PCR-based strategy for concurrent detection of major toxigenic *Fusarium* species. Lane *1*—1 kb DNA marker. Lane *2*—Mixed DNA. Lane *3*—Sample contaminated with zearalenone-positive *Fusarium*. Lane *4*—Sample contaminated with trichothecene-positive *Fusarium*. Lane *5*—Sample contaminated with Fumonisin-positive *Fusarium*. Lane *6*—Non-toxigenic *Fusarium* species. Lanes *7* and *8*—Negative controls

1.3 Multiplex PCR Detection and Identification of Moniliformin- and Enniatins-Producing Fusarium avenaceum, F. arthrosporioides, and F. tricinctum Isolates

In the third approach the *F. avenaceum, F. arthrosporioides*, and *F. tricinctum* isolates could be divided to those producing a PCR product with none, one (either 606 or 1071 bp fragment) or both primer pairs (Table 1, Fig. 3).

1.4 Multilocus Genotyping

The species and trichothecene chemotype composition of *F. graminearum species* complex and closely related species, such as *F. culmorum* and *F. cerealis*, can be analyzed by using multiplex PCR with several primer pairs followed by probes of the gene sequences produced by PCR in the MLGT assay. The products of the first PCR are used as templates in the multilocus genotyping assay with Luminex 100 flow cytometer (Luminex corporation) [29]. We have studied the species and trichothecene chemotype composition of *F. graminearum, F. ussurianum, F. vorosii, F. culmorum*, and *F. cerealis* isolates by using multiplex PCR with six primer pairs followed by a 37 probe version (Table 2, Fig. 4) of six gene sequences of the MLGT assay [30]. Later it has been possible to increase the number of probes, when the number of species in the *F. graminearum* species complex has increased (e.g., [31, 32]). Similar MLGT analyses should also be developed for other *Fusarium* species complexes.

1.5 Multiplex qPCR

Finally, we demonstrate the use of multiplex TaqMan real-time qPCR for the simultaneous quantification of *F. sporotrichioides/F. langsethiae/F. sibiricum* and *F. poae* DNA and multiplex PCR in multiplex genotype (MLGT) analyses, in which, e.g., SNPs of

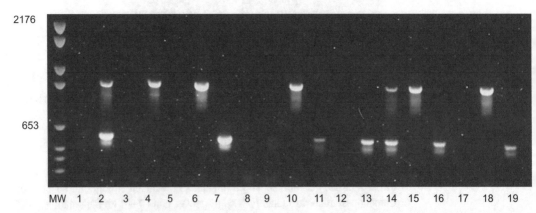

Fig. 3 PCR products obtained with specific primer pairs Fa5f,r (*upper band*, 1071 bp) and Fa8f,r (*lower band*, 604 bp) from 19 isolates of *F. avenaceum*. MW is the molecular marker VI (Boehringer/Mannheim), whose markers are 2176, 1766, 1230, 1033, 653, 517, 473, 394, 298, 234, 220, and 154 bp

Table 2
Primers used in multiplex MLGT [29]

Primer	Sequence	Size of the PCR product	ΔG for hairpin structure
Red-f	AGACTCATTCCAGCCAAG	702	0.83
Red-r	TCGTGTTGAAGAGTTTGG	702	0.75
Tri110-f	CAAGATACAGCTCGACACC	911	1.56
Tri110-r	CTGGGTAGTTGTTCGAGA	911	1.25
EF-1f	CGACCACTGTGAGTACCA	456	0.19
EF-1r	GTCAAGAACCCAGGCGTA	456	0.98
MAT-f	TTCTCAGGAACGACTCAAC	1040	0.23
MAT-r	TGTCGGTTCAGAACGATCA	1040	−0.5
TRI13-f	AACCTGAGCCCTCCAGT	912	−0.14
TRI13-r	TGGCAAAGACTGGTTCAC	912	0.21
TRI12-f	CATGAGCATGGTGATGTC	1163	−0.43
TRI12-r	AAGCATCAGCCTCTGCTC	1163	−1.25

several loci are used for genotyping and species identification. We have used this method for quantification of *Fusarium* DNA from grain samples [25].

2   Materials

Use reagents of molecular biology or HPLC[1] grade. Prepare all solutions by using ultrapure water at room temperature. However, PCR mixture could be assembled on chilled ice. Store solutions at room temperature unless it is mentioned that the temperature is indicated as +4 °C or −20 °C. Use disposable pipette tips, autoclavable micropipettes, microtubes of 1.5 mL sterilized by autoclaving, sterile 15 mL Falcon tubes, and remaining DNase/RNase-free plastic material. Use glass material including Erlenmeyer flask, beaker, and scalpel sterilized by dry heat sterilization.

Bioinformatic primer check: PCR primer secondary structure formation test by OligoAnalyzer of Integrated DNA Technologies. Nucleotide sequence of each primer and primer sets were tested for secondary structure formation (including dimers and hairpin) possibility/potential. Primers with ΔG value between +2 and −9 kcal/mol were used in PCR assays. *See* Table 1 for ΔG value matching of each primers for hairpin formation possibilities.

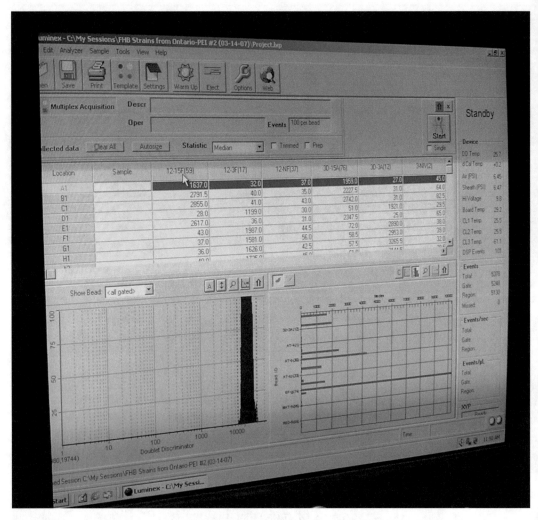

Fig. 4 An example of MLGT results of three 15ADON chemotypes and four 3ADON chemotypes

2.1 Multiplex PCR

2.1.1 PCR Components

Double-sterilized water: Autoclaved water could also be provided from commercial trademarks.

1. PCR primers: Oligonucleotides with no tags such as radioactive label or fluorescence tag. Primers could be in 15–26 bases length. They should be at least 25 nmol of amount for researches including more than 1000 tests. For 25 nmol of lyophilized primer molecule; 250 μL of water should be added to tube including primer and mixture should be kindly mixed by finger vortex or inverting tube up and down. Thus, tube including 100 pmol (*see* **Note 1**) of primer should be saved as stock solution. Store it at −20 °C. For preparing working solution, dilute 100 pmol of primers as 10 pmol to new microtubes. The number and volume of working solution could be two or more and 100 μL, respectively (*see* **Note 2**). Store it at

−20 °C. Primers could be synthesized at HPLC grade instead of PCR grade. We strongly recommend to use HPLC-grade oligonucleotides to avoid from false-positive results.

2. PCR buffer (5× buffer): The buffer could include agarose gel loading dye or not (colorless buffer). The pH of buffer is 8.5 and it provides standard PCR conditions of 50 mM Tris–Cl (pH 9.0), 50 mM NaCl. The buffer is ready to use (*see* **Note 3**). It should be saved at −20 °C. However, in routine studies one or more buffer tubes could be left at +4 °C.

3. Genomic DNA: Diluted template DNA molecules (*see* **Note 4**). DNA could be diluted as 5, 10, 25, 50 ng/μL in volume of 100 μL. General concentration range is 10–25 ng/μL. Store it at −20 °C.

4. Magnesium chloride: 25 mM of $MgCl_2$ in water. Ready-to-use solution. Store it at −20 °C. In studies with high number of samples, one or more tubes to be stored at +4 °C.

5. Deoxyribonucleoside triphosphate (dNTP) mixture: 10 mM of each dNTPs: dATP, dGTP, dCTP, dTTP. It is strongly recommended that dNTP mixtures to be aliquoted into small volumes as 250 μL. Store it at −20 °C.

6. *Taq* DNA polymerase: 5 U/μL DNA polymerase enzyme from *Thermus aquaticus* in glycerol. Store it at −20 °C (*see* **Note 5**).

7. PCR device: Thermal cycler machine heats samples between +4 and 115 °C temperature ranges.

2.1.2 Agarose Gel Electrophoresis Components

1. Gel comb with 12 teeth in size of proper with horizontal gel system.

2. Horizontal gel system: Gel system of 24.5 × 18 × 9.5 cm in sizes. System is proper to work for general city voltage values 220 V. Maximum voltage and amper values are 220 V and 500 mA, respectively. Vessel buffer volume is 600–800 mL.

3. Tris-acetic acid-ethylenediamine tetraacetic acid (EDTA) (TAE) buffer of 50×: 2 M Tris—Acetate, 0.05 EDTA. It used for electrophoresis. Weigh 242 g trisma base, 100 mL EDTA (0.5 M, pH 8.0) and measure 57.1 mL glacial acetic acids. Transfer them to beaker including 500 mL of water. Mix well by magnetic mixer. Adjust pH to 8.8 value. Volume up to 1 L and transfer solution lab bottle wrapped with aluminum foil.

4. Electrophoresis work buffer: 1× TAE solution. Dilute 50× TAE buffer in water as 1× concentration. Store it as mentioned before.

5. Ethidium bromide (EtBr) solution: 10 mg/mL EtBr (in water). It can also be commercially provided (*see* **Note 6**). Weigh 500 mg EtBr and dissolve it in 50 mL water. Mix it well, transfer to lab bottle covered by aluminum foil. Store it at +4 °C.

6. Agarose gel loading dye 6×: 10 mM Tris–Cl, 0.03 % bromophenol blue, 0.003 % xylene cyanol FF, 60 % glycerol, 60 mM EDTA. Store it at −20 °C.

7. DNA size marker 1: 100 bp DNA ladder of 50 µg/500 mL. It includes ten DNA bands ranging from 100 to 1000 bp. Mix 100 µL 6× loading dye, 100 µL DNA ladder and 400 µL of water well. Store it at −20 °C.

8. DNA size marker 2: Lambda/*HindIII* (and/or 1 kb) DNA size marker. Recommendations are same as for marker 1. These markers for analyzing genomic DNA molecules are optionally used since they do not correspond to real sizes of template DNA molecules.

9. Agarose gels: 1% of agarose gel matrix. Weigh 0.8 g agarose and transfer it to erlenmeyer flask including 40 mL (*see* **Note 7**) of 1× TAE buffer. Incubate flask in microwave oven at medium level of temperature for 2 min with occasional mixing, when it starts to boil. Then, wait until the time that flask could be hold by hand (30–35 °C) or that the temperature of the agarose is below 60 °C. Add 2 µL of EtBr solution (10 mg/mL) flask by directly touching pipette tip to liquid agarose-TAE mixture. Mix flask well. Spread liquid phase to electrophoresis gel cassettes including 12-teeth comb. When agarose becomes solid phase, remove plastic liner from up and down sites of cassette and introduce gel plate including solid gel into vessel system including 600 mL 1× TAE solution.

10. Ultraviolet (UV) transilluminator: Transilluminator system suitable for detecting DNA bands under UV light (220–365 nm wavelengths). Integrated camera system of 2 megapixel resolution is used to capture pictures under UV light.

2.1.3 Internal Amplification Control

An internal amplification control (IAC) was constructed by targeting pUC19 DNA with 5′ overhanging ends of Tri6 primer pair.

IAC F (5′- GATCTAAAGCACTATGAATCACCACATCG AACTGGATCTCAACAGC-3′) and IAC R (5′- GCCTATAGTG A T C T C G C A T G T C T A C G G G G T C T G A CGCTCAGT-3′) followed by the protocol of Kumar et al. [33]. Designing of IAC is shown in Fig. 5.

2.1.4 Amplification Mixture for the Second and Third Approach

The amplification mixture consists of template DNA (1.0 µL), MgCl$_2$ (2.0 mM), 1× PCR buffer, dNTP mix (200 µM), *Taq* DNA polymerase (1 unit), and primer pairs specific to the targeted genes *tri6*, *pks13*, and *fum1* are added at a concentration of 100 nM, 150 nM, 200 nM, and 50 nM, respectively. IAC is added at a concentration of 1000 copies per reaction. Total reaction volume is 30 µL.

2.2 Multiplex Genotyping (MLGT)

2.2.1 Primers (Table 2)

1. REDf/REDr for reductase.

2. TRI101-f/TRI101-r for Tri-101.

3. EF1-f/EF1-r for elongation factor 1.

4. TRI3-f/TRI3-r for Tri-3.

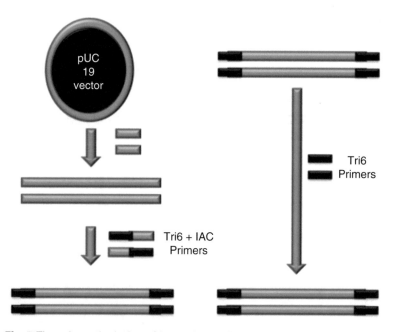

Fig. 5 The schematic design of internal amplification control for multiplex PCR assays

5. TRI12-f/TRI12-r for Tri-12.

6. MAT-f/MAT-r for MAT.

2.2.2 PCR

1. ddH$_2$O.

2. 10× PCR buffer—MgSO$_4$ (50 mM).

3. dNTP mixture (10 mM each).

4. Each primer (10 pmol μL$^-$) High Fidelity Taq DNA polymerase (5 U μL^{-1}) (*see* **Note 8**).

5. Dilute genomic DNA (about 100 ng).

6. = >vortex master mix and spin down.

7. 96-Well polycarbonate PCR microplate (*see* **Note 9**).

8. Microseal "A" Film (*see* **Note 10**).

Instead of separate PCR buffer, MgSO$_4$, polymerase, and nucleotides it is also possible to use Multiplex PCR master mix [31].

2.2.3 Gel Electrophoresis

– Use either 1× TAE or TBE (Tris-borate-EDTA) buffer containing 20 μL of ethidium bromide (1 mg/mL) in 100 mL of agarose gel and running buffer. Run ca. 1 h at 100 V to separate the PCR products. Place the gel on UV transilluminator to scan in image.

– Tris-borate-EDTA 10×: 1 M boric acid, 1 M Tris, and 0.02 M EDTA in 1000 mL ddH$_2$O.

2.2.4 PCR Cleanup		1. ddH$_2$O.

2. Millipore plate with a suitable pressure.

Alternatively multiplex PCR products can be purified using, e.g., Sephadex column [31].

2.2.5 Multiplex ASPE (Allele-Specific Primer Extension) Reaction

1. 10× ASPE buffer (200 mM Tris–HCl, pH 8.4, 500 mM KCl).
2. MgCl$_2$ (50 mM).
3. dATP, dCTP, and dGTP (100 μM of each).
4. Biotin-CTP (400 μM) (*see* **Note 11**).
5. Probe mix (all probes, 500 nmol each, Table 3).

Table 3
ASPE probes used in multiplex MLGT [29, 30, 34, 35]

ASPE probe[a]	Target	Probe sequence[b]
T12-15(59)	15ADON	TCATCAATCAATCTTTTTCACTTTtacagcggtcgcaacttc
T3-15(76)	15ADON	AATCTAACAAACTCATCTAAATACactgacccaagctgccatc
T12-3(17)	3ADON	CTTAATCCTTTATCACTTTATCActtttggcaagcccgtgca
T3-3(12)	3ADON	TACACTTTCTTTCTTTCTTTCTTTcgcattggctaacacatg
T12-N(37)	NIV	CTTTTCATCTTTTCATCTTTCAAATtggtctcctcgttgtatctgg
T3-N(2)	NIV	CTTTATCAATACATACTACAATCAgacaagtgcacagaatatacg
EFsp(31)	*Fusarium* sp.	TTCACTTTTCAATCAACTTTAATCgtagtttcacatttccgatgta
ATsp(33)	*Fusarium* sp.	TCAATTACTTCACTTTAATCCTTTtgttcctcgtcttgtagc
ATb(30)	B-FHB clade	TTACCTTTATACCTTTCTTTTTACacggtgctatggatatgg
ATce(24)	*F. cerealis*	TCAATTACCTTTTCAATACAATACgaggtagatcatcagattgtt
REDce(5)	*F. cerealis*	CAATTCAAATCACAATAATCAATCgttgcagacactacacaa
REDcu(18)	*F. culmorum*	TCAAAATCTCAAATACTCAAATCAgaagaaacgcttgtatcgaa
ATcu(19)	*F. culmorum*	TCAATCAATTACTTACTCAAATACaggacgttcctcgtgtta
EFl(94)	*F. lunulosporum*	CTTTCTATCTTTCTACTCAATAATccctcttcccacaaaccattt
MATl(49)	*F. lunulosporum*	TCATCAATCTTTCAATTACTTACgccctattcggtcctgattat
ATp(44)	*F. pseudogrami nearum*	TCATTTACCAATCTTTCTTTATACtgcagctcaacttcatcg
REDp(6)	*F. pseudogrami nearum*	TCAACAATCTTTTACAATCAAATCcaagccgatgccaagtcc
EFg(14)	*Fg* complex	CTACTATACATCTTACTATACTTTtcatcatcacgtgtcaac

(continued)

Table 3 (continued)

ASPE probe[a]	Target	Probe sequence[b]
ATg(10)	*Fg* complex	ATCATACATACATACAAATCTACAccattcaccgaagaggaaat
EF5(54)	*F. acaciae-mearnsii*	CTTTTTCAATCACTTTCAATTCATgtctcattttcctcgatcgcc
AT5(29)	*F. acaciae-mearnsii*	AATCTTACTACAAATCCTTTCTTTggtcttaagcgcttctc
MAT6(65)	*F. asiaticum*	CTTTTCATCAATAATCTTACCTTTggctacttttctgagtactct
AT6(85)	*F. asiaticum*	ATACTACATCATAATCAAACATCAaagctgggcgttcttcaa
EF1(60)	*F. austroamericanum*	AATCTACAAATCCAATAATCTCATgtcaaccagtcactaact
AT1(77)	*F. austroamericanum*	CAATTAACTACATACAATACATACatccctctcaatatcccg
RED3(3)	*F. boothii*	TACACTTTATCAAATCTTACAATCattggtgttgccttcgcc
AT3(51)	*F. boothii*	TCATTTCAATCAATCATCAACAATaaggtcttaagcgcttcg
MAT9(95)	*F. brasilicum*	TACACTTTAAACTTACTACACTAAcagatttcgatcgctgatgaa
RED9(64)	*F. brasilicum*	CTACATATTCAAATTACTACTTACggcttacaaaggtgagtg
EF8(39)	*F. cortaderiae*	TACACAATCTTTTCATTACATCATgacacttggcggggtagttt
AT8(80)	*F. cortaderiae*	CTAACTAACAATAATCTAACTAACgggtatgagaaaggcgga
ATae(9)	*F. aethiopicum*	TAATCTTCTATATCAACATCTTACggccagtaccaggcctg
REDae(66)	*F. aethiopicum*	TAACATTACAACTATACTATCTACcggaagaaacatgatgggt
RED10(55)	*F. gerlachii*	TATATACACTTCTCAATAACTAACgaactagaactagtcaatgcc
AT10(40)	*F. gerlachii*	CTTTCTACATTATTCACAACATTAtgacgatgctctttcggcc
EF7(28)	*F. graminearum*	CTACAAACAAACAAACATTATCAAactcgagcgacaggcgtc
AT7(26)	*F. graminearum*	TTACTCAAAATCTACACTTTTTCAatagttccttaccttgaaaactat
EFLa(41)	*F. louisianense*	TTACTACACAATATACTCATCAATgccctctcccacaaaccac
ATLa(48)	*F. louisianense*	AAACAAACTTCACATCTCAATAATaaacttcatcaagggcggactt
AT2(50)	*F. meridionale*	CAATATACCAATATCATCATTTACctcgtgttgtagtgaaagat
RED2(69)	*F. meridionale*	CTATAAACATATTACATTCACATCcagtattgatcatgaggcta
RED4(35)	*F. mesoamericanum*	CAATTTCATCATTCATTCATTTCAgttgtcattacgggtggt
AT4(1)	*F. mesoamericanum*	CTTTAATCTCAATCAATACAAATCcgagggaaacacaggaat
EFnep(8)	*F. nepalense*	ATTCCTTTTACATTCATTACTTACcacgacgactcgatacgt
ATu(25)	*F. ussurianum*	CTTTTCAATTACTTCAAATCTTCAgatgtagctggtggtgat
REDu(27)	*F. ussurianum*	CTTTTCAAATCAATACTCAACTTTcatcacgtgtcaaccagc
RED11(16)	*F. vorosii*	AATCAATCTTCATTCAAATCATCAcaaaggtgagtatgagtat
MAT11(100)	*F. vorosii*	CTATCTTTAAACTACAAATCTAACtaagtccgaatgaagccccgg

[a]Probe locations in the PCR products produced by primers of Table 2 are indicated by two or three marks at the beginning of the probe name. Luminex microsphere sets used for hybridization are indicated in parentheses
[b]The 5′ sequence tag portions of extension probes are capitalized, while the 3′-target-specific sequences are in lower case

6. Platinum GenoTYPE Tsp Taq (5 U/μL) (*see* **Note 12**).

7. Multiplex product from PCR.

8. 96-Well polycarbonate PCR microplate.

9. Microseal "A" Film.

2.2.6 Hybridization

1. Carboxylated fluorescent microspheres with covalently attached anti-TAG sequences.

2. Biotin-labeled targets with appropriate TAG sequence modification.

3. 2× Tm hybridization buffer—0.4 M MaCl, 0.2 M Tri, 0.16, and Triton X-100, pH 8.0.

4. 96-Well PCR plate and cover.

2.2.7 Hybridization Cleanup

1. 1× Tm hybridization buffer.

2. 1× Tm hybridization buffer containing 2 μg/mL streptavidin-R-phycoerythin (*see* **Note 13**).

3. dH$_2$O.

2.3 Multiplex TaqMan Real-Time PCR [25]

We have studied the presence and amount of T-2/HT-2-producing *Fusarium* species (*F. sporotrichioides, F. langsethia,* and *F. sibiricum*) and NIV-producing *F. poae* in cereals by using multiplex TaqMan real-time PCR with ABI Prism® 7700 cycler. The R^2 (= coefficient of determination) values of the standard curves are at the same level as when the qPCR reactions were performed separately, but more cycles are required for the threshold value with multiplex PCR.

2.3.1 TaqMan Real-Time qPCR Components

1. Plates: 96-Well plate suitable for measurements processed at 492/521 nm excitation/emission values.

2. Real-time PCR device: Real-time machine works proper with monocolor hydrolyze-dual-color probes, Sybr Green-Eva Green dyes, melting curve-ending point genotyping.

3. Genomic DNA templates. Genomic DNA solutions from barley and *Fusarium* samples diluted as standard series with 5 logs.

4. Ultrapure water: PCR-grade water included in Sybr Green I PCR master mix.

5. Primer and probe molecules: Oligonucleotide primers and probes designed for TaqMan real-time qPCR.

6. qPCR mix: 2× real-time PCR master mix including PCR buffer, MgCl$_2$, dNTP mix and *Taq* DNA polymerase. 10 μL of mix provides 1× buffer, 2.5 mM MgCl$_2$, 0.1 mM of dNTPs, and 1 U of enzyme to 20 μL PCR mix. It can also include passive control dye, Rox.

7. Centrifuge device: Centrifuge device with rotors for 96-well plates.

8. Sealing foil: Foil with proper sizes for 96-well plates. It covers the surface of plate. Optical compression pad can be used for improved film sealing.

2.3.2 Primers and Probes

F. poae-specific primers and probes [25]:

- 100 nM TMpoaef (5′-GCTGAGGGTAAGCCGTCCTT-3′) and TMpoaer (5′-TCTGTCCCCCCTACCAAGCT-3′) primers (stock solution 50 pmol μL^{-1}).

- 100 nM Tmpoae probe (5′ATTTCCCCAACTTC GACTCTCCGAGGA-3′) labelled at the 5′end with TET (tetrachloro-6-carboxy-fluorescein) and at the 3′end with 3′Eclipse Dark Quencher (stock solution 20 pmol μL^{-1}).

F. sporotrichioides/F. langsethiae/F. sibiricum-specific primers and probe [25, 36, 37]:

- 300 nM TMLANf (5′-GAGCGTCATTTCAACCCTCAA-3′) and TMLANr (5′-GACCGCCAATCAATTTGGG-3′) primers (300 nM, stock solution 50 pmol μL^{-1}).

- 100 nM TMLAN probe (5′-AGCTTGGTGTTGGGATC TGTCCTTACCG-3′) (stock solution 50 pmol μL^{-1}) labeled at the 5′ends with 6-FAM (6-carboxy-fluorescein) and at the 3′end with TAMRA (5-carboxytetramethylrhodamine) for the quencher.

DNA sample containing 1–6 ng DNA.
Genomic DNA standards containing 1, 0.1, 0.01, 0.001, 0.0001 ng/μL DNA of the *F. poae* or *F. sporotrichioides/F. langsethiae/F. sibiricum* DNA.

3 Methods

3.1 Multiplex PCR for Trichothecene-Producing Species

3.1.1 Polymerase Chain Reaction

1. Take out PCR components but *Taq* DNA polymerase from −20 °C and chill out components on ice up to 30 min.

2. All PCRs are conducted in reaction volume of 25 μL.

3. Before starting the PCR process, dissolve all components. Mix PCR buffer and MgCl$_2$ by using vortex device for 5–10 s. Finger vortex dNTP mix, primers, and DNA tubes gently. The order of PCR compounds is not significant but in general process is started by adding sterilized water and adding PCR buffer, MgCl$_2$ solution, dNTP mix, primers, DNA, and enzyme follows it.

4. For uniplex PCR of monosporic *Fusarium* spp. identification with primers UBC85F/R, OPT18F/R, FPGF/R, FP82F/R

and Tox5-1/2, PCRs were conducted as it follows: 50 ng of template DNA (5 μL), 1.5 mM of MgCl$_2$ (1.5 μL), 1× PCR Buffer (5 μL), 0.25 mM (1 μL) of each primer, 0.25 mM of each dNTPs (1 μL), 0.5 U Taq DNA Polymerase (0.1 μL) and 11.4 μL of water. PCR conditions were performed at 30 cycles including 95 °C for 1 min and 57–65 °C for 1 min (Table 1). Pre-denaturation and final extraction steps are maintained at 94 °C for 2 min and 72 °C for 5 min, respectively.

5. For multiplex PCR of barley samples with primers given above, PCRs were conducted as follows: 200 ng of template DNA (4 μL), 2.5 mM of MgCl$_2$ (2.5 μL), 1× PCR buffer (5 μL, *see* **Note 14**), 0.125 mM (0.5 μL) of each primer, 0.5 mM of each dNTP (2 μL) and 1 U (*see* **Note 15**) of Taq DNA polymerase (0.2 μL). The volume was completed to 25 μL by adding 6.3 μL of water. PCRs were carried out by pre-denaturation at 95 °C for 1 min, 59 °C for 1 min and 72 °C for 2 min and final extension at 72 °C for 10 min. For multiplex of monosporic fungal samples conditions were used as 10 ng for each species in PCR performing (Fig. 1). For negative control assays, no template control samples are used.

6. In PCR assays, mineral oils or heated lids are used generally to avoid from loss in volume. Also, 5–10 % volume of DMSO, glycerol, and BSA could be used in increasing the efficiency of PCR. The volume corresponds to 1.25–2.5 μL (5–10 % of percentage).

7. Store PCR tubes at 4 °C if samples would be analyzed in 1–3 days. On contrary, samples could be stored at −20 °C for 6 months.

3.1.2 Agarose Gel Electrophoresis

1. Transfer PCR and genomic DNA tubes to chilled ice for 30 min.

2. Mix 10 μL DNA samples and 2 μL of agarose gel loading dye in separate microtubes or on parafilm surface.

3. Load 12 μL of mixed samples including loading dye to agarose gel lines with micropipette.

4. Load 2 μL of DNA size markers to lines.

5. Close vessel system and bind positive (black) and negative (red) cables to power supply.

6. Initiate the electrophoresis at 70 V, 200 mA for 1 h and 30 min to separate all DNA bands well.

7. Remove vessel buffer from agarose gel after finishing electrophoresis. Transfer gel to UV transilluminator and capture PCR profiles under UV light.

Table 4
Primers used for group-specific detection of trichothecene- and fumonisin-producing *Fusarium* species by mPCR [10, 18]

Primer name	Primer sequence (5′ to 3′)	Gene targeted	Tm (°C)	Amplicon size (bp)	ΔG for hairpin structure
Its 1	GCA TGC CTG TTC GAG CGT	rDNA	58	300	0.46
Its 2	CTG TTG CCG CTT CAC TCG C				0.59
Tri5 f	GAG AAC TTT CCC ACC GAA TAT	tri5	56	450	0.3
Tri5 r	GAT AAG GTT CAA TGA GCA GAG				1.24
Tri6 f	GAT CTA AAC GAC TAT GAA TCA CC	tri6	58	546	0.8
Tri6 r	GCC TAT AGT GAT CTC GCA TGT				0.09
Fum1 f	ATT ATG GGC ATC TTA CCT GGA T	Fum1	58	798	0.12
Fum1 r	ACG CAA GCT CCT GTG ACA GA				−0.54
Fum 13 f	AGT CGG GGT CAA GAG CTT GT	fum13	58	988	−1.05
Fum13 r	TGC TGA GCC GAC ATC ATA ATC				0.31
ZEN F	CATTCTTGGTCTTGTGAGGA	PKS4	56	400	0.48
ZEN R	GCAGCCGCCAACCGGAAAGT				−2.95
Fcu F	GATGCCAGACCAAGACGAAG	rDNA	58	302	1.48
Fcu R	GGTTAGAATCATGCCGACC				0.57
Tri7 F	ATAGGTACCGGATCGCAGG	tri7	58	794	0.22
Tri7 R	CCGAAAGCCTCTAATAGTGT				0.53
Tri13 F	GTTGCAGTTCGCTTGATTTCG	tri13	58	1000	−1.03
Tri13 R	GTTGCAGTTCGCTTGATTCAG				−1.03

3.2 Multiplex PCR for Trichothecene-, Zearalanone-, and Fumonisin-Producing Species

3.2.1 Specificity Determination for Multiplex PCR

The specificity of the mPCR primers can be determined against different standard cultures shown in Table 4 by taking 1 μL of genomic DNA from each of the organism. A minimum quantity of 6 pg of DNA for genus specific recognition, 150 pg for the trichothecene producing *Fusarium graminearum* and 100 pg for the fumonisin producing *Fusarium verticillioides* is needed to produce a good visible band on ethidium bromide stained agarose gel [19].

The PCR cycling conditions are carried out with an initial denaturation at 94 °C for 4 min, followed by 30 cycles of 94 °C for 1 min, 58 °C for 1 min and 72 °C for 1.5 min, with a final extension of 72 °C for 8 min. Optimized mPCR representing agarose gel image is shown in Fig. 2.

3.2.2 Multiplex PCR Assay on Artificially Inoculated Food Grains

To determine the practical use of multiplex PCR assay, sterile rice grains (5 g) can be experimentally spiked with the individual spore suspensions at concentrations of 1×10^7, 1×10^6 and 1×10^5 CFU/g of *F. graminearum*, *F. culmorum*, *F. sporotrichioides*, *F. verticillioides*, and *F. solani*. One uninoculated sample is used as negative control. All the samples are incubated for 2 days. DNA is isolated from all the samples and subjected to multiplex PCR assay (Fig. 2). This method also works with naturally contaminated grain samples.

3.3 Multiplex PCR for Moniliformin- and Enniatins-Producing Fusarium Species

For multiplex PCR with monosporic *F. avenaceum*, *F. arthrosporioides*, and *F. tricinctum* isolates. PCRs were conducted as follows: 0.5–1 ng of template DNA (1 µL), 10× buffer (2 µL), 60 ng of each primer Fa5f/r amd Fa8f/r (4×1 µL), 150 µM of each dNTP (0.8 µL), and 10× polymerase. The volume was completed to 20 µL by adding water. During the first cycle DNA was denatured at 94 °C. The subsequent 30 cycles included 94 °C for 3 min, 58 °C for 1:30 min and 72 °C for 1:20 min, followed by an extension at 72 °C for 3 min [24].

3.4 Multiplex Genotyping

3.4.1 PCR for One Sample

1.2 µL ddH$_2$O.

1.0 µL 10× PCR buffer 0.4 µL MgSO$_4$ (50 mM).

0.2 µL dNTP (10 mM each).

0.2 µL each primer (10 pmol µL^{-1})

0.1 µL High-fidelity polymerase (5 U µL^{-1}).

4.7 µL dilute genomic DNA (about 100 ng).

Instead of separate PCR buffer, MgSO$_4$, polymerase, and nucleotides it is also possible to use, e.g., Qiagen Multiplex PCR master mix [31].

3.4.2 PCR Parameters

94 °C for 1:30 min.

94 °C for 30 s.

50 °C for 30 s $\geq$40 cycles $\geq$4 °C hold.

68 °C for 1:30.

3.4.3 Gel Electrophoresis

After PCR the PCR products (2 µL + 10 µL 6× loading buffer) are run on the 1.5 % agarose gel in order to check for the sizes of the multiplex product together with 100 bp ladder (2 µL). Use either 1× TAE or TBE buffer containing 20 µL of ethidium bromide (1 mg/mL) in 100 mL of agarose gel and running buffer. Run ca. 1 h at 100 V to separate the PCR products. Place the gel on UV transilluminator to scan in image.

3.4.4 PCR Clean Up

Add 150 µL of ddH$_2$O to each sampled to be cleaned. Transfer the samples to a Millipore plate and place on the manifold for 10 min with a suitable pressure. After 10 min add another 150 μL of ddH$_2$O to the plate and place it back to the manifold for another 10 min. Next, add 60 μL ddH$_2$O to the plate and place it to the plate shaker for 20 min. Then transfer contents into a clean plate. If you are not using these samples in the near future, put a seal over the top and place the plate in the freezer.

Alternatively multiplex PCR products can be purified using, e.g., Sephadex column [31].

3.4.5 Multiplex ASPE Reaction

Each reaction with the volume of 20 μL:

- 2 μL 10× ASPE buffer.
- 0.5 μL MgCl$_2$.
- 1.0 μL ATP, dCTP, and dGTP.
- 0.25 μL biotin-CTP.
- 1.0 μL probe mix (500 nmol each) (Table 3).
- 0.15 μL Platinum GenoTYPE Tsp Taq (5 U/μL).
- 15.1 μL Multiplex product from PCR.

Extension Reaction:

94 °C for 1:30 min.

94 °C for 30 s.

55 °C for 1:00 min ≥40 cycles ≥4 °C hold.

74 °C for 2:00 min.

Place the seal over the plate and put in the freezer.

3.4.6 Hybridization

Thaw and vortex the microspheres. Remember that they are light sensitive and keep them in ice after thawing.

Combine the different microspheres into a mixture with 2× Tm hybridization buffer (pH 8.0).

Add 20 μL ddH$_2$O and 5 μL of the extension reaction to the plate per reaction.

When the microspheres (*see* **Note 16**) are thawed, vortex briefly and then combine all sets into the same tube. Next, spin at 2250 × *g* for 3 min after balancing. A pellet should be visible in the bottom. Remove the supernatant without disturbing the pellet and add 2× Tm hybridization buffer to aliquot 25 μL of the mix per reaction. After 25 μL of the microsphere mix is added to the ddH$_2$O + extension reaction (total volume 50 μL), mix thoroughly.

Cover the plate to prevent evaporation and denature at 96 °C for 1:30 min and then hybridize at 37 °C for 45 min. Do not spin down the plate.

During hybridization the Luminex 100 is warmed up and set up (*see* **Notes 17** and **18**) according to the user manual.

3.4.7 Hybridization Clean-Up

After hybridization is complete, centrifuge the samples at $2250 \times g$ for 3 min. Remember to balance. Remove the supernatant and wash the microspheres in 70 μL 1× Tm hybridization buffer. Mix thoroughly with pipette. Centrifuge again at $2250 \times g$ for 3 min. Remove supernatant and resuspend the microspheres in 70 μL 1× Tm hybridization buffer containing 2 μg/mL streptavidin-R-phycoerythin. Mix thoroughly with pipette. Incubate at 37 °C for 10 min inside the Luminex 100.

When the incubation is complete, the Luminex 100 (*see* **Note 19**) is started and the sampled are read.

3.5 Multiplex TaqMan Real-Time PCR

3.5.1 PCR Mix for One Sample

- 10.88 μL ddH$_2$O.
- 0.15 μL TMlanf (50 pmol μL^{-1}).
- 0.15 μL TMlanr (50 pmol μL^{-1}).
- 0.125 μL TMlanp (20 pmol μL^{-1}) labeled with 6-FAM (6-carboxy-fluorescein)   and   TAMRA (5-carboxytetramethylrhodamine).
- 0.05 μL TMpoaef (50 pmol μL^{-1}).
- 0.05 μL TMpoaer (50 pmol μL^{-1}).
- 0.0625 μL TMpoae probe (20 pmol μL^{-1}) labeled with TET (tetrachloro-6-carboxy-fluorescein)   and   3′Eclipse   Dark Quencher—12.5 μL 2× real-time qPCR master mix.
- 1 μL dilute DNA (about 1 ng total DNA for grain sample).

PCR parameters

1. 50 °C for 2:00.
2. 95 °C for 10:00.
3. 95 °C 0:15 ≥40 cycles.
4. 60 °C 1:00.

Initiate the qPCR assay and save qPCR results as new file. According to standard series, compare DNA amount of *Fusarium* in samples. Record Ct or Cp (cycle threshold or crossing point), draw standard graphic, find your samples' location at this graphic, and then calculate the concentration value according to standard series.

4 Notes

1. If 25 nmol of lyophilized primers are diluted with 250 μL of water, stock concentration of primers could also be defined/calculated as 100 μM instead of 100 pmol μL^{-1}. It depends on researcher's choice.

2. More than one primer working solutions could increase the success in PCR studies since transferring stock solutions from −20 °C to +4 °C and/or 25 °C for several times would adversely affect the storage time of primer stocks.

3. In general, more than two PCR buffers are involved in PCR kits. They can be aliquoted (1:1 volume) to new microtubes. Also you can stop to use the buffer tube when you consume the half of it.

4. Isolated DNA molecules should be quantitatively measured by spectrophotometer and Δ 260/280 ratio should be approximately 1.8 for pure DNA molecules.

5. Since *Taq* DNA polymerase enzyme is involved in glycerol solution, it does not become freeze at −20 °C. Thus, just take it out from −20 °C refrigerator when you would use it and immediately transfer it to −20 °C again (to avoid from shortened half life process).

6. Since EtBr has potential carcinogenic effects, always use it by separate gloves, and after electrophoresis work, do not use these gloves again.

7. The volume of Erlenmeyer used in agarose gel preparation should be at least two times more than the volume of agarose gel.

8. Platinum Taq DNA Polymerase High Fidelity 5 U/μL.

9. Thermowell 96-well polycarbonate pcr microplate model P, nonsteril 25/case.

10. Microseal "A" Film, Microseal "B" film or Wax seals 50/box to seal the plate.

11. Biotin-14-dCTP (0.4 nM).

12. Platinume Geno Type TSP DNA polymerase 5 U/μL, 250 Units.

13. Streptavidin, R-phycoerythrin conjugate (SAPE)-1 mg/mL.

14. Volume of PCR buffer could be increased as 2× concentration in multiplex PCR assays.

15. Concentration of *Taq* DNA polymerase could be adjusted as 2 U in multiplex PCR.

16. Luminex microsphere beads (1 mL/bottle).

17. Luminex Classification Calibration Beads.

18. Luminex Reporter Calibration Beads.

19. Instead of Luminex 100 it is also possible to use, e.g., FLEXMAP 3D.

Acknowledgements

This work was supported by the Academy of Finland (no. 52104), National Technology Agency of Finland (No. 40168/03), and the Nordic Research Board (No. 040291).

References

1. Chandler EA, Simpson DR, Thomsett MA et al (2003) Development of PCR assays to *tri7* and *tri13* trichothecene biosynthetic and characterisation of chemotypes of *Fusarium graminearum, Fusarium culmorum* and *Fusarium cerealis*. Physiol Mol Plant Pathol 62:355–367

2. Nicholson P, Simpson DR, Wilson AH et al (2004) Detection and differentiation of trichothecene and enniatin-producing *Fusarium* species on small-grain cereals. Eur J Plant Pathol 110:503–514

3. Ward E, Foster SJ, Fraaije BA et al (2004) Plant pathogen diagnostics: immunological and nucleic acid-based approaches. Ann Appl Biol 145:1–16

4. Miedaner T, Cumagun CJR, Chakraborty S (2008) Population genetics of three important head blight pathogens *Fusarium graminearum, F. pseudograminearum* and *F. culmorum*. J Phytopathol 156:129–139

5. Yli-Mattila T, Paavanen-Huhtala S, Parikka P et al (2004) Toxigenic fungi and mycotoxins in Finnish cereals. In: Logrieco A, Visconti A (eds) An overview on toxigenic fungi and mycotoxins in Europe. Kluwer Academic Publishers, The Netherlands, pp 83–100

6. Chamberlain JS, Gibbs RA, Rainer JE et al (1998) Deletion screening of the Duchenne muscular dystrophy locus via multiplex DNA amplification. Nucleic Acids Res 16(23):11141–11156

7. Edwards MC, Gibbs RA (1994) Multiplex PCR: advantages, development, and applications. Genome Res 3:65–75

8. Henegariu O, Heerema NA, Dlouhy SR et al (1997) Multiplex PCR: critical parameters and step-by-step protocol. Biotechniques 23(3):504–511

9. Chandra NS, Wulff EG, Udayashankar AC et al (2011) Prospects of molecular markers in *Fusarium* species diversity. Appl Microbiol Biotechnol 90:1625–1639

10. Ramana MV, Shilpa P, Balakrishna K et al (2013) Incidence and multiplex PCR based detection of trichothecene producing Fusarium culmorum isolated from maize and Paddy samples collected from India. Braz J Microbiol 44:401–406

11. Hohn TM, McCormick SP, Desjardin AE (1993) Evidence of a gene cluster involving trichothecene-pathway biosynthetic genes in *Fusarium sporotrichioides*. Curr Genet 24:291–295

12. Kimura M, Tokai T, Donnell KO et al (2003) The trichothecene biosynthesis gene cluster of Fusarium graminearum F15 contains a limited number of essential pathway genes and expressed non-essential genes. FEB Lett 27:105–110

13. Niessen ML, Vogel RF (1998) Group specific PCR-detection of potential trichothecene-producing Fusarium-species in pure cultures and cereal samples. Syst Appl Microbiol 21:618–631

14. Agodi A, Barchitta M, Ferrante M et al (2005) Detection of trichothecene producing *Fusarium* spp. by PCR: adaptation, validation and application to fast food. Ital J Public Health 3:7–11

15. Demeke T, Clear RM, Patrick SK et al (2005) Species specific PCR based assays for the detection of Fusarium species and a comparison with the whole seed agar plate method and trichothecene analysis. Int J Food Microbiol 103:271–284

16. Bluhm BH, Flaherty JE, Cousin MA et al (2002) Multiplex polymerase chain reaction assay for the differential detection of trichothecene- and fumonisin-producing species of *Fusarium* in cornmeal. J Food Prot 65:1955–1961

17. Lee T, Oh DW, Kim HS et al (2001) Identification of deoxynivelenol- and nivalenol-producing chemotypes of *Gibberella zeae* by using PCR. Appl Environ Microbiol 67:2966–2972

18. Waalwijk C, van der Lee T, de Vries I et al (2004) Synteny in toxigenic *Fusarium* species: the fumonisin gene cluster and the mating type region as examples. Eur J Plant Pathol 110:533–544

19. Ramana MV, Balakrishna K, Murali HS et al (2011) Multiplex PCR-based strategy to detect contamination with mycotoxigenic *Fusarium* species in rice and finger millet collected from southern India. J Sci Food Agric 91:1666–1673

20. Waalwijk C, Kastelein P, de Vries I et al (2003) Major changes in *Fusarium* spp. in wheat in the Netherlands. Eur J Plant Pathol 109:743–754

21. Kerényi ZA, Moretti C, Waalwijk B et al (2004) Mating type sequences in asexually reproducing *Fusarium* species. Appl Environ Microbiol 70:4419–4423

22. Brandfass C, Karlovsky P (2006) Simultaneous detection of *Fusarium culmorum* and *Fusarium graminearum* in plant material by duplex PCR with melting curve analysis. BMC Microbiol 6:1–10

23. Quarta A, Mita G, Haidukowski M et al (2006) Multiplex PCR assay for the identification of nivalenol, 3- and 15-acetyl-deoxynivalenol chemotypes in *Fusarium*. FEMS Microbiol Lett 259(1):7–13

24. Yli-Mattila T, Paavanen-Huhtala S, Parikka P et al (2004) Molecular and morphological diversity of *Fusarium* species in Finland and northwestern Russia. Eur J Plant Pathol 110:573–585

25. Yli-Mattila T, Paavanen-Huhtala S, Jestoi M et al (2008) Real-time PCR detection and quantification of *Fusarium poae, F. graminearum, F. sporotrichioides* and *F. langsethiae* in cereal grains in Finland and Russia. Arch Phytopathol Plant Protect 41:243–260

26. Schilling AG, Möller EM, Geiger HH (1996) Polymerase chain reaction-based assays for species-specific detection of *Fusarium culmorum, F. graminearum* and *F. aveneceaum*. Mol Plant Pathol 86(5):515–522

27. Hue FX, Huerre M, Rouffault MA et al (1999) Specific detection of *Fusarium* species in blood and tissues by a PCR technique. J Clin Microbiol 37:2434–2438

28. Jennings P, Coates ME, Turner JA et al (2004) Determination of deoxynivalenol and nivalenol chemotypes of *Fusarium culmorum* isolates from England and Wales by PCR assay. Plant Pathol 53:182–190

29. Ward TJ, Clear R, Rooney A et al (2008) An adaptive evolutionary shift in *Fusarium* head blight pathogen populations is driving the rapid spread of more toxigenic *Fusarium graminearum* in North America. Fungal Genet Biol 45:473–484

30. Yli-Mattila T, Gagkaeva T, Ward TJ et al (2009) A novel Asian clade within the *Fusarium graminearum* species complex includes a newly discovered cereal head blight pathogen from the Russian far east. Mycologia 101:841–852

31. Davari M, Wei SH, Babay-Ahari A et al (2013) Geographic differences in trichothecene chemotypes of *Fusarium graminearum* in the Northwest and North of Iran. World Mycotoxin J 6:137–150

32. Boutigny A-L, Ward TJ, Ballois N et al (2014) Diversity of the *Fusarium graminearum* species complex on French cereals. Eur J Plant Pathol 138:133–138

33. Kumar S, Balakrishna K, Batra HV (2006) Detection of Salmonella enterica serovar Typhi (S. typhi) by selective amplification of invA, viaB, fliC-d and prt genes by polymerase chain reaction in multiplex format. Lett Appl Microbiol 42:149–154

34. O'Donnell K, Ward TJ, Aberra D et al (2008) Multilocus genotyping and molecular phylogenetics resolve a novel head blight pathogen within the *Fusarium graminearum* species complex from Ethiopia. Fungal Genet Biol 45:1514–1522

35. Sarver BAJ, Ward TJ, Gale LR et al (2011) Novel Fusarium head blight pathogens from Nepal and Louisiana revealed by multilocus genealogical concordance. Fungal Genet Biol 48:1096–1107

36. Halstensen AS, Nordby KC, Eduard W et al (2006) Real-time PCR detection of toxigenic Fusarium in airborne and settled grain dust and associations with trichothecene mycotoxins. J Environ Monit 8(12):1235–1241

37. Yli-Mattila T, Gavrilova O, Hussien T et al (2015) Identification of the first *Fusarium sibiricum* isolate in Iran and *Fusarium langsethiae* isolate in Siberia by morphology and species-specific primers. J Plant Pathol 97:183–187

Chapter 19

Multiplex Detection of Toxigenic *Penicillium* Species

Alicia Rodríguez, Juan J. Córdoba, Mar Rodríguez, and María J. Andrade

Abstract

Multiplex PCR-based methods for simultaneous detection and quantification of different mycotoxin-producing *Penicillia* are useful tools to be used in food safety programs. These rapid and sensitive techniques allow taking corrective actions during food processing or storage for avoiding accumulation of mycotoxins in them. In this chapter, three multiplex PCR-based methods to detect at least patulin- and ochratoxin A-producing *Penicillia* are detailed. Two of them are different multiplex real-time PCR suitable for monitoring and quantifying toxigenic *Penicillium* using the nonspecific dye SYBR Green and specific hydrolysis probes (TaqMan). All of them successfully use the same target genes involved in the biosynthesis of such mycotoxins for designing primers and/or probes.

Key words Multiplex, PCR, qPCR, *Penicillium*, Patulin, Ochratoxin A, *idh*, *otanps*PN

1 Introduction

Ochratoxin A, patulin, citrinin, or cyclopiazonic acid are some of the most important mycotoxins produced by *Penicillium* species [1–3]. These mycotoxins are contaminants of a wide variety of foodstuffs such as cereals, fruits, cheeses, or cured meats as consequence of toxigenic *Penicillia* growth [3–9]. The early detection and quantification of toxigenic *Penicillium* strains in foods seem to be critical for the production of safe foods, since mycotoxins cannot be removed from foods. For this purpose, PCR-based methods appear to be an efficient tool compared to traditional culturing methods to control the presence of mycotoxin-producing *Penicillium* in foods [10–12]. For detection of potentially toxigenic molds by PCR, unique DNA sequences must be selected as primer binding sites [3]. Most PCR methods for detecting or quantifying of mycotoxin-producing *Penicillium* are relied on target genes involved in the mycotoxin biosynthetic pathways since such genes are exclusively present in the toxigenic molds. Although some individual conventional PCR and real-time PCR (qPCR) methods for detection of *Penicillium* producers of different mycotoxins have been published [5, 13–21], the use of multiplex

Antonio Moretti and Antonia Susca (eds.), *Mycotoxigenic Fungi: Methods and Protocols*, Methods in Molecular Biology, vol. 1542, DOI 10.1007/978-1-4939-6707-0_19, © Springer Science+Business Media LLC 2017

PCR and qPCR methods capable of monitoring such toxigenic molds in a single reaction simplifies the detection procedure. Since in this kind of PCR a simultaneous amplification of several target sequences is performed in a single PCR reaction by using more than one primer pair, less reagents and time of analysis are required [11]. Thus, multiplex PCR-based methods seem to be the best alternative in the early detection and quantification of toxigenic *Penicillium* in foods.

In this chapter, one multiplex PCR and two multiplex qPCR methods for detecting and/or quantifying patulin- and ochratoxin A-producing *Penicillium* are detailed. The multiplex conventional PCR assay and one of the multiplex qPCR protocols are able to detect aflatoxigenic molds too [22–24]. In all the above methods, the *isoepoxydon dehydrogenase* (*idh*) and the *non-ribosomal peptide synthetase* (*otanps*PN) genes involved in the patulin and ochratoxin A biosynthesis, respectively, were used as target genes. In the case of procedures capable of detecting aflatoxin-producing molds, the *sterigmatocystin O-methyltransferase* (*aflP*) gene was also used as target. The triplex TaqMan-based qPCR described in this chapter is the only one which allows simultaneous quantification of each desired target gene separately. However, the duplex SYBR Green-based qPCR is also of great utility since allows a quantification of both target genes jointly. For routine analysis the duplex SYBR Green-based qPCR and the multiplex conventional PCR methods may be more appropriate than the triplex qPCR since they are cheaper. All multiplex procedures described here have been validated in foods and they are useful tools for taking preventive and/or correctives actions to avoid any hazard associated with accumulation of the above mycotoxins in foods. On the other hand, it should be noted that the success of detection of toxigenic *Penicillium* species by PCR-based methods depends on the mold DNA extraction from foods. An efficient mold DNA extraction useful for extracting high-quality DNA from pure cultures and mold-contaminated foods to be used in PCR methods is described in Chapter 14 of this book (*Targeting other mycotoxin biosynthetic genes*).

2   Materials

Prepare all reagents and stock solutions using ultrapure water and analytical grade chemicals. Unless indicated otherwise, store at room temperature after the preparation. Follow the official waste disposal regulations and consult the safety data sheet provided for each supplier.

2.1   Multiplex Conventional PCR

1. Oligonucleotides primers (Table 1). Store at −20 °C (*see* **Note 1**).
2. Deoxynucleotide triphosphates (dNTPs: dATP, dGTP, dCTP, dTTP) mix (10 mM each dNTP). Store at −20 °C.

3. Taq DNA Polymerase (2 U/μL). Store at −20 °C (*see* **Note 2**).

4. 10× Mg free PCR buffer (usually supplied with the enzyme). Store at −20 °C.

5. 50 mM MgCl₂ solution (usually supplied with the enzyme). Store at −20 °C (*see* **Notes 3** and **4**).

6. Agarose for routine use.

7. Sterile ultrapure water.

8. TAE buffer 50×: 2 M Tris base, 1 M glacial acetic acid, 0.05 M EDTA, pH 8.0. Store at 4 °C (*see* **Notes 5** and **6**).

9. TAE buffer 1×: Prepare it from TAE buffer 50× by diluting in ultrapure water. Store at 4 °C.

10. Loading dye buffer: 50% glycerol, 0.25% bromophenol blue, 25 mM EDTA (*see* **Notes 7** and **8**). Store at 4 °C.

Table 1
Nucleotide sequences of primers and probes used for the multiplex PCR-based methods described in this chapter

Methods	Gene	Primer/ probe name	Sequence (5′-3′)	PCR product size (bp)	References
Multiplex conventional PCR	*idh*	FC2	CGATGTTGCTAGCAAAGACG	496	[22]
		IDH2	ACCTTCAGTCGCTGTTCCTC		
	*otanps*PN	F2OT	GTGACTGGGTTGAACTTCTCGCC	373	
		R2OT	GGCGGTGGACCCCTCTCC		
	aflP	AFF2	ATTCATGCCTTGGTTGGATT	289	
		AFR3	CGAACCTCGTCCACAGTGC		
Multiplex Real-Time PCR	*idh*	F-idhtrb	GGCATCCATCATCGT	229	[23, 24]
		R-idhtrb	CTGTTCCTCCACCCA		
		IDHprobe	[FAM]-CCGAAGGGCATCCG-[TAMRA]		
	*otanps*PN	F-npstr	GCCGCCCTCTGTCATTCCAAG	117	
		R-npstr	GCCATCTCCAAACTCAAGCGTG		
		NPSprobe	[Cy5]-CGGCCGACCTCGGGAGAGA [BHQ2]		
	aflP	F-omt	GGCCGCCGCTTTGATCTAGG	123	
		R-omt	ACCACGACCGCCGCC		
		OMTprobe	[HEX]-CCACTGGTAGAGGAGATGT-[BHQ1]		

11. DNA molecular marker. Store at −20 °C (*see* **Note 9**).

12. Ethidium bromide (or equivalent; *see* **Note 10**). Store at the temperature specified by the supplier.

2.2 Multiplex qPCR

1. SYBR Green Master Mix. Store at −80 °C for long storage or 4 °C for short storage (less than 6 months) (*see* **Note 11**).

2. TaqMan Master Mix. Store at −20 °C (*see* **Note 11**).

3. Oligonucleotide primers and probes (Table 1). Store at −20 °C (*see* **Notes 1, 12, 13**).

4. Sterile ultrapure water.

3 Methods

3.1 Multiplex PCR

A multiplex PCR method reported by Luque et al. [22] for the simultaneous detection of potentially patulin- and ochratoxin A-producing *Penicillium* is below detailed as example of this methodology. Besides, such protocol allows detecting aflatoxigenic molds in the same reaction. The multiplex PCR method uses three primer sets based on the *idh* (FC2-IDH2 primers), the *otanps*PN (F2OT-R2OT primers), and the *aflP* (AFF2-AFR3 primers) genes able to amplify patulin-, ochratoxin A-, and aflatoxin-producing molds, respectively [22]. All of such genes are involved in the biosynthesis pathway of the corresponding mycotoxin. The procedure described below has been reported as suitable to detect the above mycotoxin-producing molds in pure cultures and in food matrices.

3.1.1 Reaction Setup

1. Thaw all the PCR components (10× Mg free PCR buffer, MgCl$_2$, dNTPs, primers, Taq DNA Polymerase, and DNA) on ice (*see* **Notes 14** and **15**). Vortex and centrifuge briefly all the reagents. They have to be prepared according to the manufacturer's instructions.

2. Prepare the PCR mixture in a sterile safe-lock microcentrifuge tube adding all the components of the reaction (Table 2; *see* **Notes 16–18**). Each sample should be performed by triplicate. Negative as well as positive controls should be included in each reaction (*see* **Notes 19–21**). These samples should be also prepared by triplicate. The PCR mixture should be performed on ice and optionally in a PCR workstation for avoiding cross contamination. Always it should be performed in an area separated from nucleic acid preparation and PCR product analysis.

3. Vortex the PCR mixture and dispense equal aliquots into PCR tubes (19 μL). Next, add 6 μL of DNA template into the corresponding PCR tubes. Centrifuge briefly the PCR tubes. Check that the mixture is collected at the bottom of the tubes. If not, centrifuge again for a longer time (*see* **Note 22**).

Table 2
PCR mixture composition of the multiplex PCR protocol for detecting patulin-, ochratoxin A-, and aflatoxin-producing molds [22]

PCR components	Volume (µL) per reaction
10× Mg free PCR buffer	2.5
MgCl$_2$ (50 mM)	0.75
dNTPs (10 mM)	2.5
FC2 (10 µM)	2.0
RC2 (10 µM)	2.0
F2OT (10 µM)	2.0
R2OT (10 µM)	2.0
AFF2 (10 µM)	2.0
AFR3 (10 µM)	2.0
Taq DNA Polymerase (2 U/µL)	1.25
DNA	6
Total volume	25

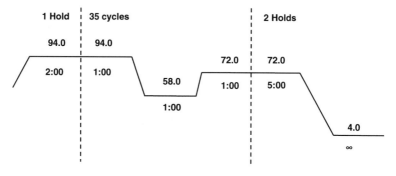

Fig. 1 Amplification conditions of the multiplex PCR protocol for detecting patulin-, ochratoxin A-, and aflatoxin-producing molds [22]

4. Place the PCR tubes on the thermal cycler and run the thermal cycling program shown in Fig. 1 (*see* **Note 23**).

3.1.2 Result Analysis

After PCR, keep amplification products at 4 °C if you are going to visualize the amplification products the same day. If not, keep them at −20 °C until use.

1. Prepare 2 % (w/v) agarose gel by mixing 2 g of agarose with 100 mL of TAE buffer 1× on a flask (*see* **Note 24**). Melt the

agarose in a microwave oven until it dissolves (*see* **Note 25**). The gel should be pour into the gel casting tray when cooled at 55–60 °C. Insert the comb immediately and wait until gel is solidified (*see* **Note 26**). Pour enough TAE buffer 1× into the electrophoresis tank (the surface should be higher than the top of the gel and not overflow).

2. For preparing samples to be loaded into the gel, mix 5 μL of each amplification product with 3 μL of a loading buffer by passing the mixed solution up and down several times through the pipette tip (*see* **Notes 27** and **28**). Carefully load all the volume of the PCR product/loading dye mixture into a well of the gel. Make sure you keep track of what sample is being loaded into each well.

3. For preparing DNA molecular size marker to be loaded into the gel, mix 1–2 μL of the marker with the same volume of a loading buffer (*see* **Note 29**). Carefully load all the volume of the DNA molecular size marker/loading dye mixture into a well of the gel.

4. When all samples are loaded, attach the electrodes from the gel box to the power supply. Run at 85 V for about 45 min to 1 h using 1×TAE as electrophoresis running buffer (*see* **Notes 30–32**).

5. After electrophoresis the gel is ready to be stained by immersion in an ethidium bromide aqueous solution (0.5 mg/mL) for about 20 min at room temperature with shaking (*see* **Note 33**).

6. Visualize the stained agarose gel under UV light box and photograph. DNA bands are visible upon exposure to UV light.

7. Calculate the sizes of the PCR products by comparison against the molecular marker. Confirm the presence of 496 and 373 bp amplification products for patulin- and ochratoxin A-producing *Penicillium*, respectively (Fig. 2). The 289 bp product because of amplification of aflatoxigenic molds can be also observed.

3.2 Multiplex qPCR

Two multiplex qPCR protocols which use SYBR Green and TaqMan methodologies are below described as examples of such methodologies. On the one hand, the duplex SYBR Green-based qPCR method allows detecting patulin- and ochratoxin A-producing *Penicillia* at the same time [24]. This method is designed on the basis of the *idh* (F-idhtrb/R-idhtrb primers) and the *otanps*PN (F-npstr/R-npstr primers) genes involved in the patulin and ochratoxin A biosynthesis, respectively. On the other hand, the triplex TaqMan-based qPCR method detects and quantifies simultaneously patulin- and ochratoxin A-producing *Penicillia*, and in addition, aflatoxin-producing molds [23]. The target genes are the *idh* (F-idhtrb/R-idhtrb primers, IDH probe), the *otanps*PN (F-npstr/R-npstr primers, NPSprobe), and the *aflP* (F-omt/R-omt primers, OMTprobe) genes involved in patulin, ochratoxin A, and aflatoxins production, respectively. Both multiplex qPCR methods have been

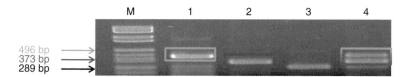

Fig. 2 Examples of PCR products generated by using the multiplex conventional PCR developed by Luque et al. [22]. *Line M*: DNA molecular weight marker of 2.1–0.15 kbp (Roche Diagnostics, Indianapolis, USA); *line 1*: PCR product obtained by using DNA from a patulin-producing *Penicillium* strain; *line 2*: PCR product obtained by using DNA from an ochratoxigenic *Penicillium* strain; *line 3*: PCR products obtained by using DNA from an aflatoxigenic *Aspergillus* strain; *line 4*: PCR product obtained by using DNA from the three types of toxigenic strains

reported as suitable to detect the above mycotoxin-producing molds in pure cultures and in food matrices.

3.2.1 Reaction Setup

1. Switch on the qPCR equipment and open the software of qPCR. Make sure that the instrument is calibrated for the dyes (e.g. SYBR Green, FAM, HEX or Cy5). The run method ought to be set up according to the user guide's instrument (*see* **Notes 34–36**). Program the thermal cycler with the specific temperature and time conditions of the multiplex qPCR protocols shown in Fig. 3 (*see* **Notes 37** and **38**).

2. Thaw all the PCR components [SYBR Green or TaqMan Master Mix, primers, probes (when TaqMan used), and DNA] on ice (*see* **Notes 14, 39–41**). Gently vortex and briefly centrifuge all the reagents. They have to be prepared according to the manufacturer's instructions.

3. Prepare DNA template (DNA template or non-template) by diluting its initial concentration up to approx. 1 ng by using sterile ultrapure water (*see* **Note 42**). It should be performed on ice.

4. Prepare tenfold dilutions of a PCR fragment (larger than the expected qPCR product) of each *idh* and *otanps*PN genes (standards), ranging from 10 to 1 log number of copies of each gene. It should be performed on ice (*see* **Notes 43–46**).

5. Prepare the PCR mix in a sterile safe-lock microcentrifuge tube adding all the components of the reaction except DNA (standards, DNA template, or non-template) (Table 3; *see* **Notes 16** and **47**). Briefly centrifuge components before adding to the mix tube to force the solution to the bottom of the tubes and to remove any bubbles. Each sample is performed by triplicate. Positive and negative controls should also be prepared by triplicate (*see* **Notes 18–21, 48** and **49**). Pipet all components on ice.

6. Vortex the PCR mixture and aliquot the mixture into PCR tubes or plates (*see* **Note 50**) and next add DNA (standards, DNA template, or non-template) (*see* **Notes 51** and **52**).

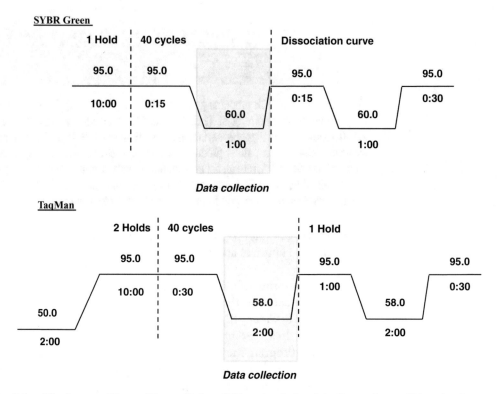

Fig. 3 Amplification conditions of the multiplex qPCR protocols for detecting and quantifying simultaneously patulin- and ochratoxin A-producing *Penicillia* [23, 24]

7. Centrifuge briefly the PCR tubes. Check that the mixture is at the bottom of the tubes and any bubble remains. If not, centrifuge again for a longer time.

8. Place the PCR tubes or plate into the qPCR system (*see* **Notes 53** and **54**) and run the corresponding qPCR protocol.

3.2.2 Results Analysis

1. Analyze the data by using the software of the qPCR instrument according to the user guide (*see* **Note 55**).

Threshold values (C_t) values are automatically determined by the software of the qPCR instrument. The amplification of the DNA (standards, DNA template, or non-template) is observed at the specific channel where each dye emits. In the case of the duplex SYBR Green-based qPCR method, the amplification of both target genes is observed at the channel whose emission wavelength is about 520 nm (fluorophore SYBR). When such methodology is used, it is also necessary to check the generated melting curve (Fig. 4, *see* **Note 56**). In case of using the triplex TaqMan-based qPCR method, the amplification of desired target genes (*idh* and *otanps*PN) is observed at the channels whose emission wavelengths are about 520 nm (fluorophore FAM; *idh*) and 596 nm (fluoro-

Table 3
PCR mixture composition of the multiplex qPCR protocols for detecting and quantifying patulin- and ochratoxin A-producing *Penicillia* [23, 24]

PCR components	Volume (µL) per reaction	
	Duplex SYBR Green-based qPCR	**Triplex TaqMan-based qPCR**
SYBR Green Master Mix (2×)	12.5	–
TaqMan Master Mix (2×)	–	12.5
ROX Reference Dye solution	–	0.1
F-idhtrb (10 µM)	0.8	1.8
R-idhtrb (10 µM)	1.4	1.8
IDHprobe (10 µM)	–	1.8
F-npstr (10 µM)	0.8	1.2
R-npstr (10 µM)	0.8	1.2
NPSprobe (10 µM)	–	1.5
F-omt (10 µM)	–	0.2
R-omt (10 µM)	–	0.4
OMTprobe (10 µM)	–	0.4
DNA	4	2.1
Sterile ultrapure water	4.7	–
Total volume	25	25

phore Cy5; *otanps*PN) (*see* **Note 57**). The amplification of the *aflP* gene can also be observed at the channel whose emission wavelength is about 556 nm (fluorophore HEX).

2. Build a standard curve relating C_t values (standards) and log number of copies of each gene. Extrapolate from this curve the quantity of the unknown samples (*see* **Note 58**).

4   Notes

1. Store the primers in small aliquots of working solutions to prevent contamination and many freeze/thaw cycles.

2. Other new enzymes are being developed. Some of them are chemically modified forms of the thermostable recombinant Taq DNA Polymerase for increasing the amplification efficiency (AmpliTaq Gold, Hot Start, etc.).

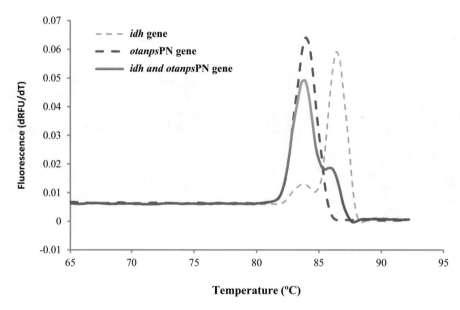

Fig. 4 Fluorescence melting curve resulting from the simultaneous amplification of the *idh* and *otanps*PN genes in the duplex SYBR Green-based qPCR method described in this chapter using DNA from a patulin-producing *Penicillium* strain and an ochratoxigenic *Penicillium* strain

3. $MgSO_4$ may be added in the PCR reaction as supplier of the Mg^{2+} divalent cation required as a cofactor for Type II enzymes. The choice of reagent depends on manufacturer's instructions of the enzyme.

4. Commercial PCR master mix containing *Taq* DNA polymerase, dNTPs, and all the other components required for PCR, except DNA template and primers, is available.

5. Mix Tris base with stir bar to dissolve in about 600 mL of ultrapure water. Add the EDTA and glacial acetic acid. Bring final volume to 1 L with ultrapure water. TAE buffer 50× is also commercially available.

6. TAE is the most commonly used electrophoresis buffer. However, other electrophoresis buffer can be used such as TBE buffer [25].

7. Several commercial loading dye buffers are also available.

8. Other loading dye buffers with different composition can be prepared [25].

9. Some DNA molecular markers are supplied with loading dye for DNA sample (or PCR product).

10. The ethidium bromide is stored at room temperature and shielded from light. Special precaution has to be taken when working with ethidium bromide since it is a hazardous compound (consult the safety data sheet provided by the supplier). Safer alternatives to this are currently available and the gels

stained with them can be visualized and photographed by using the same equipment for those stained with ethidium bromide.

11. A Core kit could be used instead of Master Mix. This kind of kit contains all components in separate tubes, so you have to mix them yourself. This allows optimizing the concentration of each component of the assay as, e.g., it has been done by Mayer et al. [26].

12. Design of primers and probes for multiplex detection of mycotoxin-producing *Penicillium* species could be also based on structural genes, e.g., *β-tubulin* gene or ITS region. However, other non-toxigenic *Penicillium* species could be detected too. Normally, this kind of primers could be very useful to amplify a non-competitive internal amplification control [19].

13. Design of primers and probes meets the criteria described by Rodríguez et al. [27].

14. Be sure that all components are completely thawed before using. Partial defrosting of PCR components could lead to a further inefficient PCR amplification since compounds remain in the frozen section.

15. An internal amplification control could be added in the reaction to avoid false-negative results [28]. An extra optimization step must be then performed.

16. Sterile gloves and filter pipette tips should be used for avoiding the introduction of contaminating nucleases or DNA. The gloves must be changed whenever there is a suspicion of being contaminated. Additionally, all sample and reagent tubes should be carefully opened and closed and kept capped as much as possible for preventing contamination.

17. Add, if possible, *Taq* Polymerase enzyme at the end, just before PCR mix is dispensed into the PCR tubes.

18. Prepare sufficient PCR mixture, excepting DNA template, for all the PCR reactions required. Adding at least two extra of each PCR component to compensate for pipetting errors is advisable. The replicates of each sample should be taken into account for the calculations too.

19. When preparing the PCR mixture, negative and positive controls must be also considered for the calculations.

20. It is recommended to include the following negative controls: (1) non-template control (water + PCR Master mix + primers), (2) non-primer/non-template control (water + PCR Master mix), and (3) DNA non-template control (water + PCR Master mix + primers + DNA non-template).

21. It is recommended to include a positive control based on a known DNA containing target which will be detected in the PCR.

22. When using a thermal cycler without a heated lid, overlay the reaction mixture with 25 µL of mineral oil before placing the PCR tubes on it.

23. Switch on the PCR machine at least 10–15 min before starting the PCR run.

24. The agarose solvent should be the same as used in the electrophoresis tank in order to avoid interferences in the mobility of the amplification products. For this, the use of the same batch for the gel and the tank is advisable.

25. Normally it takes 1–2 min, depending on the power of the microwave. You should cover the flask containing the mix with a small piece of tissue before heating in the microwave to avoid any spillage when agarose starts to boil. Once the mix is see-through you can take with care the flask using heat resistant gloves.

26. Normally it takes around 20–30 min. The agarose gel is opaque when solidified.

27. You can prepare the PCR product/loading dye mix in fresh PCR tubes or on a piece of Parafilm®, being the last one the most economic.

28. When quite faint PCR products are expected, 10–15 µL of each one should be loaded into the gel. The amount of loading dye should be increased too.

29. The selection of the DNA molecular weight marker should be based on the size of the expected PCR products. A DNA molecular size marker of 2.1–0.15 kbp was used by Luque et al. [22].

30. Be sure that electrodes are attached in the proper place. Since DNA is negatively charged, it will migrate in an electric field towards the positive electrode (from the negative to positive electrode).

31. The voltage and the running time depend on the percentage of agarose in the gel as well as the size of the expected PCR products. When higher is the percentage of agarose, the pores created in the gel matrix are smaller, and it is more difficult for large PCR products to move through the matrix. In general, shorter PCR products move at a faster rate than longer ones.

32. Run the gel until the dye line is approximately 75–80% of the way down the gel.

33. The ethidium bromide (or other alternatives safe-view dyes) could be also added to the gels before running. In this case, before pouring agarose solution into the gel casting tray (Subheading 3.1.2, **step 1**), add 5 µL of ethidium bromide (0.5 mg/mL). Take care that agarose solution cool down up

to 60 °C before the addition of ethidium bromide to prevent its volatilization.

34. A quantification assay must be selected. Normally, when DNA is used as target, absolute quantification using the standard curve method is performed.

35. In most qPCR software at least three types of samples can be selected: (a) Non-template control *(NTC)* consisting of a sample which does not contain template; (b) standard which consists of a sample with known concentration used to construct a standard curve. By running standards of several concentrations, a standard curve is created for extrapolating the quantity of an unknown sample; (c) unknown sample consisting of a sample containing an unknown quantity of template to be quantified.

36. Make sure that the correct emission channel for each dye has been properly selected. If not, you will not see your results although the PCR is running.

37. If the used SYBR Green (or TaqMan) mix contains Uracil DNA glycosylase (UDG) for preventing non-template DNA amplification an incubation step at 50 °C should be performed prior to the PCR cycling to activate the enzyme [29]. After this step necessary for the cleaving of the enzyme to the uracil residues from any contaminating DNA, the UDG is inactivated when temperature increases up to 95 °C [29].

38. If SYBR Green methodology is used, a melting curve, also called dissociation curve, must be included after thermal cycling conditions. Specificity of amplification products must be checked by analyzing the generated melting curves, since the double-stranded DNA-binding SYBR Green dye can bind to primer-dimers and other reaction artifacts producing a fluorescent signal [30, 31].

39. The SYBR Green Master Mix and TaqMan probes (and the later prepared PCR mixture) are light sensitive. They have to be stored and processed away from light.

40. If the SYBR Green Mix or TaqMan Master Mix does not have passive reference dye (normally ROX), it should be added as additional PCR component. ROX reference dye normalizes signal and ensures data integrity.

41. Other multiplex qPCR protocols for detecting toxigenic *Penicillium* species use internal amplification controls (e.g., [19, 20, 32]).

42. DNA can also be diluted using the elution buffer provided by commercial extraction kits or TE buffer.

43. The standard dilutions may also be prepared using tenfold dilutions from a DNA template aliquot of a known concentration.

44. For preparing standards for patulin-producing *Penicillium* strains, a 600 bp fragment of the *idh* gene amplified with primers IDH1 and IDH2 and the conventional PCR protocol reported by Paterson [33] may be used as standard stock solution. After, the concentration of the PCR product in the stock solutions is determined by a biophotometer and the number of copies is then calculated as follows: Number of copies $= w_{PCR}/(660 \text{ g/mol}) \times (L_{PCR}) \times (6.023 \times 10^{23})$, where w_{PCR}: weight or amount of PCR product (g/μL); 660 g/mol: average weight of a base pair (dsDNA); L_{PCR}: length of PCR product (bp); and 6.023×10^{23}: Avogadro's number.

    Later, the stock solutions are serially diluted by a factor of 10 and aliquots of these dilutions are used as a copy number standard during each setup of the qPCR reaction.

45. For preparing standards for ochratoxin A-producing *Penicillium* strains, a 750 bp fragment of the *otanps*PN gene amplified with primers otanps_for and otanps_rev and the conventional PCR protocol reported by Bogs et al. [5] may be used as standard stock solution. To obtain copy number standards of such gene, proceed as indicated in **Note 44**.

46. The quantification of aflatoxigenic molds by using the triplex TaqMan method could be performed by preparing the standard stock solution for aflatoxin-producing molds amplifying a 1254 bp fragment of the *aflP* gene by using primers OMT-forward and OMT-reverse and the conventional PCR method reported by Richard et al. [34]. To obtain copy number standards of such gene, proceed as indicated in **Note 44**.

47. The PCR mixture must be prepared in a UV-equipped PCR workstation in order to prevent cross contamination of samples. Before and after use clean the surface of workstation with 70 % (v/v) ethanol. In addition, UV light should be switched on for decontamination before using for 10–15 min.

48. Reaction volumes can be scaled up or down as long as the final concentrations of the reaction components remain the same.

49. When SYBR Green is used, a no amplification control (NAC) tube that contains sample, but not the enzyme could be necessary. If the absolute fluorescence of the NAC is greater than that of the NTC after PCR, fluorescent contaminants may be present in the sample or in the heat block of the thermal cycler [35].

50. The PCR mixture can be loaded in different formats: PCR plates (48-, 96-, or 384-well plates depending on the necessity) or PCR tubes (single tubes or eight-strip tubes). When using the plates, they have to be sealed by using adhesive films after filling in the wells with the PCR components.

51. Avoid open the PCR tubes containing DNA (or standards or DNA non-template control) once it has been added.

52. When qPCR plate or PCR tubes are loaded, prevent cross contamination. Try to keep reactions and components capped as much as possible. Open and close tubes carefully too.

53. If the plate or tubes are not going to be loaded into the qPCR instrument immediately, place them at refrigeration in dark conditions.

54. Avoid touch the optical surface of the cap or sealing film without gloves, as fingerprints may interfere with fluorescence measurements.

55. Make sure that the negative controls did not amplify and the positive controls amplified to verify that the qPCR run was adequate.

56. The expected amplicons (229 and 117 bp) give two melting peaks (T_m) at 86.9 and 83.9 °C for patulin- and ochratoxin A-producing *Penicillium*, respectively.

57. The choice of dyes which emit in different wavelengths allows quantification of each target gene separately but simultaneously in the same multiplex qPCR reaction.

58. The criteria considered for reliability of the qPCR methods are the correlation coefficient (R^2) and the amplification efficiency calculated from the formula $E = [10^{(-1/S)}] - 1$ where S is the slope of the standard curve. Generally slopes between -3.1 and -3.6 with PCR efficiency values in the range of 90–110% are considered satisfactory. Furthermore, the optimal correlation coefficient (R^2) derived from the standard curve has to be between 0.99 and 0.999 [27].

Acknowledgments

This work was supported by AGL2010-21623 and Carnisenusa CSD2007-00016—ConsoliderIngenio 2010 from the Spanish Comision Interministerial de Ciencia y Tecnología projects and GR10162 of the Junta de Extremadura and FEDER. Dr. A. Rodríguez is supported by a "*Juan de la Cierva-Incorporación*" Senior Research Fellowship (IJCI-2014-20666) from the Spanish Ministry of Economy and Competitiveness.

References

1. Frisvad JC, Smedsgaard J, Larsen TO et al (2004) Mycotoxins, drugs and other extrolites produced by species in *Penicillium* subgenus *Penicillium*. Stud Mycol 49:201–241

2. Lund F, Frisvad JC (2003) *Penicillium verrucosum* in wheat and barley indicates presence of ochratoxin A. J Appl Microbiol 95:1117–1123

3. Niessen L (2007) PCR-based diagnosis and quantification of mycotoxin producing fungi. Int J Food Microbiol 119:38–46

4. Baily JD, Tabuc C, Quérin A et al (2005) Production and stability of patulin, ochratoxin A, citrinin, and cyclopiazonic acid on dry-cured ham. J Food Prot 68:1516–1520

5. Bogs C, Battilani P, Geisen R (2006) Development of a molecular detection and differentiation system for ochratoxin A producing *Penicillium* species and its application to analyse the occurrence of *Penicilliumnordicum* in cured meats. Int J Food Microbiol 107:39–47

6. Dorner JW (2002) Recent advances in analytical methodology for cyclopiazonic acid. Adv Exp Med Biol 504:107–116

7. Núñez F, Díaz MC, Rodríguez M et al (2000) Effects of substrate, water activity, and temperature on growth and verrucosidin production by *Penicillium polonicum* isolated from dry-cured ham. J Food Prot 63:232–236

8. Rodríguez A, Rodríguez M, Martín A et al (2012) Presence of ochratoxin A on the surface of dry-cured Iberian ham after initial fungal growth in the drying stage. Meat Sci 90:728–734

9. Tangni EK, Theys R, Mignolet E et al (2003) Patulin in domestic and imported apple-based drinks in Belgium: occurrence and exposure assessment. Food Addit Contam 20:482–489

10. Dao HP, Mathieu F, Lebrihi A (2005) Two primer pairs to detect OTA producers by PCR method. Int J Food Microbiol 36:215–220

11. Hayat A, Paniel N, Rhouati A et al (2012) Recent advances in ochratoxin A-producing fungi detection based on PCR methods and ochratoxin A analysis in food matrices. Food Control 26:401–415

12. Rodríguez A, Andrade MJ, Rodríguez M et al (2014) Detection of mycotoxin-producing moulds and mycotoxins in foods. In: Rai VR, Bai JA (eds) Microbial food safety and preservation techniques. CRC Press, New York, pp 191–213

13. Geisen R, Mayer Z, Karolewiez A et al (2004) Development of a Real Time PCR system for detection of *Penicillium nordicum* and for monitoring ochratoxin A production in foods by targeting the ochratoxin polyketide synthase gene. Syst Appl Microbiol 27:501–507

14. Luque MI, Rodríguez A, Andrade MJ et al (2011) Development of a PCR protocol to detect patulin producing moulds in food products. Food Control 22:1831–1838

15. Luque MI, Córdoba JJ, Rodríguez A et al (2013) Development of a PCR protocol to detect ochratoxin A producing moulds in food products. Food Control 29:270–278

16. Paterson RRM, Archer S, Kozakiewicz Z et al (2000) A gene probe for the patulin metabolic pathway with potential use in novel disease control. Biocontrol Sci Technol 10:509–512

17. Rodríguez A, Luque MI, Andrade MJ et al (2011) Development of real-time PCR methods to quantify patulin-producing molds in food products. Food Microbiol 28:1190–1199

18. Rodríguez A, Rodríguez M, Luque MI et al (2011) Quantification of ochratoxin A-producing molds in food products by SYBR Green and TaqMan real-time PCR methods. Int J Food Microbiol 149:226–235

19. Rodríguez A, Córdoba JJ, Werning ML et al (2012) Duplex real-time PCR method with internal amplification control for quantification of verrucosidin producing molds in dry-ripened foods. Int J Food Microbiol 153:85–91

20. Rodríguez A, Werning ML, Rodríguez M et al (2012) Quantitative real-time PCR method with internal amplification control to quantify cyclopiazonic acid-producing molds in foods. Food Microbiol 32:397–405

21. Schmidt-Heydt M, Richter W, Michulec M et al (2008) Comprehensive molecular system to study the presence, growth and ochratoxin A biosynthesis of *Penicillium verrucosum* in wheat. Food Addit Contam 25:989–996

22. Luque MI, Andrade MJ, Rodríguez A et al (2013) Development of a multiplex PCR method for the detection of patulin-, ochratoxin A- and aflatoxin-producing moulds in foods. Food Anal Meth 6:1113–1121

23. Rodríguez A, Rodríguez M, Andrade MJ et al (2012) Development of a multiplex real-time PCR to quantify aflatoxin, ochratoxin A and patulin producing molds in foods. Int J Food Microbiol 155:10–18

24. Rodríguez A (2012) Desarrollo de métodos de PCR en tiempo real para la detección y cuantificación de mohos productores de micotoxinas en alimentos. Doctoral thesis, University of Extremadura, Spain.

25. Sambrook J, Fritsch EF, Maniatis T (1989) Molecular cloning: a laboratory manual, 2nd edn. Cold Spring Harbor Laboratory Press, New York

26. Mayer Z, Bagnara A, Färber P et al (2003) Quantification of the copy number of *nor-1*, a gene of the aflatoxin biosynthetic pathway by real-time PCR, and its correlation to the cfu of *Aspergillus flavus* in foods. Int J Food Microbiol 82:143–151

27. Rodríguez A, Rodríguez M, Córdoba JJ et al (2015) Design of primers and probes for quantitative real-time PCR methods. In: Chhandak B (ed) PCR primer design (Series: Methods in Molecular Biology). Humana Press, New York, pp 31–56

28. Priyanka SR, Venkataramana M, Balakrishna K et al (2015) Development and evaluation of a multiplex PCR assay for simultaneous detection of major mycotoxigenic fungi from cereals. J Food Sci Technol 52:486–492

29. Invitrogen (2008) Real-Time PCR: from theory to practice. Invitrogen Corporation, Carlsbad

30. Holland PM, Abramson RD, Watson R et al (1991) Detection of specific polymerase chain reaction product by utilizing the 50–30 exonuclease activity of *Thermus aquaticus* DNA polymerase. Proc Natl Acad Sci U S A 88:7276–7280

31. Heid CA, Stevens J, Livak KJ et al (1996) Real time quantitative PCR. Genome Res 6:986–994

32. Bernáldez V, Rodríguez A, Martín A et al (2014) Development of a multiplex qPCR method for simultaneous quantification in dry cured ham of an antifungal-peptide *Penicillium chrysogenum* strain used as protective culture and aflatoxin-producing moulds. Food Control 36:257–266

33. Paterson RRM (2004) The *isoepoxydon dehydrogenase* gene of patulin biosynthesis in cultures and secondary metabolites as candidate PCR inhibitors. Mycol Res 108:1431–1437

34. Richard E, Heutte N, Bouchart V et al (2009) Evaluation of fungal contamination and mycotoxin production in maize silage. Anim Feed Sci Tech 148:309–320

35. Applied Biosystems (2010) Fast SYBR® Green Master Mix Protocol. https://tools.lifetechnologies.com/content/sfs/manuals/cms_046776.pdf. Accessed July 2010.

Part IV

**Combined PCR and Other Molecular Approaches
for Detection and Identification of Mycotoxigenic Fungi**

Chapter 20

PCR-RFLP for *Aspergillus* Species

Ali Atoui and André El Khoury

Abstract

Polymerase chain reaction-restriction fragment length polymorphism (PCR-RFLP) is the most simple method for single-nucleotide change detection. It is widely used in the detection and differentiation between mycotoxigenic species. It is based on PCR amplification of a target region containing the variant site of the studied species followed by restriction endonuclease digestion and gel electrophoresis to visualize the RFLP patterns. In this method primers are designed to flank the polymorphic site and positioned in such a way as to create unequally sized fragments upon restriction endonuclease cleavage of the PCR products. Here, we describe the protocol of PCR-RFLP developed for the detection and differentiation between *Aspergillus flavus* and *A. parasiticus* by amplifying a 674 bp fragment of the *aflR–aflJ* intergenic region followed by restriction endonuclease analysis using *BglII* to obtain RFLP patterns.

Key words *Aspergillus*, PCR- RFLP, Detection, Differentiation

1   Introduction

Molecular methods have been widely applied in the detection and differentiation of a large number of *Aspergillus* species. In this context, several molecular approaches have been developed, including species-specific diagnostic PCR, random amplified polymorphic DNA (RAPD) analysis and restriction fragment length polymorphism (RFLP) analysis [1–3]. In recent years the combined PCR-RFLP has been widely used for the detection and differentiation between mycotoxigenic species. The first step in this method is to select a target gene. This target gene must show interspecies variability between the studied species in order to have a RFLP patterns. Internal transcribed spacer (ITS), intergenic spacer (IGS), *β*-tubulin, as well as genes involved in mycotoxin biosynthesis genes have been widely used as target gene to identify and differentiate between different *Aspergillus* species using PCR-RFLP [2, 4–7]. The sequences of the target gene of each studied species are then analyzed to identify the appropriate restriction enzyme. After PCR, a portion of the reaction is subjected to restriction endonuclease digestion and gel electrophoresis to visualize the resultant

Antonio Moretti and Antonia Susca (eds.), *Mycotoxigenic Fungi: Methods and Protocols*, Methods in Molecular Biology, vol. 1542, DOI 10.1007/978-1-4939-6707-0_20, © Springer Science+Business Media LLC 2017

bands representing the RFLP patterns. This approach of differentiating these species seems to be simpler, less costly, and quicker than conventional sequencing of PCR products and/or morphological identification.

This chapter describes the approach used in order to perform a PCR-RFLP assay. We will explore an example of an assay used to discriminate *Aspergillus flavus* and *Aspergillus parasiticus* using the *aflR–aflJ* intergenic region as a target gene.

2   Materials

Prepare all solutions using ultrapure water (prepared by purifying deionized water to attain a sensitivity of 18 MΩ cm at 25 °C) and molecular grade reagents. Prepare and store all reagents at room temperature (unless indicated otherwise).

2.1   Fungal Culture

Czapek yeast extract agar (CYA) medium: 30 g/L Sucrose, 5 g/L yeast extract, 1 g/L·K$_2$HPO$_4$, 0.01 g/L·ZnSO$_4$·7H$_2$O, 0.005 g/L CuSO$_4$·5H$_2$O, 10 mL/L Czapek concentrate, 15 g/L agar.

2.2   DNA Extraction

1. Lysis buffer: 400 mM Tris–HCl [pH 8.0], 60 mM EDTA [pH 8.0], 150 mM NaCl, 1% sodium dodecyl sulfate (SDS).

2. Potassium acetate buffer (pH 4.8): 60 mL of 5 M potassium acetate, 11.5 mL of glacial acetic acid, 28.5 mL of H$_2$O.

3. Isopropyl alcohol.

4. 70% Ethanol solution: Prepare 100 mL of solution by mixing 70 mL of absolute ethanol and water to reach the 100 mL volume.

2.3   PCR-RFLP

1. *Taq* recombinant polymerase (Invitrogen, USA).

2. dNTP mix.

3. Agarose.

4. Loading dye.

5. Molecular marker.

6. Tris-acetate-EDTA buffer (TAE 1×): 40 mM Tris, 20 mM acetic acid, 1 mM EDTA.

7. Restriction enzyme: It should be selected according to the map restriction analysis described in Subheading 3.4.

3   Methods

3.1   Fungal Cultures

1. Select a number of aflatoxigenic and non-aflatoxigenic fungal strains of *Aspergillus flavus* and *A. parasiticus*.

2. Grow the isolates in Petri dishes containing CYA medium.

3. Incubate the culture at 28 °C for 3 days in order to have a young mycelium.

3.2 Identification of RFLP Enzymes and Primer Design

1. The *aflR–aflJ* intergenic region of *A. flavus* and *A. parasiticus* is the selected target gene in this method. The following steps describe the identification of the appropriate restriction enzyme allowing discrimination between *A. flavus* and *A. parasiticus* and the design of primers (*see* **Note 1**).

2. Obtain all available sequences of the selected target gene (or gene fragment) from several isolates of both species using GenBank search.

3. Align the obtained sequence for each species separately using Clustal X in order to show the absence of intraspecies variability.

4. Align then all sequences for the two species using Clustal X and show the interspecies variability (Fig. 1).

5. Go to enzyme restriction site (Example: http://bio.lundberg.gu.se/cutter2/).

6. Enter the sequence of each species and make restriction analysis by choosing all enzymes in the database.

7. Choose the best enzyme by comparing the restriction maps of the *aflR-aflJ* intergenic region sequence of both species (*see* **Note 2**, Fig. 2).

8. After choosing the best enzyme showing RFLP patterns, design a primer pair by one program such as Primer3, to amplify a fragment containing the sequences variability. In the present protocol the designed primer pair amplifies the 674 bp *aflR-aflJ* intergenic region sequence in *A. flavus* and *A. parasiticus*.

9. Examine the specificity of the primers. This can be assessed by BLAST search and PCR (*see* **Note 3**).

3.3 DNA Extraction from Pure Fungal Cultures

1. Select a rapid and suitable DNA isolation method for yielding high quality of DNA. Here we describe the method performed according to Lui et al. [8].

2. Add 500 mL of lysis buffer into 1.5 mL Eppendorf tube.

3. Take a small lump of mycelia from young culture using a sterile toothpick, with which the lump of mycelia is disrupted in to the 1.5 mL containing the lysis buffer.

4. Leave the tube at room temperature for 10 min.

5. Add 150 mL of potassium acetate buffer, vortex the tube briefly, and centrifuge at $10,000 \times g$ for 1 min.

6. Transfer the supernatant to another 1.5 mL Eppendorf tube and centrifuge again as described above.

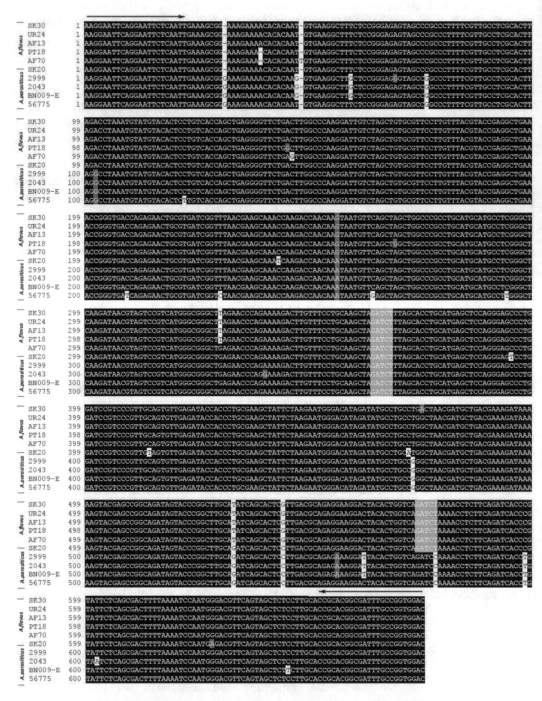

Fig. 1 Alignment of *aflR-aflJ* intergenic spacer region sequences in ten strains of *A. flavus* and *A. parasiticus* isolates. The location of selected primers is represented by *bold arrows*. The regions shadowed in *pale gray* represent the restriction site for *BglII* endonuclease enzyme

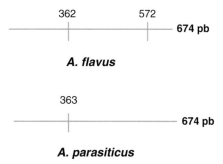

A. flavus

A. parasiticus

Fig. 2 Schematic representation of restriction sites for *BglII* on *aflR-aflJ* intergenic region sequence (674 bp) of *A. flavus* and *A. parasiticus* (not to scale). According to the sequence analysis, there are two restriction sites for *BglII* in the sequence of *A. flavus* that should cleave the PCR products into three fragments of 362, 210, and 102 bp. However, there is only one restriction site for this enzyme in the sequence of *A. parasiticus* that should produce two fragments of 363 and 311 bp. The obtained RFLP patterns revealed substantial variability

7. Take the supernatant to a new 1.5 mL Eppendorf tube and add an equal volume of isopropyl alcohol.

8. Mix the tube by inversion. Centrifuge the tube at $10,000 \times g$ for 2 min.

9. Discard the supernatant and wash the resultant DNA pellet in 300 μL of 70 % ethanol.

10. Centrifuge the pellet $16,000 \times g$ for 1 min and discard the supernatant.

11. Leave the DNA pellet to be dried and dissolve the dried pellet in 50 μL of deionised H_2O.

3.4 PCR-RFLP

1. Perform the PCR reaction for all isolated fungal DNA in duplicate with the *Taq* recombinant polymerase (Invitrogen, USA) in 50 μL reaction mixture containing 5 μL of *Taq* polymerase buffer 10×, 1.5 μL of 50 mM $MgCl_2$, 1 μL of dNTP 10 mM of each (Promega), 1 μM of each primer, 1.5 U of *Taq*, about 1 μL of genomic DNA, H_2O up to 50 μL (*see* **Note 4**). Reaction conditions were 94 °C for 4 min, (94 °C for 40 s, T °C for 40 s and 72 °C for t s) × 35 cycles followed by an incubation at 72 °C for 10 min (*see* **Note 5**).

2. Include negative controls (no DNA template) to test for the presence of DNA contamination of reagents and reaction mixtures.

3. Include positive control using *β*-tubulin gene (or actin gene) in the same PCR conditions for all examined fungal DNA. Non aflatoxigenic species should give positive result with positive control and negative result with the designed target gene primers.

4. The amplified products are examined by 1% w/v agarose (Promega) gel electrophoresis (Fig. 3)

5. In a sterile 1.5 mL microfuge tube, prepare the restriction mixture on ice (*see* **Note 4**) for digesting 10 μL of the PCR product (*see* **Note 6**) as indicated in Table 1.

6. Label sterile 0.5 mL microfuge tubes and dispense the appropriate volume of the restriction mixture in each tube (30 μL).

7. Add the adequate amount of PCR product (10 μL) and mix. Centrifuge the mixture shortly to settle down all the droplets on the walls.

8. Incubate at 37 °C for 2–3 h in water bath. Meanwhile, prepare 2 % gel agarose.

9. After incubation, mix maximal 20 μL of the digested samples with 3 μL of loading dye. This can be done by adding the loading buffer to the microfuge tubes.

10. Slowly load the sample mixture into the slots of the submerged gel under TAE 1× using a disposable micropipette.

11. Run the agarose gel at a voltage of 4–5 V/cm (110–120 V) for 1 h 45 min.

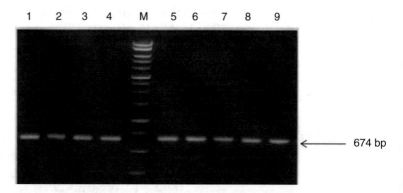

Fig. 3 0.8 % of agarose gel electrophoresis of PCR products with the designed primers of different of *A. flavus* and *A. parasiticus* isolates (*lanes 1–9*); *Lane M*—100-bp DNA marker (GeneRuler, Fermentas)

Table 1
Composition of restriction mixture for digesting 10 μL PCR product

	Volume added (μL)
10× buffer	4
H_2O	24
BglII restriction enzyme 10 U/μL (*see* **Note 7**)	2
Total volume	30

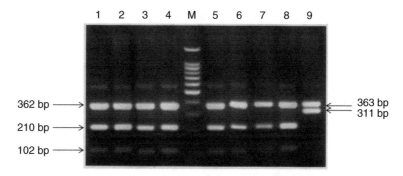

Fig. 4 Electrophoretic analysis showing the restriction profiles of the *aflR-aflJ* intergenic spacer PCR product (674 bp) digested with *BglII. Lanes 1–8 A. flavus* strains; *lane M*—100 bp DNA marker (GeneRuler, Fermentas); *lane 9—A. parasiticus strain*

12. When the dyes have migrated a sufficient distance through the gel, turn off the electricity and remove the gel from the tank. Examine the gel by UV light and photograph the gel showing the RFLP patterns (Fig. 4).

4 Notes

1. There are now several programs for the design of PCR-RFLP in which the selection of primers and restriction enzymes has been integrated [9].

2. Restriction fragment sizes lesser than 100 bp should not be taken into consideration because they are not clearly resolved by electrophoresis on 2% agarose gel. The enzyme should be reliable and inexpensive.

3. The selected primer pairs should be highly specific, amplifying only the studied species and yielding amplicons of the expected size (in this protocol 674 bp) and no additional or non-specific bands should be observed. In addition none of the other species (i.e., non-aflatoxigenic species) should give a positive result with this PCR primer set.

4. Keep all reagents on ice. Vortex the restriction mixture carefully and centrifuge it shortly to settle down all the droplets on the walls.

5. The annealing temperature (T) is specific for primer pair used and the extension time (t) is selected according to the expected amplicon length (1 min per 1 Kb is recommended).

6. Enzyme manufacturers provide lots of information about restriction digestion. This is available both in the catalogue and on the web.

7. It is always good to add the buffer and water into the tube first. Do not use more enzyme than 10 % of the final reaction volume. This is because the enzyme storage buffer contains glycerol. Excess quantity of glycerol will inhibit the digestion.

References

1. Nicholson P, Simpson DR, Weston G et al (1998) Detection and quantification of *Fusarium culmorum* and *Fusarium graminearum* in cereals using PCR assays. Physiol Mol Plant Pathol 53:17–37

2. Paterson RRM (2006) Identification and quantification of mycotoxigenic fungi by PCR. Process Biochem 71:1467–1474

3. Ferrer C, Colom F, Frases S et al (2001) Detection and identification of fungal pathogens in blood by using molecular probes. J Clin Microbiol 39:2873–2879

4. Gonzalez-Salgado A, Patino B, Vazquez C et al (2005) Discrimination of *Aspergillus niger* and other *Aspergillus* species belonging to section Nigri by PCR assays. FEMS Microbiol Lett 245:353–361

5. Martinez-Culebras PV, Ramon D (2007) An ITS-RFLP method to identify black *Aspergillus* isolates responsible for OTA contamination in grapes and wine. Int J Food Microbiol 113:147–153

6. El Khoury A, Atoui A, Rizk T et al (2011) Differentiation between *Aspergillus flavus* and *Aspergillus parasiticus* from pure culture and aflatoxin-contaminated grapes using PCR-RFLP analysis of *aflR-aflJ* intergenic spacer. J Food Sci 76:M247–M253

7. Somashekar D, Rati ER, Chandrashekar A (2004) PCR-restriction fragment length analysis of *aflR* gene for differentiation and detection of *Aspergillus flavus* and *Aspergillus parasiticus* in maize. Int J Food Microbiol 93:101–107

8. Lui D, Caloe C, Biard R et al (2000) Rapid mini preparation of fungal DNA for PCR. J Clin Microbiol 38:471

9. Rasmussen HB (2012) Restriction fragment length polymorphism analysis of PCR-amplified fragments (PCR-RFLP) and gel electrophoresis - valuable tool for genotyping and genetic fingerprinting. In: Sameh Magdeldin Dr. (ed) Gel electrophoresis - principles and basics, ISBN:978-953-51-0458-2, InTech. doi:10.5772/37724

Chapter 21

PCR ITS-RFLP for *Penicillium* Species and Other Genera

Sandrine Rousseaux and Michèle Guilloux-Bénatier

Abstract

Among numerous molecular methodologies developed for highly specific identification of filamentous fungi isolates, here we describe restriction digestion analysis of the ITS products as an easy method to identify isolates of filamentous fungi. This technique is a rapid and reliable method appropriate for routine identification of filamentous fungi. This can be used to screen large numbers of isolates from various environments in a short time. The use of different endonucleases allowed generating individual restriction profiles. The individual profiles obtained were combined into composite restriction patterns characteristic of a species. Eleven different genera can be differentiated and among them 41 different species.

Key words Filamentous fungi, PCR ITS-RFLP, Identification, Composite profiles

1 Introduction

Filamentous fungi are widely distributed in various habitats and can be isolated from soil [1, 2], food [3, 4], or humans [5]. They act as human pathogens [6], food spoilage organisms, mycotoxins, or off-flavor producers [7–9]. Culture-dependent methods with macroscopic and microscopic examination are traditionally used to identify filamentous fungi, but may fail to identify the complete diversity of fungi present. Morphological and physiological characteristics are influenced by culture conditions and consequently this approach can provide incomplete or ambiguous results. Moreover, these methods are also time consuming and laborious. Therefore, numerous molecular culture-dependent methodologies have been developed for highly specific identification. Among these techniques, random amplified polymorphisms DNA (RAPD) method was applied to identify *Penicillium* starter cultures [10, 11] or to study the intraspecific diversity of *Geotrichum candidum* isolated from cheese [12]. Restriction fragment length polymorphism (RFLP) technique was used to discriminate species among the *Aspergillus* genus [13]. Polymerase chain reaction-restriction fragment length polymorphism (PCR-RFLP) method was described to

Antonio Moretti and Antonia Susca (eds.), *Mycotoxigenic Fungi: Methods and Protocols*, Methods in Molecular Biology, vol. 1542, DOI 10.1007/978-1-4939-6707-0_21, © Springer Science+Business Media LLC 2017

identify strains of the *A. niger* aggregate [14, 15], *P. aurantiogriseum* from foods [16], medically relevant fungi [17], and *Penicillium* subgenus *Bivertilillium* [10]. Amplified fragment length polymorphism technique (AFLP) analysis was used to detect *Aspergillus flavus* in food samples [18] and was applied for differentiating between *Alternaria alternata* and *A. infectoria* isolated from wheat [19]. A multiplex PCR method has developed to identify *Aspergillus versicolor*, *Cladosporium* spp., *Penicillium purpurogenum*, and *Stachybotrys chartarum* [20].

Most of the developed methods are based on the analysis of the internal transcribed spacer (ITS) region. ITS sequences including the 5.8S rRNA gene (the coding region which is conserved) and two flanking regions ITS1 and ITS2 (noncoding and variable) (Fig. 1) show low intraspecific polymorphism and high interspecific variability and have proved useful for identification of different fungi and yeasts [14, 21, 22].

Most of the methods described in the literature have been used for the identification of specific fungi present in different environments, but few techniques have been developed for describing and identifying in a same time the different genera of filamentous fungi.

Here, we describe a PCR ITS-RFLP method that we have developed to be a fast and easy method for identifying species of fungal genera. By this method, we can differentiate eleven genera: *Acremonium*, *Alternaria*, *Aspergillus*, *Botrytis*, *Cladosporium*, *Fusarium*, *Epicoccum*, *Penicillium*, *Pilidiella*, *Thanatephorus*, and *Trichoderma*. Among the genus *Penicillium*, 22 different species were differentiated: *P. aurantiogriseum*, *P. bilaiae*, *P. brevicompactum*, *P. chrysogenum*, *P. commune*, *P. corylophilum*, *P. crustosum*, *P. expansum*, *P. fellutanum*, *P. herquei*, *P. italicum*, *P. islandicum*, *P. minioluteum*, *P. oxalicum*, *P. paneum*, *P. paxilli*, *P. raistrickii*, *P. roqueforti*, *P. simplicissimum*, *P. spinulosum*, *P. verrucosum*, and *P. verruculosum*. Only the species *P. thomii* and *P. glabrum* were not differentiating.

For the genus *Aspergillus*, all seven different species analyzed were discriminated: *A. aculeatus*, *A. carbonarius*, *A. fumigatus*, *A.

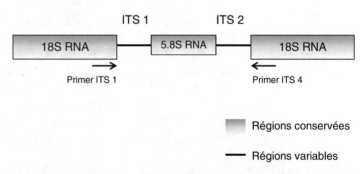

Fig. 1 Schematic of the nuclear ribosomal RNA repeat. *S* svedberg sedimentation coefficient, *ITS* internal transcribed spacer; universal primers [24]

japonicus, A. niger, A. terreus, and *A. wentii.* Moreover, we can discriminate *Cladosporium cladosporioides* of *Cladiosporium herbarum* and the three species *Trichoderma harzianum, Trichoderma koningiopsis* and *Trichoderma longibrachiatsum.*

The PCR ITS-RFLP method we describe facilitates rapid and easy identification of fungal species isolated without sequencing. This assay is a routine, sensitive, and reliable compared to morphological identification and can be used to screen vast numbers of isolates in a short time. Consequently, it could be very useful for studies comparing large samples of isolates where sequencing cannot reasonably be undertaken.

2   Materials

Prepare all media using sterile distilled water and all solutions using ultrapure water.

2.1   Culture Media

1. Dichloran rose-bengal chloramphenicol (DRBC) agar medium: 1.5 % agar, 1 % d-glucose, 0.5 % bacteriological peptone, 0.1 % KH_2PO_4, 0.05 % $MgSO_4 \cdot 7H_2O$, 0.5 mL of rose-bengal solution, 1 mL of dichloran solution, 0.01 % chloramphenicol. Final pH 5.6 ±0.2 at 25 °C. Sterilize (*see* **Note 1**).

2. Malt extract agar (MEA) medium: 2 % Agar, 2 % malt extract, 2 % d-glucose, 0.1 % bacteriological peptone. Sterilize (*see* **Note 2**).

3. Potato dextrose agar (PDA): 1.7 % Agar, 1 % d-glucose, 0.5 % potato extract. Sterilize.

4. Potato dextrose broth (PDB): 1 % d-Glucose, 0.5 % potato extract. Sterilize.

These two media are used for the culture of filamentous fungi (*see* **Note 3**).

2.2   Washing and Dilution Suspensions for Sampling Procedure

1. Sterile bags and sterile disposable inoculating.

2. Plates of DRBC and MEA media.

3. Washing suspension: 0.9 % (w/v) NaCl; 0.2 % (v/v) Tween 80. Make up to 1 L with distilled water and sterilize (*see* **Note 4**).

4. Dilution suspension: 0.9 % (w/v) NaCl. Make up to 1 L with distilled water and sterilize (*see* **Note 5**).

5. Ultrasonic cleaner (Cleaning bath Ultrasons-Digit, Dutscher Scientific).

6. An orbital shaker.

7. Rinse solution: 0.05 % (v/v) Tween 80. Make up to 1 L with distilled water and sterilize.

8. Solution for conservation of spores at –80 °C: 40 % (v/v) glycerol. Make up to 200 mL with distilled water and sterilize (*see* **Note 6**).

9. Cryovial (2 mL).

2.3 DNA Preparation

1. Plates of PDA medium or 6 mL of PDB.

2. Sterile disposable inoculating.

3. Rinse solution: 0.05 % (v/v) Tween 80. Make up to 1 L with distilled water and sterilize.

4. Sterile water (aliquot of 2 mL).

5. Malassez hematimeter cell: consists of a thick glass microscope slide with a grid of perpendicular lines etched in the middle. The grid has specified dimensions (1 mm³) so that the area covered by the lines is known, which makes it possible to count the number of cells in a specific volume of solution.

   Total cells/mL = (Total cells counted × 10000 cells/mL)/(dilution factor × number of squares).

6. Liquid nitrogen.

2.4 DNA Extraction

1. Commercial EZNA Fungal DNA Kit (Omega bio-tek, Doraville, USA).

2. Cold absolute ethanol.

3. Ultrapure water (aliquot 200 μL).

4. Biophotometer (Eppendorf, Le Pecq, France).

2.5 PCR

PCR reactions were performed using Taq polymerase (Promega Corp., USA) (*see* **Note 7**).

1. MyClycler thermal cycler (Bio-Rad, Hercules, USA).

2. PCR program:
   - Initial step: 94 °C, 3 min.
   - Annealing step (34 cycles).
   - Denaturation: 94 °C, 1 min 30 s.
   - Hybridization: 55 °C, 1 min 30 s.
   - Elongation: 72 °C, 2 min.
   - Final elongation: 72 °C, 15 min (34 cycles).

3. Submerged horizontal electrophoresis cells (Wide Mini-Sub Cell GT Cell, 30 samples per comb, Bio-Rad, Hercules, USA).

4. TAE: Tris-Acetate-EDTA (1×) (TAE): 40 mM Tris, 20 mM acetic acid, and 1 mM EDTA. It is used as electrophoresis buffer.

5. DNA size markers (GeneRuler 100 bp DNA ladder, Fermentas, France).

6. Agarose gel at 1.5 % is recommended to control the amplication. Gels were stained with ethidium bromide (5 μL of ethidium bromide at 10 mM to a final concentration 0.5 mM).

7. Migration in TAE 1×; 45 min at 90 V.

8. DNA band visualization: computer program Quantity One 4.6.5 Bio-Rad.

2.6 DNA Digestion

1. Endonucleases (*Hae*III, *Hin*fI, *Mse*I *Sdu*I, *Bfm*I, *Mae*II, and *Cfr*9I, (Fermentas, France), *Hpy*188I and *Psp*GI (New England Biolabs, UK) (Fig. 2).

2. Restriction mix (final volume 20 μL): 10 μL of PCR product, 7 μL of sterile ultrapure water, reaction buffer 2 μL and 1 μL of endonuclease (10 U/μL) (*see* **Note 8**).

3. Submerged horizontal electrophoresis cells (Wide Mini-Sub Cell GT Cell, 30 samples per comb, Bio-Rad, Hercules, USA).

4. Electrophoresis buffer: Tris-acetate EDTA, 1×.

5. DNA size markers (GeneRuler 100 bp DNA ladder, Fermentas, France).

6. Agarose gel at 3 % is recommended to obtain restriction profiles. Gels were stained with ethidium bromide (5 μL of ethidium bromide at 10 mM to a final concentration 0.5 mM) (*see* **Note 9**).

7. Migration in TAE 1×; 1 h–1 h 30 min at 90–100 V.

8. DNA band visualization: computer program Quantity One 4.6.5 Bio-Rad.

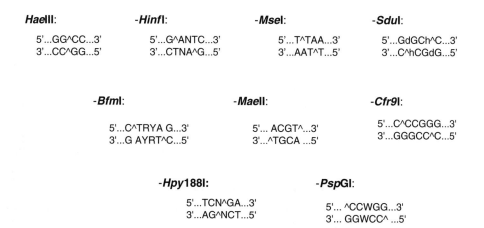

Fig. 2 Endonucleases used and their restriction sites

3 Methods

3.1 Sampling Procedure and Isolation of Filamentous Fungi

The three first steps are realized to release spores and/or mycelium from the surface of the sample.

1. Collect samples aseptically (for example sterile bag) and conserve them at 4 °C during the transport and before use.

2. Place the studied sample (for example 200 grape berries) in a flask containing 200 mL of sterile washing suspension.

3. Sonicate this mixture for 1 min and then stir on an orbital shaker for 30 min to put the microorganisms in suspension. Hold washing suspension in a sterile flask at 4 °C before use.

4. Then serially dilute (1/10) washing suspension and plate each dilution (100 µL) on DRBC medium (three repetitions by dilution). Plates were returned and incubated for 4–7 days at 25 °C.

5. Pick and isolate on MEA medium from each plate single fungal colony, considered representing different genera and species (by macroscopic observations). Return and incubate plates for 4–7 days at 25 °C.

6. After incubation, deposit 1 mL of rinse solution on the surface of each MEA plate which is scraped using a sterile disposable inoculating. Collect and deposit in a sterile cryovial the spore suspension (≈500 µL). Add 500 µL of glycerol 40 % (for a final concentration 20 %) and keep the cryovial at −80 °C.

3.2 DNA Preparation

1. From the frozen stock, inoculate a PDA medium for each fungal isolate.

2. After a culture of 7 days or more (in function of the isolates), add 5 mL of rinse solution and scrap gently the medium surface using a sterile disposable inoculating. This step is repeated to recover as much as possible spores.

3. Count the spore using a Malassez cell.

4. Collect the spores in a tube containing 2 mL of sterile water.

5. Centrifuge the resulting suspension (10,000 xg, 10 min).

6. Take up the pellet of spores containing between 150 and 200 mg in 2 mL of sterile water (*see* **Notes 10** and **11**).

3.3 DNA Extraction

1. Extract DNA from 2 mL of a suspension of spores or 2 mL of crushed mycelium using DNA extraction kit fungi Miniprep EZNA® (Omega Bio-tek, Doraville, USA) according to the manufacturer's instructions.

2. Specifically, centrifuge 2 mL of suspension (10,000 xg for 20 min).

3. Incubate the pellet with 600 μL of buffer FG1 and 5 μL RNase (20 mg/mL) for 1 min.

4. Add 10 μL of 2-mercaptoethanol and incubate the mixture at 65 °C for 5 min.

5. Add then 140 μL of FG2 buffer and incubate on ice for 5 min.

6. Centrifuge for 10 min at 10,000 xg, transfer the supernatant and add a ½ volume of buffer FG3 and a volume of absolute ethanol.

7. Elute the DNA with 100 μL of ultrapure water H_2O using Hi-bond®spin columns.

8. Control the quality of the DNA extracts before each use. Determine the reports of the OD (260 nm/280 nm and 260 nm/230 nm) to assess the quality of the samples. The DNA concentration should be from 100 to 150 ng/μL.

9. Store the DNA at –20 °C before analysis (*see* **Note 12**).

3.4   PCR ITS

1. PCR reactions were performed as described in **Note 7** and Table 1.

2. To control amplifications, deposit 15 μL of PCR products in each well of an agarose gel at 1.5 % and 5–7 μL of DNA size markers in the two most outer wells.

3. The size of the amplified PCR products differs between some genera and allows to distinguish, for example, a strain of the genus *Aspergillus* (580 bp) from a strain of the genus *Botrytis* (540 bp) (Fig. 3a). However, a PCR ITS was unable to discriminate different species among a same genera (Fig. 3b).

4. Thus generate restriction profiles to allow discrimination between species.

Table 1
PCR reaction (final volume 50 μL)

	Concentration of the stock solution	Volume	Final concentration in each PCR reaction
Ultrapure water		26.35 μL	
Buffer		5 μL	
$MgCl_2 \cdot 6H_2O$	25 mM	3 μL	1.5 mM
dNTP	25 mM (each)	0.4 μL	0.2 mM
Primer ITS1	10 μM	2.5 μL	0.5 μM
Primer ITS4	10 μM	2.5 μL	0.5 μM
Taq polymerase	5 U/μL	0.25 μL	0.025 U/μL
DNA	100–150 ng/μL	10 μL	20–30 ng/μL

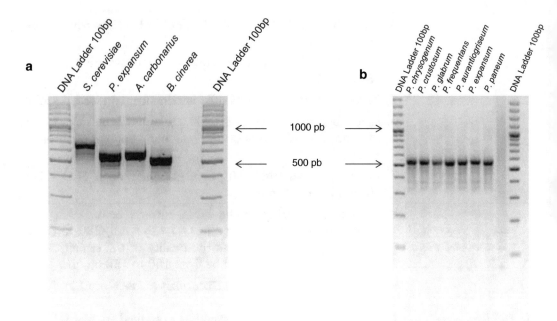

Fig. 3 PCR ITS amplification of DNA from different genera and species with primers ITS1 and ITS4 [24]. (**a**) *Lanes 1* and *7*: DNA ladder 100 bp. *Lane 2*: *Saccharomyces cerevisiae* (840 bp); *lane 3*: *Penicillium expansum* (570 bp); *lane 4*: *Aspergillus carbonarius* (580 bp); *lane 5*: *Botrytis cinerea* (540 bp). (**b**) *Lanes 1* and *10*: DNA ladder 100 bp. *Lane 2*: *P. chrysogenum; lane 3*: *P. crustosum; lane 4*: *P. glabrum; lane 5*: *P. frequentans; lane 6*: *P. aurantiogriseum; lane 7*: *Penicillium expansum;* and *lane 8*: *P. paneum*

3.5   PCR ITS-RFLP

1. Digest PCR products with the restriction enzymes *Sdu*I, *Hinf*I, *Mse*I, and *Hae*III, separately.

2. Use an additional endonuclease in some cases to complete discrimination: *Mae*II, *Cfr*9I, *Hpy*188I or *Psp*GI. After electrophoresis visualize, DNA bands under UV light (*see* **Notes 8** and **9**).

3. Designate the profiles obtained with one restriction endonuclease with letters as follows: *Sdu*I: 16 different profiles from "a" to "r"; *Hinf*I: 6 different profiles from "a" to "f"; *Mse*I: 9 different profiles from "a" to "i"; *Hae*III: 9 different profiles from "a" to "i" (Table 2). The endonuclease *Sdu*I is especially discriminant, which generates 16 different profiles.

4. Combine the individual profiles obtained into composite restriction patterns characteristic of a species (Table 2).

5. By this method, 41 different species among 11 genera can discriminate. Only the species *P. thomii* and *P. glabrum* gave the same composite profile. Only four endonucleases are necessary to discriminate the majority of species [23].

Table 2
ITS individual restriction patterns and composite restriction patterns for different species of filamentous fungi

Species	Individual restriction patterns by endonuclease				Composite restriction pattern
	*Sdu*I (pb)	*Hinf*I (pb)	*Mse*I (pb)	Other enzymes (pb)	
Acremonium alternatum	576[a] (l)	–	–	290+168+85 (d, *Hae*III)	l-d
Alternaria alternata	570[a] (l)	–	–	433+137 (e, *Hae*III)	l-e
Aspergillus aculeatus	357+99 (j)	269+185+112 (f)	–	–	jf
Aspergillus carbonarius	285+175 (b)	284+195+113 (e)	–	–	be
Aspergillus fumigatus	265+171 (a)	289+289 (b)	–	–	ab
Aspergillus japonicus	362+99 (j)	283+283 (b)	–	–	jb
Aspergillus niger	291+175 (b)	289+289 (b)	–	–	bb
Aspergillus terreus	287+174+98 (g)	296+199+105 (e)	371+215 (b)	–	geb
Aspergillus wentii	289+167+97 (g)	285+195+108 (e)	252+200+65 (h)	–	geh
Botrytis cinerea	539[a] (k)	–	–	433+111 (a, *Hae*III)	k-a
Cladosporium cladosporioides	418+126 (o)	–	–	544[a] (c, *Hae*III)	o-c
Cladosporium herbarum	421+123 (o)	–	–	510[a] (f, *Hae*III)	o-f
Epicoccum nigrum	545[a] (k)	–	–	545[a] (c, *Hae*III)	k-c
Fusarium oxysporum	540[a] (k)	–	–	340+110+90 (b, *Hae*III)	k-b
Pilidiella diplodiella	299+138+96 (n)	–	–	440+172 (i, *Hae*III)	n-i
Penicillium aurantiogriseum	262+168 (a)	289+173+ 113 (a)	366+204 (b)	586[a] (a, *Bfm*I)	aab-a
Penicillium bilaiae	279 +171 (b)	287 +287 (b)	368+207 (b)	328 +276 (d, *Psp*GI)	bbb-d
Penicillium brevicompactum	348+164 (e)	287+287 (b)	352+186 (c)	–	ebc

(continued)

Table 2
(continued)

Species	Individual restriction patterns by endonuclease				Composite restriction pattern
	*Sdu*I (pb)	*Hin*fI (pb)	*Mse*I (pb)	Other enzymes (pb)	
Penicillium corylophilum	292+164 (b)	282+282 (b)	361+207 (b)	584[a] (c, *Psp*GI)	bbb-c
Penicillium chrysogenum	257+167 (a)	285+285 (b)	362+97+97 (a)	277+180+130 (a, *Cfr*9I)	aba-a
Penicillium crustosum	263+169 (a)	293+293 (b)	362+98+98 (a)	191+174+174 (a, *Hpy*188I)	aba-a
Penicillium commune	260+164 (a)	285+285 (b)	353+101+101 (a)	178+178+143 (b, *Hpy*188I)	aba-b
Penicillium expansum	257+164 (a)	285 +181+106 (a)	364+101+101 (a)	328+ 271 (a, *Mae*II)	aaa-a
Penicillium fellutanum	294+173 (b)	272 +272 (b)	353+197 (b)	268+ 223+110 (a, *Psp*GI)	bbb-a
P. glabrum/P. thomii	358+194 (c)	290+139+139 (c)	366+206 (b)	–	ccb
Penicillium herquei	295+170+95 (g)	290+ 290 (b)	293+203+72 (i)	–	gbi
Penicillium islandicum	320+72 (h)	185+185+120+95 (d)	365+211 (b)	–	hdb
Penicillium italicum	262+170 (a)	290+180+110 (a)	352+102+102 (a)	322+195+83 (b *Mae*II)	aaa-b
Penicillium minioluteum	453 (f)	290+204+104 (e)	573 (f)	–	fef
Penicillium oxalicum	264+170 (a)	313 +313 (b)	316 +207 (e)	–	abe
Penicillium paneum	290+121 (d)	284+180+110 (a)	368+203 (b)	–	dab
Penicillium paxilli	259+170 (a)	297+ 297 (b)	247+133+74 (g)	–	abg
Penicillium raistrickii	259+164 (a)	306 +306 (b)	359+206 (b)	–	abb
Penicillium roqueforti	285+165 (b)	273+171+101 (a)	359+99+99 (a)	–	baa
Penicillium simplicissimum	290+172 (b)	278+278 (b)	361+203 (b)	336+213 (b, *Psp*GI)	bbb-b
Penicillium spinulosum	350+165 (e)	290+136+136 (c)	354+200 (b)	–	ecb

(continued)

Table 2
(continued)

Species	Individual restriction patterns by endonuclease				Composite restriction pattern
	SduI (pb)	*HinfI* (pb)	*MseI* (pb)	Other enzymes (pb)	
Penicillium verrucosum	260+168 (a)	297+180+110 (a)	360+205 (b)	480+114 (b, *Bfm*I)	aab-b
Penicillium verruculosum	450[a] (f)	300+300 (b)	550 (d)	–	fbd
Thanatephorus cucumeris	477+150 (p)	–	–	507+119 (h, *Hae*III)	p-h
Trichoderma harzianum	620[a] (m)	–	–	434+180 (i, *Hae*III)	m-i
Trichoderma koningiopsis	600[a] (r)	–	–	437+168 (i, *Hae*III)	r-i
Trichoderma longibrachiatum	623[a] (m)	–	–	461+90 (g, *Hae*III)	m-g

[a]No enzyme restriction site

4 Notes

1. The DRBC medium was used to isolate filamentous fungi. The use of the antifungal agent, dichloran, restricts spreading of mucoraceous fungi and restricts the colony size of other genera. Rose bengal also assists in the reduction of colony sizes and is selective against bacteria. Additional selectivity against bacterial growth is achieved by the incorporation of the heat-stable antibiotic Chloramphenicol. Plates were returned and incubated for 4–7 days at 25 °C.

2. The MEA medium was used to isolate and to conserve filamentous fungi. For fungi count, a pH value between 3.5 and 5.6 is recommended depending of the microorganisms. It is recommended to adjust the pH more acidic with addition of 10 % lactic acid and 5 % tartaric acid. Plates were returned and incubated for 4–7 days at 25 °C.

3. Plates were returned and incubated for 5 days at 28 °C.

4. 200 mL of washing solution were used to wash the surface of 200 grape berries. The volume must be adapted in function of the studied sample.

5. After sterilization, prepare sterile vials containing 900 µL of dilution suspension.

6. Glycerol is viscous: to collect more easily, cut the tip to have a wider opening. Take very gently.

7. The 5.8S-ITS region was amplified by PCR using universal fungal primers ITS1 and ITS4 [24]. ITS1 (5′-TCCGTAGGTGAA CCTGCGG-3′) and ITS4 (5′-TCCTCCGCTTATTGATATGC-3′).

8. The reaction mixture was placed in a water bath for 4 h at optimum temperature: at 37 °C to *Sdu*I, *Hin*fI, *Hae*III, *Cfr*9I, *Bfm*I, and *Hpy*188I, 65 °C for *Mse*I; *Mae*II and at 75 °C for *Psp*GI.

9. Put the agarose into a flask, then add TAE 1×, and mix thoroughly. For a Wide Mini-Sub Cell, 90 mL of 3% agarose gel must be used (2.7 g of agarose and 90 mL of 1× TAE). Weigh the flask. Heat gently. Stop heating when the mixture begins to boil. Mix thoroughly. Weigh and add water to reach the initial weight. Repeat the operation until obtain a clear solution. Cool 5–10 min before pouring the gel into the electrophoresis cell.

10. The spore suspension can be stored at 4 °C if the DNA extraction is carried out on the same day and at −20 °C if the DNA extraction was performed on another day.

11. DNA preparation may also be performed from cultures in liquid medium. From the frozen stock, sample was inoculated for 4 days at 28 °C in liquid medium (6 mL of culture medium PDB) for each fungal isolate. After incubation, cell suspension was then centrifuged ($10,000 \times g$, 10 min) to collect the mycelium which was frozen in liquid nitrogen and ground to a fine powder. The ground material (between 150 and 200 mg fresh weight) is then placed in sterile water (2 mL).

12. The quality of extracted DNA is also controlled by electrophoresis on agarose gel at 1% (migration 15 min at 110 V). The DNAs are then stored at −20 °C before analysis.

References

1. van Elsas JD, Duarte GF, Keijzer-Wolters A et al (2000) Analysis of the dynamics of fungal communities in soil via fungal-specific PCR of soil DNA followed by denaturing gradient gel electrophoresis. J Microbiol Methods 43:133–151

2. Anderson IC, Cairney JW (2004) Diversity and ecology of soil fungal communities: increased understanding through the application of molecular techniques. Environ Microbiol 6:769–779

3. Haasum I, Nielsen PV (1998) Physiological characterization of common fungi associated with cheese. J Food Sci 63:157–161

4. Alborch L, Bragulat MR, Castellá G et al (2012) Mycobiota and mycotoxin contamination of maize flours and popcorn kernels for human consumption commercialized in Spain. Food Microbiol 32:97–103

5. Peleg AY, Hogan DA, Mylonakis E (2010) Medically important bacterial-fungal interactions. Nat Rev Microbiol 8:340–349

6. Li Q, Wang C, Tang C et al (2014) Dysbiosis of gut fungal microbiota is associated with mucosal inflammation in Crohn's disease. J Clin Gastroenterol 48:513–523

7. Araguas C, Gonzalez-Penas E, Lopez de Cerain A (2005) Study on ochratoxin A in cereal-derived products from Spain. Food Chem 92:459–464

8. Copetti MV, Iamanaka BT, Pereira JL et al (2011) Aflatoxigenic fungi and aflatoxin in cocoa. Int J Food Microbiol 148:141–144

9. Rousseaux S, Diguta CF, Radoï-Matei F et al (2014) Non-*Botrytis* grape-rotting fungi responsible for earthy and moldy off-flavors and mycotoxins. Food Microbiol 38:104–121

10. Dupont J, Magnin S, Marti A et al (1999) Molecular tools for identification of *Penicillium* starter cultures used in the food industry. Int J Food Microbiol 49:109–118

11. Geisen R, Cantor MD, Hansen TK et al (2001) Charac-terization of *Penicillium roqueforti* strains used as cheese starter cultures by RAPD typing. Int J Food Microbiol 65:183–191

12. Sacristán N, Mayo B, Fernández E et al (2013) Molecular study of *Geotrichum* strains isolated from Armada cheese. Food Microbiol 36:481–487

13. De Valk HA, Klaassen CHW, Meis JFGM (2008) Molecular typing of *Aspergillus* species. Mycoses 51:463–476

14. Accensi F, Cano J, Figuera L et al (1999) New PCR method to differentiate species in the *Aspergillus niger* aggregate. FEMS Microbiol Lett 180:191–196

15. Bau M, Castella G, Bragulat MR et al (2006) RFLP characterization of *Aspergillus niger* aggregate species from grapes from Europe and Israel. Int J Food Microbiol 111:18–21

16. Colombo F, Vallone L, Giaretti M et al (2003) Identification of *Penicillium aurantiogriseum* species with a method of polymerase chain reaction-restriction fragment length polymorphism. Food Control 14:137–140

17. Dean TR, Kohan M, Betancourt D et al (2005) A simple polymerase chain reaction/restriction fragment length polymorphism assay capable of identifying medically relevant filamentous fungi. Mol Biotechnol 31:21–27

18. Liu P, Li B, Yin R et al (2014) Development and evaluation of ITS- and aflP-based LAMP assays for rapid detection of *Aspergillus flavus* in food samples. Can J Microbiol 60:579–584

19. Oviedo MS, Sturm ME, Reynoso MM et al (2013) Toxigenic profile and AFLP variability of *Alternaria alternata* and *Alternaria infectoria* occurring on wheat. Braz J Microbiol 44:447–455

20. Dean TR, Ropp B, Betancourt D et al (2005) A simple multiplex polymerase chain reaction assay for the identification of four environmentally fungal contaminants. J Microbiol Methods 61:9–16

21. Lee SL, Taylor JW (1992) Phylogeny of five like protoctisan *Phytophtora* species inferred from the internal transcribed spacers of ribosomal DNA. Mol Biol Evol 9:636–653

22. Esteve-Zarzoso B, Belloch C, Uruburu F et al (1999) Identification of yeasts by RFLP analysis of the 5.8S rRNA gene and the two ribosomal internal transcribed spacers. Int J Syst Bacteriol 49:329–337

23. Diguta CF, Vincent B, Guilloux-Benatier M et al (2011) PCR ITS-RFLP: a useful method for identifying filamentous fungi on grapes. Food Microbiol 28:1145–1154

24. White TJ, Bruns T, Lee S et al (1990) Amplification and direct sequencing of fungi ribosomal RNA genes for phylogenetics. In: Innis MA, Gelfand DH, Ninsky JJ, White TJ (eds) PCR protocols. a guide to methods and applications. Academic Press, San Diego, pp 315–322

Part V

New Methodologies for Detection and Identification of Mycotoxigenic Fungi

Chapter 22

Identification of Ochratoxin A-Producing Black Aspergilli from Grapes Using Loop-Mediated Isothermal Amplification (LAMP) Assays

Michelangelo Storari and Giovanni A.L. Broggini

Abstract

The loop-mediated isothermal amplification (LAMP) allows the rapid and specific amplification of target DNA under isothermal conditions without a prior DNA purification step. Moreover, successful amplifications can be directly evaluated through a color change of the reaction solutions. Here, we describe two LAMP assays for the detection of ochratoxin-A producing black aspergilli isolated from grapes. The two assays can detect DNA of OTA-producing black aspergilli following a very simple sample preparation and have the potential to significantly speed up the routine monitoring of these toxigenic molds in vineyards.

Key words Mycotoxins, Hydroxynaphthol blue, Polyketide synthase, Climate change, Wine

1 Introduction

The loop-mediated isothermal amplification (LAMP) was developed by [1]. In a LAMP reaction, six primers (FIP, BIP, LF, LB, F3, and B3) mediate the continuous amplification of the target DNA under isothermal conditions through the formation of DNA loops. This leads to the production of a great quantity of polymerized DNA (up to 10^9 copies of the target DNA). The advantages of the LAMP reaction compared to a conventional PCR are multiple. Firstly, the LAMP reaction is less sensitive to inhibitors and this permits the amplification of DNA without a prior purification step [2]. Secondly, the successful amplification of the target DNA can be visualized in several ways in addition to gel electrophoresis. These are turbidity through the precipitation of magnesium pyrophosphate during DNA amplification and the addition of DNA intercalating dyes to the reagent tubes [3, 4]. Another interesting way to detect positive reactions is based on the addition of complexometric dyes such as hydroxynaphthol blue (HNB) to the master mix [4]. Before amplification, HNB in the presence of Mg^{2+}

Antonio Moretti and Antonia Susca (eds.), *Mycotoxigenic Fungi: Methods and Protocols*, Methods in Molecular Biology, vol. 1542,
DOI 10.1007/978-1-4939-6707-0_22, © Springer Science+Business Media LLC 2017

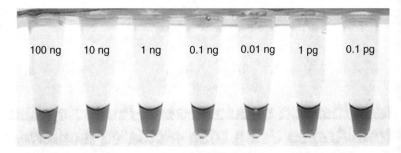

Fig. 1 Amplification of different amounts of *Aspergillus carbonarius* DNA by loop-mediated isothermal amplification (LAMP). A color change from *violet* to *sky blue* indicates successful amplification of the target DNA. The limit of detection of the LAMP assay was found to be between 0.1 and 0.01 ng DNA. Adapted from ref. [15].

ions gives to the master mix solution a violet color. However, DNA amplification and the subsequent precipitation of magnesium pyrophosphate induce the color of the solution to shift towards sky blue (Fig. 1). This allows to confirm the occurred amplification simply through a colour change without any post-reaction manipulation. In summary, the LAMP reaction is a rapid and specific detection tool and a valid alternative to conventional PCR and qPCR since it does not require complex DNA extraction procedures, gel electrophoresis, or expensive real-time machines.

Since its first publication, the LAMP reaction has found a huge amount of applications in medicine, food safety, agriculture, and environmental microbiology [5–7]. In the last years, the first applications of LAMP in the detection of mycotoxin-producing molds have come to light. These include assays for the detection of *Fusarium graminearum* [8] and *Aspergillus flavus* [9]. Here we present a protocol based on the LAMP reaction to identify ochratoxin A-producing black aspergilli isolated from grapes.

Black aspergilli are responsible for the contamination of grapes and wines with ochratoxin A (OTA), a major mycotoxin [10]. Survey of their presence in vineyards is part of the strategy to reduce the risk of OTA contamination of these products [11, 12]. However, identification of OTA-producing strains is a tedious process involving the isolation of black aspergilli from grapes, their identification through DNA extraction and molecular techniques such as gene sequencing or the use of specific primers and, finally, the detection of OTA production on synthetic media [13, 14]. To speed up this process, we developed two LAMP assays for the specific detection of OTA-producing *A. carbonarius* and *A. niger/A. awamori* strains [15]. The two primer sets were designed on two PKS genes whose presence in the genome of *A. carbonarius* and *A. niger/A. awamori* (also called *A. niger clade*) has been shown to correlate positively with the ability of these black aspergilli to produce OTA [16, 17]. The two LAMP assays were shown to work properly with samples processed by simply heating the fungal mycelium in water. Using the protocol presented here it is therefore possible to greatly simplify the procedure of identification of OTA-producing black aspergilli isolated from grapes.

2   Materials

2.1   DNA Extraction from Pure Cultures

1. Potato dextrose broth (PDB; BD Difco, USA): 4 g L^{-1} potato starch (from infusion), 20 g L^{-1} dextrose. Pour 10 mL of autoclaved PDB in glass or plastic tubes.

2. Horizontal shaker.

3. Lyophilizator.

4. Commercial DNA extraction kit for plant or fungal tissues (e.g., DNeasy® Plant Mini Kit, Qiagen, Germany).

5. NanoDrop (NanoDrop Technologies, USA).

2.2   Loop-Mediated Isothermal Reaction

All stock and working solutions are aliquoted and stored at –20 °C.

1. LAMP primers:

    – Primer sequences of the two LAMP assays are listed in Table 1. These primers were designed using the PrimerExplorer V3 software (Eiken Chemical Co., Ltd., Tokyo, Japan) on the PKS genes *acps14* (Accession

Table 1
LAMP primer sets for the detection of OTA-producing *A. carbonarius* and *A. niger*

Target	Primers	Sequence (5′–3′)
A. carbonarius–AcPKS	FIP	GCTGCAATGCACCCGGTAGTTGAAGACGTGGAGGGCTTCT
	BIP	AAGTCACGTCAAAAGCCCTGGTGATTCCTTGGGAGGTTGGTC
	LF	GCTGGCCGAAAAGATCGCTAA
	LB	CCCCTCTGCTATGAAGTCCG
	F3	TGGTGGTACGAATGCACAC
	B3	CGAAATGACAAACAGGCGGT
A. niger–AnPKS	FIP	CCTGCGCCACCTTCCAAGTGCGATTCGCCCCTCTATGTTG
	BIP	CTGATCTCAGCCACACTGGCTGATCTTGGGGTTCAAGCTCTG
	LF	CCAGCACGGATTTTACCGATC
	LB	TAGAGTCGAAGATGATACCCCCAGT
	F3	TGCATTAGGTGTTGCCCG
	B3	AGGTCATCGCGTTGAGGA

Number GU001531) for *A. carbonarius* and *an15g07920* for *A. niger* and *A. awamori*.

- FIP and BIP primers should be of HPLC-purified grade (dissolve lyophilized primers in nucleotide-free water to a final concentration of 100 μM).

- Prepare primer working solutions by diluting the stock solutions in nuclease-free water to a final concentration of 10 μM for LF, LB, F3, and B3 and at 20 μM for FIP and BIP, respectively.

2. Deoxyribonucleotides (dNTPs): 10 mM for each dNTP.

3. 10× Thermopol buffer (New England Biolabs, USA).

4. 100 mM $MgSO_4$.

5. 5 M Betaine working solution.

6. 3 mM Hydroxynaphthol Blue (HNB) working solution.

7. *Bst* DNA polymerase (New England Biolabs, USA).

8. Nuclease-free water.

9. Conventional Thermocycler for PCR reactions.

2.3 Fungal Pure Culture

1. Potato dextrose agar (PDA; BD Difco, USA): 4 g L^{-1} potato starch (from infusion), 20 g L^{-1} dextrose, 15 g L^{-1} agar. Pour autoclaved PDA in 90 mm Petri dishes.

3 Methods

3.1 DNA Extraction

1. Grow black aspergilli strains in 10 mL of PDB for 7–10 days at 24 °C on an orbital shaker (*see* **Notes 1** and **2**).

2. Separate the grown mycelia from the broth by centrifugation ($4000 \times g$; 30 min) and lyophilize them.

3. Extract DNA from lyophilized mycelia using a commercial DNA extraction kit (if use DNeasy® Plant Mini Kit start from about 100–200 mg of lyophilized mycelium).

4. Quantify DNA using NanoDrop.

5. Check the integrity of DNA by PCR amplification using universal primers for fungi (e.g., CL1 and CL2A) [18].

3.2 Loop-Mediated Isothermal Amplification Reaction

LAMP master mix preparation should be carried out on ice.

LAMP assays should be able to amplify DNA concentrations as small as 0.05 ng $μL^{-1}$.

1. Prepare the following master mix for each LAMP assay (*see* **Note 3**):

- 1.6 μmol L^{-1} each FIP and BIP.

- 0.2 μmol L^{-1} each F3 and B3.

- 0.8 µmol L^{-1} each LF and LB.
- 1.2 mmol L^{-1} each deoxyribonucleotide (dNTP).
- 100 µmol L^{-1} hydroxynaphthol blue (HNB).
- 8 mmol L^{-1} MgSO$_4$.
- 1× Thermopol buffer.
- 0.8 mol L^{-1} betaine.
- 0.320 U µL^{-1} *Bst* DNA polymerase.
- 2 µL DNA.
- Nuclease-free water up to 25 µL.

2. Include in your reaction one or more positive controls (pure DNA of *A. niger* or *A. carbonarius* depending on the assay you are carrying on) and one or more negative controls (nuclease-free water instead of extracted DNA).

3. Incubate the reaction mixtures at 65 °C for 60 min in a thermocycler with the lid preheated at 99 °C.

4. Raise the incubation temperature to 80 °C for 2 min to terminate the reaction.

5. Check visually the output of the LAMP reaction: a colour shift from violet to sky blue indicates the presence of the target DNA in your sample (Fig. 1, *see* **Notes 4** and **5**).

3.3 Identification of Ochratoxin A-Producing Black Aspergilli in Pure Culture

Each reaction should be carried out twice to ensure consistency of the results.

1. Inoculate potato dextrose agar (PDA) with strains of black aspergilli under a laminar flow cabinet.

2. Incubate the strains at 24 °C for 2–3 days until the white mycelium is clearly visible but still not covered by black conidia.

3. Scrape a visible piece of young mycelium from the agar plate using a sterile pipette tip.

4. Mix the mycelium in 50 µL of sterile water.

5. Incubate the mycelium/water mixture for 10 min at 95 °C in a thermocycler (lid at 99 °C).

6. Use 2 µL of the mycelium/water mixture in the LAMP assay (*see* Subheading 3.2).

4   Notes

1. *Aspergillus* spp. are classified as Biosafety Level 2 (BSL-2) microorganisms. Proper procedures should be followed to minimize their dispersion in the working environment.

2. *Aspergillus* spp. can produce mycotoxins during growth. Avoid contact of the skin with the broth where black aspergilli were grown.

3. LAMP master mix preparation and DNA addition should be carried out in different rooms with different lab coats and sets of pipettes to reduce the risk of contamination of the master mixes.

4. The LAMP reaction produces a huge amount of DNA amplicons of different sizes. To avoid the risk of contamination and appearance of false-positives we recommend not to open the reactions tubes once the reaction is completed and not to accumulate them in the trash close to the working bench.

5. The output of LAMP reaction can also be visualized on an agarose gel. However, because of the risk of contamination of the working zone and instruments with LAMP amplicons (*see* **Note 4**), we recommend to use always an indirect indicator of the amplification like HNB or DNA-intercalating dyes.

Acknowledgments

This work was funded by the Autonomous Province of Trento, project ENVIROCHANGE, Call for proposal Major Projects 2006.

References

1. Notomi T, Okayama H, Masubuchi H et al (2000) Loop-mediated isothermal amplification of DNA. Nucleic Acids Res 28:e63

2. Kaneko H, Kawana T, Fukushima E, Suzutani T (2007) Tolerance of loop-mediated isothermal amplification to a culture medium and biological substances. J Biochem Bioph Meth 70:499–501

3. Tomita N, Mori Y, Kanda H, Notomi T (2008) Loop-mediated isothermal amplification (LAMP) of gene sequences and simple visual detection of products. Nat Protoc 3:877–882

4. Goto M, Honda E, Ogura A et al (2009) Colorimetric detection of loop-mediated isothermal amplification reaction by using hydroxynaphthol blue. Biotechniques 46:167–172

5. Parida M, Sannarangaiah S, Dash PK et al (2008) Loop mediated isothermal amplification (LAMP): a new generation of innovative gene amplification technique; perspective in clinical diagnosis of infectious diseases. Rev Med Virol 18:407–421

6. Niessen L, Luo J, Denschlag C, Vogel RF (2013) The application of loop-mediated isothermal amplification (LAMP) in food testing for bacterial pathogens and fungal contaminants. Food Microbiol 36:191–206

7. Kuan CP, Wu MT, Lu YI, Huang HC (2010) Rapid detection of squash leaf curl virus by loop mediated isothermal amplification. J Virol Methods 169:61–65

8. Niessen L, Vogel RF (2010) Detection of *Fusarium graminearum* DNA using a loop-mediated isothermal amplification (LAMP) assay. Int J Food Microbiol 140:183–191

9. Luo J, Vogel RF, Niessen L (2012) Development and application of a loop-mediated isothermal amplification assay for rapid identification of aflatoxigenic molds and their detection in food samples. Int J Food Microbiol 159:214–224

10. Perrone G, Susca A, Cozzi G et al (2007) Biodiversity of Aspergillus species in some important agricultural products. Stud Mycol 59:53–66

11. Hocking AD, Leong SL, Kazi BA et al (2007) Fungi and mycotoxins in vineyards and grape products. Int J Food Microbiol 119:84–88

12. Visconti A, Perrone G, Cozzi G, Solfrizzo M (2008) Managing ochratoxin A risk in the grape-wine food chain. Food Add Contam 25:193–202

13. Battilani P, Giorni P, Bertuzzi T et al (2006) Black aspergilli and ochratoxin A in grapes in Italy. Int J Food Microbiol 111:S53–S60

14. Storari M, Broggini GAL, Bigler L et al (2012) Risk assessment of the occurrence of black aspergilli on grapes grown in an alpine region under a climate change scenario. Eur J Plant Pathol 134:631–645

15. Storari M, von Rohr R, Pertot I et al (2013) Identification of ochratoxin A producing *Aspergillus carbonarius* and *A. niger* clade isolated from grapes using the loop-mediated isothermal amplification (LAMP) reaction. J Appl Microbiol 114:1193–1200

16. Storari M, Pertot I, Gessler C, Broggini GAL (2010) Amplification of polyketide synthase gene fragments in ochratoxigenic and nonochratoxigenic black aspergilli in grapevine. Phytopathol Mediterr 49:393–405

17. Castellá G, Cabañes FJ (2011) Development of a real time PCR system for detection of ochratoxin A-producing strains of the Aspergillus niger aggregate. Food Control 22:1367–1372

18. O'Donnell K, Nirenberg HI, Aoki T, Cigelnik E (2000) A multigene phylogeny of the Gibberella fujikuroi species complex: detection of additional phylogenetically distinct species. Mycoscience 41:61–78

Chapter 23

Detection of Transcriptionally Active Mycotoxin Gene Clusters: DNA Microarray

Tamás Emri, Anna Zalka, and István Pócsi

Abstract

Various bioanalytical tools including DNA microarrays are frequently used to map global transcriptional changes in mycotoxin producer filamentous fungi. This effective hybridization-based transcriptomics technology helps researchers to identify genes of secondary metabolite gene clusters and record concomitant gene expression changes in these clusters initiated by versatile environmental conditions and/or gene deletions. Such transcriptional data are of great value when future mycotoxin control technologies are considered and elaborated. Giving the readers insights into RNA extraction and DNA microarray hybridization steps routinely used in our laboratories and also into the normalization and evaluation of primary gene expression data, we would like to contribute to the interlaboratory standardization of DNA microarray based transcriptomics studies being carried out in many laboratories worldwide in this important field of fungal biology.

Key words DNA microarray, RNA extraction, Acidic guanidinium thiocyanate-phenol reagent, Transcriptomics, Secondary metabolism, Mycotoxins, Gene clusters, Culture conditions, Environmental stress, Gene deletion mutants

1 Introduction

Although the number of DNA microarray-based transcriptomics studies carried out in filamentous fungi exceeded 50 already by 2007 the majority of these early studies were performed on coated glass slides carrying PCR-amplified cDNA sequences as probes [1]. A most recent study by Emri et al. [2] found a poor correlation between gene expression data collected by expressed sequence tag (EST)-based microarrays and whole-genome-based Agilent 60-mer oligonucleotide high-density arrays, which was explained, at least in part, by potential cross-hybridization of cDNAs of paralogue gene pairs. Nevertheless, even early EST-based chips were successful to shed light on the transcriptional fine-tuning of numerous basically important cell biological processes related to primary and secondary metabolisms, sexual and asexual developments as well as pathogenesis and symbiosis [1]. Owing to the next-generation

Antonio Moretti and Antonia Susca (eds.), *Mycotoxigenic Fungi: Methods and Protocols*, Methods in Molecular Biology, vol. 1542, DOI 10.1007/978-1-4939-6707-0_23, © Springer Science+Business Media LLC 2017

sequencing platforms, which are spreading wide and fast, the number of publically available fungal genomes exceeded 180 in 2014 [3]. The good quality of the assembled genomes makes the construction of whole-genome-based DNA microarrays now as part of the daily routine. Excellent platforms are available to gain myriads of global transcriptional data and versatile easy-to-use and time-saving bioinformatics tools help researchers to mine these data to catch and evaluate the transcriptional changes they are interested in [3]. It is important to note that, in addition to hybridization-based methods like DNA microarrays, sequence-based transcriptomics tools (RNA-seq) are also employed now frequently to map global changes in the fungal gene expression levels [3].

Considering their outstanding biomedical, agricultural, industrial, and economic importance, deciphering secondary metabolite (drugs, antibiotics, mycotoxins) biosynthetic pathways was always among the high priorities of fungal transcriptomics studies in both the aspergilli [4–6] and the fusaria [7–10]. Elimination of transcriptional regulators and subsequent comparative transcriptomics are exceptionally powerful approaches, when the target genes of these factors, e.g., among secondary metabolite biosynthetic clusters, are screened [5, 11–17]. Environmental conditions facilitating or, just the opposite, hindering secondary metabolite productions are easy to identify and discuss when global transcriptional data sets recorded under various circumstances, e.g., in cultures exposed to various types of environmental stress, cultivated in the presence of various nutrients or grown under conductive or non-conductive conditions for mycotoxin production, are available [2, 4, 18–29]. The significance of this kind of information is almost impossible to underestimate when future biocontrol strategies to limit the growth and toxin production of toxinogenic fungi are considered and elaborated [30–37].

The genomes of toxinogenic fungi like the aspergilli and fusaria contain several dozens or even more secondary metabolite gene clusters [38–43]. Some of them are responsible for the production of well-known and harmful mycotoxins, e.g., aflatoxins, ochratoxins, and trichothecenes, but the products and regulations of many other "untapped" secondary metabolite gene clusters have remained yet to be elucidated [44–48]. Recent studies demonstrated that the regulation of secondary metabolite production is not uniform at all, and conditions which induce the activity of one cluster may even repress others [2]. DNA microarray based transcriptome analyses provide us with valuable tools to follow the activities of all secondary metabolite gene clusters concomitantly in a given species and also allow us to correlate these transcriptional fingerprints with the culture conditions tested. Here we present methods for the isolation of RNA for transcriptome analysis as well as a technical overview on DNA microarray based transcriptome analysis. Our aim is to let the reader have a deeper insight into today's global gene expression studies and also to facilitate inter-laboratory standardization in this field as much as possible.

1.1 RNA Isolation

Isolation of high-quality RNA from filamentous fungi is sometimes difficult due to their complex cell wall, their high intracellular RNase activity and also because of the presence of polysaccharides which may co-precipitate with RNA. Not surprisingly, several papers addressing these technical difficulties have been published thus far [49–51], and companies provide customers with several kits and ready-to-use reagents to make RNA extraction faster and easier. We present here a simple method based on acid guanidinium thiocyanate-phenol-chloroform extraction [19, 52–55], which is a typical phase separation system employing aqueous sample and water-saturated phenol-chloroform solvent [52–54]. Guanidinium thiocyanate is incorporated as a strong chaotropic agent to inactivate RNases and to denaturate RNA-binding proteins. The acidic pH (pH 4–6) is crucial since it helps RNA partitioning into the aqueous phase while denaturated proteins and DNA partition into the organic phase.

1.2 Microarray-Based Gene Expression Analysis (Agilent DNA Microarray System)

As summarized nicely by Aguilar-Pontes et al. [3], a number of excellent DNA microarray platforms are available for transcriptome analyses in fungi. In our global gene expression investigations, we routinely use the Agilent platform (http://microarrayservice.jp/download/pdfs/Agilent_DNA_Microarray_Platform.pdf) and, hence, we focus our attention here on this technology.

The Agilent One-Color Microarray-based Gene Expression Analysis (http://www.chem.agilent.com/library/usermanuals/Public/G4140-90040_GeneExpression_OneColor_6.7.pdf) uses cyanine three-labeled targets to measure gene expression in experimental and control samples. The presence and quantity of targets are detected by hybridization to the hundreds of thousands of immobilized single-stranded DNA probes, in length of 60 bp. This probe size is appropriate to achieve excellent specificity (e.g., one probe/gene) and to minimize the risk of cross hybridization between probes. The probes are synthetized in situ into nano-spots (features), by printing by nucleotides into a defined feature position onto a specific glass surface, using phosphoramidite chemistry. Precision jetting enables high levels of spatial multiplexing and flexible designs. One feature contains thousands of homologous single-stranded probes, which allow the quantitation of the target nucleic acid over 5 logs of dynamic range. The level of the fluorescent signal, detected after microarray processing, depends on number of target nucleic acid-probe hybrids within a feature. The on-array position of every features and sequence specific information are delivered in a data file with array slide and readable for the Feature Extraction software. The Agilent produces the $1 \times 3'$

Table 1
Agilent array formats

Array format	Array/slide	Features/array	Sample/slide
1×1 M	1	1,000,000	1
1×244 K	1	244,000	1
2×400 K	2	400,000	2
2×105 K	2	105,000	2
4×180 K	4	180,000	4
4×44 K	4	44,000	4
8×60 K	8	60,000	8
8×15 K	8	15,000	8

standard glass slide size microarrays in different array formats with different number of features/array (Table 1).

Experimental design is an important step when planning a microarray experiment. Replicates are essential to perform reliable statistical analysis of results. Biological replicates (multiple cases per group) are necessary to find variation between subjects within the same treatment or between specimens from the same subject. Technical replicates (repeatedly processed RNA samples from one case) provide information about the variability of the labeling, hybridization and quantification processes. The number of biological and technical replicates will be determined mostly by the budget. The actual number should be determined case by case, during consultation with bioinformaticians.

In the case of Agilent microarray slides the array content could be designed by the factory or by the researcher via a free array design tool—eArray. Use existing design or target specific chromosomal regions of interest. The Web-based interface makes it easy to create, search, and share designs (http://www.genomics.agilent.com/en/Custom-Design-Tools/eArray/?cid=AG-PT-122&tabId=AG-PR-1047).

2   Materials

2.1   RNA Isolations

Standard molecular biology laboratory equipment including refrigerated Eppendorf centrifuge and thermo block or equivalents is suitable for RNA preparation. For quantification of RNA samples and for quality control, a NanoDrop spectrophotometer (Thermos Scientific) and Agilent 2100 Bioanalyzer (Agilent Technologies) or equivalents are recommended.

1. Acidic guanidinium thiocyanate-phenol reagent: 4.3 M phenol, 0.8 M guanidine thiocyanate, 0.4 M ammonium thiocyanate, 0.1 M sodium acetate and 0.68 M glycerol. Alternatively, ready to use solutions are available, as TRIzol (Life Technologies) or TRI Reagent (Sigma-Aldrich). Preparing acidic guanidinium thiocyanate-phenol reagent is a real alternative but it is recommended for advanced users only.

2. Standard saline citrate solution (SSC, NaCl-Na-citrate solution): 150 mM NaCl and 15 mM Na_3-citrate (pH 7.0). Prepare 20× SSC stock solution as follows (or buy it from suppliers): Dissolve 17.5 g NaCl and 8.8 g Na-citrate in 90 mL distilled water. Set the pH to 7.0 (using HCl and NaOH) and set the final volume to 100 mL. Use molecular biology-grade powders and nuclease-free or DEPC-treated water. If you wish to treat the 20× SSC with DEPC (it is recommended) you can simply use distilled water. Store the stock solution at 4 °C. Prepare (1×) SSC solution as follows: Dilute 1 volume of 20× SSC with DEPC-treated or nuclease-free water to 20 volumes just before RNA preparation.

3. 70 v/v % ethanol: For preparing 70 v/v % ethanol dilute 7 volume absolute ethanol (molecular biology or ACS grade or equivalents) with DEPC-treated or nuclease-free water to 10 volumes. Store at –20 °C or cool them down to –20 °C before use. Do not use 70 % v/v ethanol sold for disinfection.

4. Chloroform (molecular biology or ACS grade or equivalents). (Do not use isoamylalcohol-chloroform mixture.) Chloroform can be replaced with the less toxic 1-bromo-3-chloropropane.

5. 2-Propanol (isopropanol, molecular biology or ACS grade or equivalents).

6. DEPC-treated water or nuclease-free water: If you do not wish to buy DEPC-treated water from suppliers do the DEPC treatment as follows: Add 0.5 mL of diethylpyrocarbonate (DEPC; molecular biology or ACS grade or equivalents) to each 500 mL distilled water or solution (e.g., 20× SSC). Mix it overnight by magnetic stirrer, and then heat-inactivate the remained DEPC in an autoclave (20 min, 121 °C).

2.2 Microarray Procedure

The appropriate array format/resolution should be selected according to the experiment design. For a general gene expression experiment the 4×44 K or 8×60 K array formats are usually used, for a focused experiment the 8×15 K array format could be the right choice. For the exon expression experiment array formats 4×180 K or 2×400 K are appropriate. Human and nonhuman catalog microarrays or custom microarrays can be ordered from the factory. The custom arrays can one design via Agilent eArray software easily by uploading own sequences, or by selection of pre-designed validated probes from Agilent databases.

1. Low Input Quick Amp Labeling Kit/One-Color (Agilent).

2. RNA Spike-In Kit/One-Color (Agilent).

3. Gene Expression Hybridization Kit (Agilent), Gene Expression Wash Buffer Kit (Agilent).

4. RNeasy Mini Kit (Qiagen).

5. Optional: 2100 Bioanalyzer, RNA 6000 Nano Assay Kit (RNA Series II Kit).

6. Isopropyl alcohol (molecular biology grade).

7. Ethanol (95–100% molecular biology grade).

8. DNase/RNase-free distilled water, Milli-Q water or equivalent.

9. Optional: Stabilization and drying solution (Agilent), acetonitrile, sulfolane.

10. Microarray scanner (Agilent SureScan D or other compatible) (Note, the microarray slides can be read by compatible scanners. The compatibility and quality of reads are limited by the scanner specification.)

11. Hybridization chamber (Agilent).

12. Hybridization chamber gasket slides (Agilent).

13. Hybridization oven (65 °C) (Agilent).

14. Circulating water baths or heat blocks set to 37, 40, 60, 65, 70, and 80 °C.

15. Optional: Ozone-barrier slide cover (Agilent).

16. Feature extraction software 10.7.1 or later (Agilent), Agilent Scan Control software.

17. Optional: GeneSpring GX 9.0 software or higher (Agilent).

3   Methods

3.1   Isolation of RNA

3.1.1   Sample Preparation

Harvest mycelia from the studied cultures. Wash them briefly with ice-cold water and then dry them between filter papers if necessary. (The intact mycelia can be stored at –70 °C for several days.) Lyophilize the frozen mycelia in Eppendorf tube and use sterile toothpick to make fine powder. (Alternatively, grind the harvested mycelia into fine powder in liquid nitrogen with pestle and mortar.)

3.1.2   Phase Separation

1. Add 1 mL TRIzol/TRI reagent to approximately 100 mg mycelial powder (*see* **Notes 1–4**). Using mycelial powder which fills half portion of the conical part of a 1.5 Eppendorf tube is generally good. However, the more mycelia is used, the more chance is to get low-quality RNA.

2. Mix it thoroughly with a toothpick and let it stand for 5 min at room temperature (*see* **Note 5**). (Samples at this stage can be stored at –70 °C for weeks.)

3. Centrifuge out the cell debris from the samples ($12,000 \times g$, 10 min, 4 °C) and transfer the supernatant into a fresh Eppendorf tube. Note the volume of the transferred supernatant (*see* **Note 6**).

4. Add 0.2× volume chloroform to the samples (*see* **Note 7**). Mix them intensively for 15 s and allow standing for 10–15 min at room temperature.

5. Centrifuge the samples ($12,000 \times g$, 10 min, 4 °C) and transfer the colorless upper phase, containing the RNA, into a fresh Eppendorf tube (*see* **Note 8**).

3.1.3   Precipitation of RNA

1. Add 0.5 mL 1× SSC and 0.5 mL 2-propanol to the samples. Mix them by inversion and incubate them at room temperature for 10–15 min or longer (*see* **Note 9**). Spin down the precipitated RNA ($12,000 \times g$, 10 min, 4–20 °C).

2. Remove the supernatant by pipetting or simply pour it out. (Be sure that RNA pellet is stuck to the wall of the Eppendorf tube.)

3. Add 1 mL 70 v/v % ethanol to the tubes. Vortex it until the pellet is removed from the wall of the tube. Spin down the pellet ($12,000 \times g$ 10 min 4 °C).

4. Repeat washing with 70 v/v % ethanol to remove all the salts and isopropanol from the samples.

5. Remove all the supernatant by pipetting or with vacuum and dry the tube and the pellet gently. (Avoid overdrying.) You can continue the process with LiCl precipitation [55] to enrich mRNA. However, total RNA is suitable for transcriptome analysis.

3.1.4   Quantification and Quality Control of RNA

1. Dissolve the pellet with 10–50 μL DEPC-treated or nuclease-free water. Incubate the samples at 68 °C for 5 min. (Always check that the pellet has completely dissolved. If not, continue to dissolve it by pipetting and/or add more water and/or incubate them for another 5 min at 68 °C.) Allow the samples standing for 30 min at room temperature.

2. Use a NanoDrop spectrophotometer (Thermos Scientific) or equivalent for quantification of RNA samples. For quality control, Agilent 2100 Bioanalyzer (Agilent Technologies) or equivalents are recommended. (Checking the quality of RNA by TAE/formamide agarose electrophoresis is possible but not always reliable when RNA is used for DNA microarray.) RNA samples can be stored at –70 °C for weeks.

3.1.5 DNase Treatment   RNA samples always contain some DNA and, therefore, the DNase treatment steps cannot be omitted. Any kits developed for DNase treatment of RNA samples are suitable for this step. If you do not wish to use kits:

1. Mix appropriate amount of RNA sample, DEPC-treated or nuclease-free water (if necessary), 1 μL 10× reaction solution (*see* **Note 10**), and 1 unit DNase I (*see* **Note 11**) (in this order). The final volume should be 10 μL and the final RNA concentration should be less than 200 ng/μL.

2. Incubate the samples at room temperature (or at 30–37 °C) for 10–15 min.

3. Add 1 μL stop solution (*see* **Note 12**), mix thoroughly, incubate at 65–70 °C for 10 min, and chill on ice.

3.2 The Microarray Procedure   Before starting read carefully all procedural descriptions (follow the manufacturer's instruction). Use a dedicated/separated area for RNA related microarray experiment procedure. Make decontamination of the work area prior to and after the workflow application. Always use reagents equilibrated on the room temperature. To prevent the contamination of the reagents and samples by nucleases always wear powder-free laboratory gloves and use dedicated solutions and pipettors with nuclease-free aerosol-resistant tips. Prior to the pipetting always mix samples and reagents on vortex mixer, and then spin down in a centrifuge for 5–10 s, to collect all material from the lid and from the tube wall, to avoid loss of material and contamination by opening. Follow—at least— Biosafety Level (BSL1) safety rules, use at least BSL1 hood for sample handling. Wear appropriate personal protective equipment as Cyanine dye is potential carcinogenic, LiCl is toxic and potential teratogen, lithium dodecyl sulfate and triton are irritating.

The following protocol contains procedure for 25 ng total RNA as starting material. This amount is ideal for four-pack and eight-pack microarray slide formats. (50 ng total RNA is necessary for one-pack and for two-pack microarray slide formats. The amount of reagents should be adapted to the amount of starting material and the microarray format, according to the manufacturer protocol.) This example uses four-pack microarray slide. The workflow is summarized in Fig. 1.

3.2.1 Sample Preparation   The Agilent one-color microarray-based gene expression analysis uses Cy3-labeled targets, to measure gene expression in experimental and control samples. For optimal performance, use high quality, intact RNA template (total or poly A+ RNA) (*see* **Notes 13–15**).

Check the RNA purity by A260/A230 ratio measurement. Determine the RNA integrity (RIN) with Agilent 2100 Bioanalyzer. The ideal RIN value for the microarray experiment is 9-10.

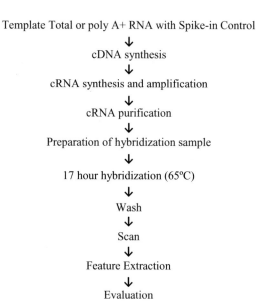

Template Total or poly A+ RNA with Spike-in Control
↓
cDNA synthesis
↓
cRNA synthesis and amplification
↓
cRNA purification
↓
Preparation of hybridization sample
↓
17 hour hybridization (65°C)
↓
Wash
↓
Scan
↓
Feature Extraction
↓
Evaluation

Fig. 1 Workflow of the microarray procedure

Application of not pure or degraded RNA could lead to poor results. If necessary, additional RNA purification steps or new RNA isolation must be performed.

The one color LIQA—low input quick amp—labeling kit generates fluorescent cRNA (complementary RNA) (*see* **Note 16**). The sample input range for one-color processing is 10–200 ng of total RNA, or minimum 5 ng of poly A+ RNA. The method uses T7 RNA polymerase blend, which simultaneously amplifies target material and incorporates Cy3-CTP.

3.2.2 Spike-In Control: Internal Control (~0.5 h)

The internal control kit—a set of artificial transcripts, with no complimentary sequences to the biological samples—is used for monitoring the microarray workflow from the sample amplification and labeling to the microarray processing. The application of the internal control kit is essential for optimizing and troubleshooting the microarray experiment. The diluted RNA controls are spiked directly into the RNA sample prior to amplification and labeling. The final amount of the spiked internal control depends on the amount of the RNA starting material.

Preparation of spike-in control work solution for 25 ng of total RNA starting sample:

1. Vortex vigorously spike-in control stock solution, heat at 37 °C for 5 min, and vortex again.

2. Spin down. Prepare a serial dilution in the following steps: 1:20, 1:25, and 1:4. Mix well and spin down during dilution steps.

1. Add 25 ng of total RNA to a 1.5 mL microcentrifuge tube in a final volume of 1.5 μL.

2. Add 2 μL of diluted spike-in control work solution to each tube.

3. Add 1.8 μL of T7 primer mix into each tube.

4. Denature the template, control, and primer mix at 65 °C in a circulating water bath for 10 min.

5. Place the tubes on ice and incubate for 5 min, then spin them down.

6. Add 4.7 μL of cDNA Master Mix to each sample tube (→cDNA), and mix by pipetting up and down.

7. Incubate samples at 40 °C in circulating water bath for 2 h.

8. Move samples to a 70 °C circulating water bath and incubate for 15 min.

9. Move samples to ice, incubate for 5 min, and spin them down.

10. Add 6 μL of Transcription Master Mix to each tube (→cRNA). Gently mix by pipetting.

11. Incubate samples in a circulating water bath at 40 °C for 2 h.

Use RNeasy Mini Kit to purify the amplified cRNA samples.

1. Add 84 μL of nuclease-free water to the cRNA sample (total volume 10 μL).

2. Add 350 μL of Buffer RLT and mix well by pipetting.

3. Add 250 μL of ethanol (96–100 % purity) and mix thoroughly by pipetting. Do not spin in a centrifuge.

4. Transfer the 700 μL of the cRNA sample to an RNeasy Mini Spin Column in a Collection Tube. Spin the sample in a centrifuge at 4 °C for 30 s at $10,000 \times g$. Discard the flow-through and collection tube.

5. Transfer the RNeasy column to a new collection tube (2 mL) and add 500 μL of Buffer RPE (containing ethanol) to the column. Spin the sample in a centrifuge at 4 °C for 30 s at $10,000 \times g$. Discard the flow-through. Reuse the collection tube.

6. Add another 500 μL of Buffer RPE to the column. Centrifuge the sample at 4 °C for 60 s at $10,000 \times g$. Discard the flow-through and the collection tube.

7. Elute the purified cRNA sample by transferring the RNeasy column to a new collection tube (1.5 mL). Add 30 μL RNase-free water directly onto the RNeasy filter membrane. Wait for 60 s, and then centrifuge at 4 °C for 30 s at $10,000 \times g$.

8. Maintain the cRNA sample-containing flow-through on ice. Discard the RNeasy column.

3.2.5 cRNA Quantification

Use the NanoDrop ND-1000 UV–VIS Spectrophotometer version 3.2.1 (or higher) to quantify the cRNA. Record the following results:

1. Cyanine 3 dye concentration (pmol/mL).
2. RNA absorbance ratio (260 nm/280 nm).
3. cRNA concentration (ng/μL).
4. Determine the cRNA yield (μg).
5. Determine the specific activity (pmol Cy3/μg cRNA).

The yield and specific activity should reach a recommended value for successful hybridization. If the specific activity does not meet the requirements, repeat the cRNA preparation. The required value differs by different array formats.

Recommended value for the four-pack microarray in this example:

Yield: 5 μg; specific activity: 6 pmol Cy3/μg cRNA.

3.2.6 Hybridization (18 h)

Video: http://www.genomics.agilent.com/article.jsp?pageId=1200039.

1. Prepare the 10× Blocking Agent work solution.
2. Add 500 μL of nuclease-free water to the vial containing lyophilized 10× Gene Expression Blocking Agent supplied with the Gene Expression Hybridization Kit.
3. Gently vortex. To resuspend the pellet, if necessary, heat the mix for 5 min at 37 °C.
4. Collect all material from the tube walls or from cap by spinning down for 10 s.

Prepare hybridization of samples (amounts for a four-pack microarray slide/array).

5. Add to a 1.5 mL nuclease-free microfuge tube 1.65 μg Cy3-labeled linearly amplified cRNA.
6. Add 11 μL 10× Gene Expression Blocking Agent work solution.
7. Add nuclease-free water—bring volume to 52.8 μL.
8. Add 25× fragmentation buffer (provided with the Hybridization Reagent kit).
9. Incubate at 60 °C for *exactly* 30 min to fragment RNA.
10. Immediatley cool on ice for 1 min.
11. Add 55 μL 2× Hi-RPM Hybridization Buffer to stop the fragmentation reaction.
12. Mix well by careful pipetting part way up and down. Do not introduce bubbles to the mix. The surfactant in the 2× Hi-RPM Hybridization Buffer easily forms bubbles. Do not vortex!

13. Spin for 1 min at room temperature at 13,000 rpm in a micro-centrifuge to collect the sample from the wall and lid. Use it immediately; do not store!

14. Place sample on ice and load onto the array as soon as possible.

Prepare the hybridization assembly, video: http://www.genomics.agilent.com/article.jsp?pageId=1200043

15. Position the slides so that the barcode label is to your left.

16. Load the samples left to right. The output files will come out in that same order.

17. Load a clean gasket slide into the Agilent SureHyb chamber base with the label facing up and aligned with the rectangular section of the chamber base. Make sure that the gasket slide is flush with the chamber base and is not ajar.

18. Slowly dispense 100 μL of prepared hybridization sample (from total volume of 110 μL) onto the gasket well (use four-well backing gasket slide for four-pack microarray slide). Fill four prepared samples into the four gasket wells.

19. Grip the slide on either end and slowly put the slide "active side" (where microarray probes are allocated) down, parallel to the SureHyb gasket slide, so that the "Agilent"-labeled barcode is facing down and the numeric barcode is facing up. Make sure that the sandwich pair is properly aligned.

20. Place the SureHyb chamber cover onto the sandwiched slides and slide the clamp assembly onto both pieces.

21. Firmly hand-tighten the clamp onto the chamber.

22. Vertically rotate the assembled chamber to wet the gasket and assess the mobility of the bubbles. If necessary, tap the assembly on a hard surface to move stationary bubbles.

3.2.7 Hybridization of Microarray

1. Load each assembled chamber into the oven rotator rack. Start from the center of the rack (position 3 or 4 when counting from the left). Set your hybridization rotator to rotate at 10 rpm when using 2× Hi-RPM Hybridization Buffer.

2. Hybridize at 65 °C for 17 h.

3.2.8 Microarray Wash

Video: http://www.genomics.agilent.com/article.jsp?pageId=1200045.

1. Add Triton X-102 to Gene Expression Wash Buffer 1 and Wash Buffer 2.

2. Dispense 1000 mL of Gene Expression Wash Buffer 2 directly into a sterile 1000 mL bottle.

3. Tightly cap the bottle and place in a 37 °C water bath the night before washing arrays.

4. Wash staining dishes, racks, and stir bars with acetonitrile or isopropyl alcohol to avoid wash artifacts on your slides and images. Conduct solvent washes in a vented fume hood.

5. Add the slide rack and stir bar to the staining dish.

6. Transfer the staining dish with the slide rack and stir bar to a magnetic stir plate.

7. Fill the staining dish with 100 % acetonitrile or isopropyl alcohol (*see* **Note 17**).

8. Turn on the magnetic stir plate and adjust the speed to a medium speed.

9. Wash for 5 min.

10. Discard the solvent as is appropriate for your site.

11. Repeat the washing process.

12. Wash dishes, racks, and stir bars with Milli-Q water. Rinse all components five times to remove any traces of contaminating material. (Do not use any detergent!)

13. Discard the used Milli-Q water.

3.2.9 Washing the Microarray Slides

1. Completely fill slide-staining dish #1 with Gene Expression Wash Buffer 1 at room temperature.

2. Place a slide rack into slide-staining dish #2. Add a magnetic stir bar. Fill slide-staining dish #2 with enough Gene Expression Wash Buffer 1 at room temperature to cover the slide rack. Place this dish on a magnetic stir plate.

3. Place the empty dish #3 on the stir plate and add a magnetic stir bar. Do not add the pre-warmed (37 °C) Gene Expression Wash Buffer 2 until the first wash step has begun.

4. Remove one hybridization chamber from incubator and record time. Record whether bubbles formed during hybridization and if all bubbles are rotating freely.

5. Place the hybridization chamber assembly on a flat surface and loosen the thumbscrew, turning counterclockwise.

6. Slide off the clamp assembly and remove the chamber cover.

7. With gloved fingers, remove the array-gasket sandwich from the chamber base by grabbing the slides from their ends. Keep the microarray slide numeric barcode facing up as you quickly transfer the sandwich to slide-staining dish #1.

8. Without letting go of the slides, submerge the array-gasket sandwich into slide-staining dish #1 containing Gene Expression Wash Buffer 1.

9. With the sandwich completely submerged in Gene Expression Wash Buffer 1, pry the sandwich open from the barcode end only, slip one of the blunt ends of the forceps between the slides.

10. Gently turn the forceps upwards or downwards to separate the slides.

11. Let the gasket slide drop to the bottom of the staining dish.

12. Grasp the top corner of the microarray slide, remove the slide, and then put it into the slide rack in the slide-staining dish #2 that contains Gene Expression Wash Buffer 1 at room temperature. Transfer the slide quickly so avoid premature drying of the slides. Touch only the barcode portion of the microarray slide or its edges!

13. When all slides are placed into the slide rack in slide-staining dish #2, stir using setting medium for 1 min.

14. During this wash step, remove Gene Expression Wash Buffer 2 from the 37 °C water bath and pour into the slide-staining dish #3.

15. Transfer slide rack to slide-staining dish #3 containing Gene Expression Wash Buffer 2 at elevated temperature. Stir using setting 4 for 1 min.

16. Slowly remove the slide rack minimizing droplets on the slides. It should take 5–10 s to remove the slide rack. If liquid remains on the bottom edge of the slide, dab it on a cleaning tissue.

17. Discard used Gene Expression Wash Buffer 1 and Gene Expression Wash Buffer 2.

18. Put the slides in a slide holder (Agilent SureScan Scanner compatible).

19. Carefully place the end of the slide without the barcode label onto the slide ledge.

20. Gently lower the microarray slide into the slide holder. Make sure that the active microarray surface faces up, toward the slide cover.

21. Close the plastic slide cover, pushing on the tab end until you hear it click.

22. Scan slides immediately to minimize the impact of environmental oxidants on signal intensities. If necessary, store slides in slide boxes in a nitrogen purge box, in the dark.

23. Video: http://www.genomics.agilent.com/article.jsp?pageId= 1200047.

3.3 Processing and Evaluation of Data

3.3.1 Scanning and Feature Extraction

The optimal scanning and row data generation for the Agilent one-color gene expression microarray, demonstrated in this protocol, can be performed on the Agilent microarray scanner. The Agilent microarray slide is readable with other compatible non-Agilent scanners. The compatibility list is available in microarray manufacturer's protocol.

Feature extraction is the process by which information from probe features is extracted from microarray scan data, allowing researchers to measure gene expression in their experiments. The Feature Extraction (FE) software, built in the Agilent scanner software, automatically finds spots for Agilent microarrays. The software finds and places microarray grids, excludes outlier pixels, determines feature intensities and ratios, and calculates statistical confidences and provides application-specific QC reports with metrics, targeted to the experiment. The Feature Extraction software is compatible with some non-Agilent scanners (with selected models of TECAN, Innopsys, Molecular Devices scanners).

1. Put the assembled slide holders into the scanner cassette.

2. After the setting up of scanner parameters, according to the instrument user guide, select the scanner protocol, appropriate to the current experiment, in this case 4×44 K Agilent one-color gene expression. Start the scan.

3. After generating the microarray scan images, extract .tif images using the Feature Extraction software. Check and evaluate the automatic QC report, prior to the bioinformatics analysis of the results.

4. Normalizing one-color microarray data. When comparing data across a set of one-color microarrays, a simple linear scaling of the data is usually sufficient for most experimental applications. In the case of Agilent microarray slides and read by Agilent scanner the signal value of the 75th percentile of all of non-control probes on the microarray is a more robust and representative value of the overall microarray signal as compared to the median or 50th percentile signal. Therefore, use the 75th percentile signal value to normalize Agilent one-color microarray signals for inter-array comparisons. Other manufacturers can determine other normalization rules.

The optimal evaluation of Agilent gene expression microarray data can be performed by Agilent GeneSpring GX bioinformatics software. The GeneSpring GX has easy to use wizard driven workflows. It is intuitive and provides interactive visualization for built-in pathway analysis. GeneSpring GX provides powerful statistical tools to put your multi-omic data into a biological context. Several other bioinformatics software are available free of charge or for fee. The minimum requirement for an appropriate software functions or components are:

- Probe- or gene-level expression analysis on all major microarray platforms, including Agilent, Affymetrix, and Illumina.

- microRNA analysis and identification of gene targets.

- The ability to do correlative analysis on mRNA expression (and miRNA) data (and splicing, QPCR, GWAS, CNV).

- Exon splicing analysis using t-tests or multivariate splicing ANOVA and filtering for transcripts on splicing index.

- NCBI Gene Expression Omnibus Importer tool for expression datasets.

- Hierarchical clustering to visualize individual samples with their metadata information.

- Built-in pathway analysis module to promote investigation and to enable understanding of data within a biological context.

eSeminars: http://www.genomics.agilent.com/article.jsp?pageId= 1500002.

3.3.2 Processing and Evaluation of Pre-normalized Data

In most cases, pre-normalized microarray data need further normalization. Several effective and reliable methods and software are available to perform normalization steps, and the following example represents only one possibility.

- In our research team, background correction of pre-normalized microarray data is typically carried out with the normexp + offset method as suggested by [56] using the implementation described by [57], which is followed by quantile normalization between arrays [58] according to Smyth [57] using the Limma software package.

- After normalization, gene functions are extracted from the available fungal genome databases, e.g., from Aspergillus Genome Database (AspGD, http://www.aspergillusgenome.org/), the Broad Institute *Aspergillus* Comparative Database (http://www.broadinstitute.org/annotation/genome/aspergillus_group/MultiHome.html), or The Central *Aspergillus* REsource (CADRE, http://www.cadre-genomes.org.uk/index.html) for the aspergilli or from the *Fusarium* Comparative Database for the fusaria, which is also available at the Broad Institute (http://www.broadinstitute.org/annotation/genome/fusarium_group/MultiHome.html).

- To identify elements of secondary metabolite gene clusters (Fig. 2), a wealth of excellent literature is now available [38–43]. To validate gene expression data of interest, Northern blot or RT-PCR techniques are routinely used [2, 19, 59].

4   Notes

1. Wear disposable gloves. Medical gloves are good enough, however, RNase-free gloves (gloves certified to be free of contamination) are also available. A clean work place is also a basic requirement. Use special surfactants to removes RNA and RNases if the quality of isolated RNA is repeatedly low.

2. Use sterile pipette tips, Eppendorf tubes, etc.

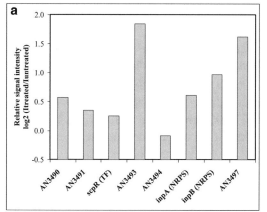
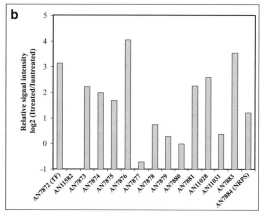
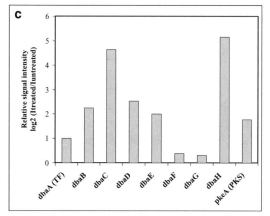
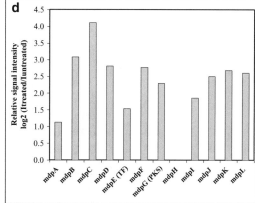

Fig. 2 Transcriptional changes recorded in selected secondary metabolite gene clusters [41] initiated by various stress conditions and/or the deletion of the *atfA* gene encoding the stress-response regulator bZip-type transcription factor AtfA in *Aspergillus nidulans*. DNA microarray data were taken from the publication of Emri et al. [2]. In this transcriptomics study, Agilent 60-mer oligonucleotide high-density arrays (4 × 44 K array format) were used [2]. Genes coding for pathway (cluster) specific transcription factors (TFs) or key biosynthetic enzymes (e.g., NRPS—non-ribosomal peptide synthase, PKS—polyketide synthase) are indicated. (**a**) Effect of 75 mM H$_2$O$_2$ on the transcription of the *inp* cluster in *A. nidulans* wild-type strain. Only few genes but no key genes (TF, NRPS) were up-regulated by H$_2$O$_2$. (**b**) Effect of 0.8 mM menadione sodium bisulfite on the gene expressions in the *AN7884* cluster in the wild-type strain. Note that the *AN11582* gene was not available on the microarray used in these experiments. Many genes including the key genes (TF, NRPS) were up-regulated by oxidative stress. (**c**) Effect of *atfA* gene deletion on the transcription of the *dba* and *F9775* hybrid cluster genes under unstressed conditions. Many genes including the key genes (TF, PKS) were up-regulated by the gene deletion. (**d**) Effect of 5 mM H$_2$O$_2$ stress treatment on the gene expression pattern in the *mdp* cluster in the *ΔatfA* gene deletion strain. Please note that the *mdpH* gene was not available on the chip microarray used in these experiments. All genes in the cluster including the key genes (TF, PKS) were induced

3. Work with phenol (TRI/TRIzol) and chloroform under the hood.

4. Addition of ribonuclease inhibitors (e.g., RNaseOUT, Invitrogen; ProtectRNA, Sigma-Aldrich) may improve the quality of RNA. Check the pH of TRIzol/TRI when using old reagents. It should be acidic.

5. Do not reduce the incubation time (dissociation of nucleoprotein complexes need time).

6. Do not try to transfer all the supernatant. Smaller but cleaner volume is better. If you work with several samples it is convenient to transfer the same volumes (e.g., 0.6 mL).

7. As an example: if the volume of the sample is 600 mL, 0.2× volume means 0.2×600 mL = 120 mL. If 1-bromo-3-chloropropane is used instead of chloroform use only 0.1× volume of it.

8. After centrifugation you will got three phases: a colorless aqueous upper phase, a white interphase and a red organic phase. Avoid aspirating the interphase and again, do not try to remove all the upper phase. Smaller, but cleaner volume is better. The interphase and the organic phase can be used for subsequent isolation of proteins and DNA [53].

9. Increase the volume of 2-propanol and 1× SSC in proportion if you started the RNA preparation with more than 1 mL TRIzol/TRI reagent. Do not cool the samples during precipitation, increase the incubation time instead if necessary. (Low temperature enhances the precipitation of salts.) Addition of 0.5 mL 1× SSC to the samples can be omitted; however it may decrease the yield.

10. 10× Reaction solution contains 20–25 mM $MgCl_2$ with 0–5 mM $MgCl_2$ in a 100 mM Tris/HCl buffer (pH 7.5–8.3), e.g., 100 mM Tris/HCl (pH 7.5) with 25 mM $MgCl_2$. Note that DEPC react easily with Tris and, therefore, DEPC treatment of the prepared Tris buffers is not efficient. Use molecular biology-grade powders and nuclease-free or DEPC-treated water for making RNase-free 10x reaction solution, instead.

11. Use RNase-free DNase only.

12. Stop solution is an 50 mM Na_2-EDTA solution (RNase free). Do not heat the samples before adding the stop solution (free Ca^{2+} and Mg^{2+} may induce RNA hydrolysis).

13. RNA is sensitive to the oxidative degradation, do not let evaporate the RNA samples.

14. RNA is sensitive to the RNase degradation effect, ensure the appropriate sample protection conditions.

15. RNA is sensitive to the heat degradation; store the samples on −80 °C in aliquots, and thaw the samples on ice, without direct heating.

16. Fluorescent dyes are photo-degradable; store dye-containing reagents in the dark until use.

17. Acetonitrile and Agilent Stabilization and Drying Solution are flammable and toxic; they must be used in a fume hood.

Acknowledgements

This work was supported by the Hungarian Scientific Research Fund(OTKAK100464,K112181)andbytheSROP-4.2.2.B-15/1/ KONV-2015-0001 project. The project has been supported by the European Union, cofinanced by the European Social Fund.

References

1. Breakspear A, Momany M (2007) The first fifty microarray studies in filamentous fungi. Microbiology 153:7–15

2. Emri T, Szarvas V, Orosz E et al (2015) Core oxidative stress response in *Aspergillus nidulans*. BMC Genomics 16:478–496

3. Aguilar-Pontes MV, de Vries RP, Zhou M (2014) (Post-)genomics approaches in fungal research. Brief Funct Genomics 13:424–439

4. Price MS, Conners SB, Tachdjian S et al (2005) Aflatoxin conducive and non-conducive growth conditions reveal new gene associations with aflatoxin production. Fungal Genet Biol 42:506–518

5. Price MS, Yu J, Nierman WC et al (2006) The aflatoxin pathway regulator AflR induces gene transcription inside and outside of the aflatoxin biosynthetic cluster. FEMS Microbiol Lett 255:275–279

6. Bok JW, Hoffmeister D, Maggio-Hall LA et al (2006) Genomic mining for Aspergillus natural products. Chem Biol 13:31–37

7. Pirttilä AM, McIntyre LM, Payne GA et al (2004) Expression profile analysis of wild-type and fcc1 mutant strains of Fusarium verticillioides during fumonisin biosynthesis. Fungal Genet Biol 41:647–656

8. Lee S, Son H, Lee J et al (2011) A putative ABC transporter gene, *ZRA1*, is required for zearalenone production in *Gibberella zeae*. Curr Genet 57:343–351

9. Brown DW, Butchko RA, Busman M et al (2012) Identification of gene clusters associated with fusaric acid, fusarin, and perithecial pigment production in *Fusarium verticillioides*. Fungal Genet Biol 49:521–532

10. Wiemann P, Sieber CM, von Bargen KW et al (2013) Deciphering the cryptic genome: genome-wide analyses of the rice pathogen Fusarium fujikuroi reveal complex regulation of secondary metabolism and novel metabolites. PLoS Pathog 9:e1003475

11. Cary JW, OBrian GR, Nielsen DM et al (2007) Elucidation of *veA*-dependent genes associated with aflatoxin and sclerotial production in *Aspergillus flavus* by functional genomics. Appl Microbiol Biotechnol 76:1107–1118

12. Perrin RM, Fedorova ND, Bok JW et al (2007) Transcriptional regulation of chemical diversity in Aspergillus fumigatus by LaeA. PLoS Pathog 3:e50

13. Seong KY, Pasquali M, Zhou X et al (2009) Global gene regulation by *Fusarium* transcription factors *Tri6* and *Tri10* reveals adaptations for toxin biosynthesis. Mol Microbiol 72:354–367

14. Wiemann P, Brown DW, Kleigrewe K et al (2010) FfVel1 and FfLae1, components of a velvet-like complex in *Fusarium fujikuroi*, affect differentiation, secondary metabolism and virulence. Mol Microbiol 77:972–994

15. Lee J, Myong K, Kim JE et al (2012) FgVelB globally regulates sexual reproduction, mycotoxin production and pathogenicity in the cereal pathogen *Fusarium graminearum*. Microbiology 158:1723–1733

16. Studt L, Schmidt FJ, Jahn L et al (2013) Two histone deacetylases, FfHda1 and FfHda2, are important for *Fusarium fujikuroi* secondary metabolism and virulence. Appl Environ Microbiol 79:7719–7734

17. Brown DW, Busman M, Proctor RH (2014) *Fusarium verticillioides SGE1* is required for full virulence and regulates expression of protein effector and secondary metabolite biosynthetic genes. Mol Plant Microbe Interact 27:809–823

18. OBrian GR, Fakhoury AM, Payne GA (2003) Identification of genes differentially expressed during aflatoxin biosynthesis in *Aspergillus flavus* and *Aspergillus parasiticus*. Fungal Genet Biol 39:118–127

19. Pócsi I, Miskei M, Karányi Z et al (2005) Comparison of gene expression signatures of diamide, H_2O_2 and menadione exposed *Aspergillus nidulans* cultures--linking genome-wide transcriptional changes to cellular physiology. BMC Genomics 6:182

20. Wilkinson JR, Yu J, Bland JM et al (2007) Amino acid supplementation reveals differential regulation of aflatoxin biosynthesis in *Aspergillus flavus* NRRL 3357 and *Aspergillus parasiticus* SRRC 143. Appl Microbiol Biotechnol 74:1308–1319

21. Wilkinson JR, Yu J, Abbas HK et al (2007) Aflatoxin formation and gene expression in response to carbon source media shift in *Aspergillus parasiticus*. Food Addit Contam 24:1051–1060

22. Wilkinson JR, Kale SP, Bhatnagar D et al (2011) Expression profiling of non-aflatoxigenic Aspergillus parasiticus mutants obtained by 5-azacytosine treatment or serial mycelial transfer. Toxins (Basel) 3:932–948

23. Yu J, Ronning CM, Wilkinson JR et al (2007) Gene profiling for studying the mechanism of aflatoxin biosynthesis in Aspergillus flavus and A. parasiticus. Food Addit Contam 24:1035–1042

24. Kim JH, Yu J, Mahoney N et al (2008) Elucidation of the functional genomics of antioxidant-based inhibition of aflatoxin biosynthesis. Int J Food Microbiol 122:49–60

25. Schmidt-Heydt M, Magan N, Geisen R (2008) Stress induction of mycotoxin biosynthesis genes by abiotic factors. FEMS Microbiol Lett 284:142–149

26. Schmidt-Heydt M, Abdel-Hadi A, Magan N et al (2009) Complex regulation of the aflatoxin biosynthesis gene cluster of Aspergillus flavus in relation to various combinations of water activity and temperature. Int J Food Microbiol 135:231–237

27. Schmidt-Heydt M, Parra R, Geisen R et al (2011) Modelling the relationship between environmental factors, transcriptional genes and deoxynivalenol mycotoxin production by strains of two Fusarium species. J R Soc Interface 8:117–126

28. Schroeckh V, Scherlach K, Nützmann HW et al (2009) Intimate bacterial-fungal interaction triggers biosynthesis of archetypal polyketides in *Aspergillus nidulans*. Proc Natl Acad Sci U S A 106:14558–14563

29. Medina A, Schmidt-Heydt M, Cárdenas-Chávez DL et al (2013) Integrating toxin gene expression, growth and fumonisin B1 and B2 production by a strain of *Fusarium verticillioides* under different environmental factors. J R Soc Interface 10:20130320

30. Edlayne G, Simone A, Felicio JD (2009) Chemical and biological approaches for mycotoxin control: a review. Recent Pat Food Nutr Agric 1:155–161

31. Reverberi M, Ricelli A, Zjalic S et al (2010) Natural functions of mycotoxins and control of their biosynthesis in fungi. Appl Microbiol Biotechnol 87:899–911

32. Atanda SA, Aina JA, Agoda SA et al (2012) Mycotoxin management in agriculture: a review. J Anim Sci Adv 2:250–260

33. de Medeiros FHV, Martins SJ, Zucchi TD et al (2012) Biological control of mycotoxin-producing molds. Ciênc Agrotec Lavras 36:483–497

34. Tsitsigiannis DI, Dimakopoulou M, Antoniou PP et al (2012) Biological control strategies of mycotoxigenic fungi and associated mycotoxins in Mediterranean basin crops. Phytopathol Mediterr 51:158–174

35. Marroquín-Cardona AG, Johnson NM, Phillips TD et al (2014) Mycotoxins in a changing global environment--a review. Food Chem Toxicol 69:220–230

36. Oliveira PM, Zannini E, Arendt EK (2014) Cereal fungal infection, mycotoxins, and lactic acid bacteria mediated bioprotection: from crop farming to cereal products. Food Microbiol 37:78–95

37. Pfliegler WP, Pusztahelyi T, Pócsi I (2015) Mycotoxins - prevention and decontamination by yeasts. J Basic Microbiol 55(7):805–818. doi:10.1002/jobm.201400833

38. von Döhren H (2009) A survey of nonribosomal peptide synthetase (NRPS) genes in *Aspergillus nidulans*. Fungal Genet Biol 46:S45–S52

39. Nielsen ML, Nielsen JB, Rank C et al (2011) A genome-wide polyketide synthase deletion library uncovers novel genetic links to polyketides and meroterpenoids in *Aspergillus nidulans*. FEMS Microbiol Lett 321:157–166

40. Andersen MR, Nielsen JB, Klitgaard A et al (2013) Accurate prediction of secondary metabolite gene clusters in filamentous fungi. Proc Natl Acad Sci U S A 110:E99–E107

41. Inglis DO, Binkley J, Skrzypek MS et al (2013) Comprehensive annotation of secondary metabolite biosynthetic genes and gene clusters of *Aspergillus nidulans, A. fumigatus, A. niger* and *A. oryzae*. BMC Microbiol 13:91

42. Sieber CM, Lee W, Wong P et al (2014) The *Fusarium graminearum* genome reveals more secondary metabolite gene clusters and hints of horizontal gene transfer. PLoS One 9:e110311

43. Hansen FT, Gardiner DM, Lysøe E et al (2015) An update to polyketide synthase and non-ribosomal synthetase genes and nomenclature in *Fusarium*. Fungal Genet Biol 75C:20–29

44. Chiang YM, Szewczyk E, Nayak T et al (2008) Molecular genetic mining of the *Aspergillus* secondary metabolome: discovery of the emericellamide biosynthetic pathway. Chem Biol 15:527–532

45. Chiang YM, Szewczyk E, Davidson AD et al (2010) Characterization of the *Aspergillus nidulans* monodictyphenone gene cluster. Appl Environ Microbiol 76:2067–2074

46. Ahuja M, Chiang YM, Chang SL et al (2012) Illuminating the diversity of aromatic polyketide synthases in *Aspergillus nidulans*. J Am Chem Soc 134:8212–8221

47. Giles SS, Soukup AA, Lauer C et al (2011) Cryptic Aspergillus nidulans antimicrobials. Appl Environ Microbiol 77:3669–3675

48. Yaegashi J, Oakley BR, Wang CCC (2014) Recent advances in genome mining of secondary metabolite biosynthetic gene clusters and the development of heterologous expression systems in *Aspergillus nidulans*. J Ind Microbiol Biotechnol 41:433–442

49. Li Y, Wang W, Du X et al (2010) An improved RNA isolation method for filamentous fungus *Blakeslea trispora* rich in polysaccharides. Appl Biochem Biotechnol 160:322–327

50. Sallau AB, Henriquez F, Nok AJ et al (2013) *Aspergillus niger* - specific ribonucleic acid extraction method. J Yeast Fungal Res 4:58–62

51. Schumann U, Smith NA, Wang MB (2013) A fast and efficient method for preparation of high-quality RNA from fungal mycelia. BMC Res Notes 6:71

52. Chomczynski P, Sacchi N (1987) Single-step method of RNA isolation by acid guanidinium thiocyanate-phenol-chloroform extraction. Anal Biochem 162:156–159

53. Chomczynski P (1993) A reagent for the single-step simultaneous isolation of RNA, DNA and proteins from cell and tissue samples. Biotechniques 15, 532–534, 536–537

54. Chomczynski P, Sacchi N (2006) The single-step method of RNA isolation by acid guanidinium thiocyanate-phenol-chloroform extraction: twenty-something years on. Nat Protoc 1:581–585

55. Hayes A, Zhang N, Wu J et al (2002) Hybridization array technology coupled with chemostat culture: tools to interrogate gene expression in *Saccharomyces cerevisiae*. Methods 26:281–290

56. Ritchie ME, Silver J, Oshlack A et al (2007) A comparison of background correction methods for two-colour microarrays. Bioinformatics 23:2700–2707

57. Smyth GK (2005) Limma: linear models for microarray data. In: Gentleman R, Carey V, Dudoit S et al (eds) Bioinformatics and computational biology solutions using R and bioconductor. Springer, New York, pp 397–420

58. Bolstad BM, Irizarry RA, Astrand M et al (2003) A comparison of normalization methods for high density oligonucleotide array data based on bias and variance. Bioinformatics 19:185–193

59. Szilágyi M, Miskei M, Karányi Z et al (2013) Transcriptome changes initiated by carbon starvation in Aspergillus nidulans. Microbiology 159:176–190

Chapter 24

Mycotoxins: A Fungal Genomics Perspective

Daren W. Brown and Scott E. Baker

Abstract

The chemical and enzymatic diversity in the fungal kingdom is staggering. Large-scale fungal genome sequencing projects are generating a massive catalog of secondary metabolite biosynthetic genes and pathways. Fungal natural products are a boon and bane to man as valuable pharmaceuticals and harmful toxins. Understanding how these chemicals are synthesized will aid the development of new strategies to limit mycotoxin contamination of food and feeds as well as expand drug discovery programs. A survey of work focused on the fumonisin family of mycotoxins highlights technological advances and provides a blueprint for future studies of other fungal natural products. Expressed sequence tags led to the discovery of new fumonisin genes (*FUM*) and hinted at a role for alternatively spliced transcripts in regulation. Phylogenetic studies of *FUM* genes uncovered a complex evolutionary history of the *FUM* cluster, as well as fungi with the potential to synthesize fumonisin or fumonisin-like chemicals. The application of new technologies (e.g., CRISPR) could substantially impact future efforts to harness fungal resources.

Key words Genomics, *Fusarium verticillioides*, *Aspergillus niger*, Fumonisins, Expressed sequenced tags (ESTs), Horizontal gene transfer

1 Introduction

The chemical and enzymatic diversity inherent in the fungal kingdom is staggering. Mycologists have only characterized the tip of the iceberg with regard to fungal species and natural product chemists have only sampled the myriad of fungal metabolites produced. Metabolites, otherwise known as natural products (NPs), play important roles in the lives of fungi and man. Fungi produce NPs for communication, development, and defense and humans utilize these chemicals as pharmaceuticals and commodity chemicals. However, there are also negative impacts for humans due to fungal NPs. Mycotoxins are defined as NPs produced by fungi that are toxic or carcinogenic to humans and other animals. Mycotoxins and other NPs, referred to as secondary metabolites (e.g., not required for growth or reproduction), are often encoded by genes that are located adjacent to each other or clustered. Common types of NPs include polyketides, non-ribosomal peptides, and terpenes.

Antonio Moretti and Antonia Susca (eds.), *Mycotoxigenic Fungi: Methods and Protocols*, Methods in Molecular Biology, vol. 1542, DOI 10.1007/978-1-4939-6707-0_24, © Springer Science+Business Media LLC 2017

Within the fungal kingdom, filamentous fungi are particularly well known for the impressive diversity of NPs that they produce. In addition to mycotoxins, other NPs include pathogen virulence factors, cell communication molecules, antibiotics, and pharmaceuticals. In most cases, an NP or family of related metabolites produced by one or more species has been characterized at the structural level but not at the genetic level. In parallel, as the number of high-quality fungal genome sequences increases, the catalog of enzymes that are predicted to produce polyketides, non-ribosomal peptides or other NPs is expanding rapidly. The combination of genomic sequence and the development of molecular genetic tools applicable to many different filamentous fungi has made the task of assigning metabolites with genes and genes with metabolites in individual fungal species much more tractable.

Understanding how fungi synthesize NPs has been motivated by a need to stop their synthesis to limit the contamination of foods and feeds with mycotoxins and a need to discover new, pharmaceutically valuable chemicals. Annual worldwide economic losses due to mycotoxins are in billions of dollars [1, 2]. Modern interest in limiting mycotoxins stem from the dramatic death of poultry in England in 1960 from "Turkey-X" disease [3]. By 1963, aflatoxin was identified as the causative toxin, produced by the common soil fungi *Aspergillus flavus* and *A. parasiticus*. The first gene spanning 1.4 kilobases (kb) involved in aflatoxin synthesis was cloned in 1992 [4], a cluster of 25 co-regulated genes spanning 60 kb involved in the synthesis of a related toxin was described in 1996 [5] and the *Aspergillus flavus* genome sequence, spanning 36.8 megabases (Mb) with approximately 12,000 predicted genes, was released in 2005 [2], http://www.aspergillusflavus.org/genomics/ and National Center for Biotechnology Information (NCBI). The trajectory of research leading to the identity of genes involved in the synthesis of other mycotoxins, like fumonisins and ochratoxin, as well as the genome sequence of other mycotoxigenic fungi, has proceeded along a similar path. Although significant progress has been made towards understanding toxin biosynthesis, progress towards developing new methods to limit mycotoxin contamination of human food and animal feeds is hampered by the slow process of identifying target genes for further study.

2   Early Genomic Research

Analysis of the first fungal genome sequence led to the discovery that they contained more genes likely involved in NP synthesis than expected [6, 7]. Many predicted NP genes are clustered or located adjacent to each other in the genome. Each cluster contains a core biosynthetic gene, modifying genes, a transcriptional regulator or two, and a gene providing protection from the NP. The core gene may encode a polyketide synthase (PKS), a non-ribosomal peptide synthase (NRPS), a terpene cyclase (TC), or dimethylallyl transferase (DMAT) while modifying genes may encode methyltransferases, oxidases, dehydrogenases, reductases, or cyclases. The *Aspergillus flavus* genome is predicted to contain 55 gene clusters of which only seven have been associated with a likely metabolite [7]. Efforts to determine the function of the unknown clusters as well as clusters in other fungi have involved a variety of approaches including gene deletion and both homologous and heterologous gene expression. Since NPs likely play a role in fitness and multiple NPs may have overlapping effects that contribute to fitness, unveiling the role of a particular NP may require the creation of fungal mutants with multiple, core NP genes deleted [7]. Studies of gene expression across the whole genome by microarray over 28 diverse culture conditions led to the identification of four patterns of expression for the predicted core gene in each cluster [7]. The development of new technologies that would allow the creation of multiple, targeted gene mutations in a timely and effective manner would substantially impact our understanding of fungal NPs.

A first approach to understand what genes are involved in a fungal NP synthesis is to look at their differential expression. Adjacent genes that share a common pattern of expression may be involved in the synthesis of the same metabolite. A common technique, referred to as Northern analysis, involved separating total RNA by electrophoreses on agarose gels followed by transfer to a membrane, hybridization with a radiolabeled DNA probe, and exposure to film. A limitation to this approach was that the expression of only one gene could be interrogated at a time per blot. A major advance in studying gene expression took advantage of improvements in sequencing technology and involved determining the nucleotide sequence of a portion of cDNA created from RNA isolated from a single biological sample after growth under specific growth conditions. The first iteration of this technology involved generating hundreds to thousands of sequences (e.g., reads) from cDNA clones and was referred to as expressed sequence tags (ESTs). The second iteration generated millions of reads and was referred to as RNA-seq. Expression levels, with statistical support by RNA-seq, are assessed by simply counting the number of reads per gene, much more precise than comparing the intensity of bands on a photographic film in different lanes relative to the total amount of RNA loaded per lane. The sequence data also provides valuable information about gene structure without any *a priori* information about the gene. The value of EST data is exemplified from studies of fumonisin gene expression synthesis by the fungus *Fusarium verticillioides* [8, 9].

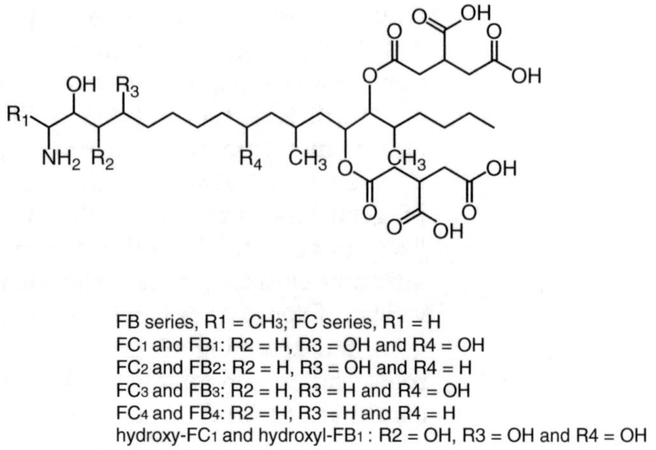

FB series, R1 = CH₃; FC series, R1 = H
FC1 and FB1: R2 = H, R3 = OH and R4 = OH
FC2 and FB2: R2 = H, R3 = OH and R4 = H
FC3 and FB3: R2 = H, R3 = H and R4 = OH
FC4 and FB4: R2 = H, R3 = H and R4 = H
hydroxy-FC1 and hydroxyl-FB1 : R2 = OH, R3 = OH and R4 = OH

Fig. 1 Structure of the polyketide-derived fumonisins

3 Fumonisin, a Case Study

Fumonisins are linear, polyketide-derived molecules with an amine, one to four hydroxyl, two methyl, and two tricarboxylic acid constituents, produced primarily by *Fusarium verticillioides* (Fig. 1). Fumonisins are common contaminants of maize and can cause multiple animal diseases, including cancer and neural tube defects in rodents [10, 11]. Consumption of fumonisin-contaminated maize is epidemiologically associated with esophageal cancer and neural tube defects in some human populations [12, 13]. Under most conditions, *F. verticillioides* infect and colonize maize without causing any symptoms. However, under some conditions, *F. verticillioides* infection can cause destructive disease at any life stage of the plant (e.g., root, stalk, and ear rot disease). Because maize is one of the world's most important food crops, and fumonisins are among the most common contaminants of maize worldwide, fumonisins are a significant safety concern to farmers, food producers, and regulatory agencies. Although resistance to insects by engineering maize to produce *Bacillus thuringiensis* (*Bt*) toxin has reduced fumonisin contamination, levels are not below recommended limits under all conditions [14]. Thus, additional strategies are needed to reduce fumonisin contamination of maize and the associated health risks to humans and other animals.

4 Using Gene Expression for Mycotoxin Characterization

Analysis of over 87,000 ESTs from 11 different *F. verticillioides* cDNA libraries identified over 700 ESTs that corresponded to genes in the fumonisin gene cluster [8]. A majority of the ESTs (586) were derived from libraries created from RNA extracted from

F. verticillioides mycelial after growth on a fumonisin production medium, GYAM. In the 24-h library, no *FUM* gene transcripts were detected while 233 transcripts were present in the combined 48- and 72-h library and 353 transcripts were present in the 96-h library (Table 1). Overall, *FUM* gene transcription increased 2.2-fold over time, consistent with previous transcriptional analysis by

Table 1
Description of *FUM* genes, total ESTs, and distribution of ESTs in selected cDNA libraries

Gene	Putative function	Predicted protein	Total # of ESTs	FvF (24 h)	FvM (48/72 h)	FvG (96 h)	Fold change in % FvM to FvG	Fold change in % NF3FvM to FvG
FUM21	C6 transcription factor	672	16	0	1 F?	10 8NF 2F?	11.0 ↑	
FUM1	Polyketide synthase	2586	41	0	1	11	12.0 ↑	
FUM6	P450 monooxygenase	1115	44	0	19	13	NC	
FUM7	Dehydrogenase	424	17	0	11	4	2.0 ↓	
FUM8	Aminotransferase	836	65	0	16	36	3.3 ↑	
FUM3	Dioxygenase	300	41	0	9	17	2.7 ↑	
FUM10	Fatty acyl-CoA synthetase	552	91	0	20	61	4.5 ↑	
FUM11	Tricarboxylate transporter	306	19	0	5 2NF	7 7NF	2.0 ↑	2.5 ↑
FUM12	P450 monooxygenase	502	85	0	17	64	5.4 ↑	
FUM20	Unknown	Na	1	0	0	1	1.0↑	
FUM13	Dehydrogenase/reductase	369	40	0	20	10	1.4 ↓	
FUM14	AA condensation domain	553	150	0	55 1NF	65 11NF	1.7 ↑	8.9 ↑
FUM15	P450 monooxygenase	596	21	0	8	10	1.8 ↑	
FUM16	Fatty acyl-CoA synthetase	676	90	0	46 10NF	37 24NF	1.1 ↑	2.7 ↑
FUM17	Longevity assurance factor	388	6	0	0	3		
FUM18	Longevity assurance factor	384	9	0	5 1NF	3 2NF	1.3 ↓	3.3 ↑
FUM19	ABC transporter	1489	1	0	0	1	1.0↑	
			737	0	233	353	2.2 ↑	4.4 ↑

NC no change, *Na* not available, and *NF* non-functional
F? functionality could not be determined with available sequence data
Bolded text highlight incease in NF transcripts overtime

Northern [15]. In addition to providing evidence for the differentially expression of the fumonisin genes, the EST collection enabled the discovery of two new *FUM* genes. The first, *FUM21*, encoded a predicted Zn(II)2Cys6 DNA-binding positive transcription factor [9]. The presence of eight introns in the gene is likely what prevented the initial identification of the ORF by BLAST analysis of genomic DNA. The second, *FUM20*, was defined by a single EST consisting of 680 nts with one intron from the 96-h library [8]. The role of *FUM20* remains unclear. BLAST analysis of the EST did not share any similarity with any previously described protein nor any DNA sequence. *FUM20* mutant strains synthesize wild-type levels of fumonisin under the growth conditions tested (unpublished). The *FUM20* transcript may be noncoding RNA. Comparative analysis of the genomic DNA located between *FUM2* and *FUM13* from multiple *Fusarium* did not identify any conserved ORF greater than 30 nucleotides (unpublished). Based on the observation that the EST overlaps *FUM2* transcript by up to 200 nts at the 5' end and likely includes a portion of the *FUM13* promoter at the 3' end, it may regulate *FUM2* or *FUM13* transcription by an unknown mechanism.

An observation we found most surprising was the number of ESTs that were presumably nonfunctional due to the presence of a stop codon in the predicted open reading frame (ORF). In every case, the presence of the stop codon in the transcript was due to the retention of an intron or the use of an alternative 3' splice border during intron excision. Alternative splicing has been extensively described in higher eukaryotes and is a process by which a single gene can code for multiple proteins. It is an essential process allowing for the production of many more proteins than expected from the number of genes in the human genome. Alternative splicing occurs during the processing of the messenger RNA generally when an exon is skipped and thus not included in the final mRNA. Upwards to 95 % of human genes with multiple exons are subject to alternative splicing of which a vast majority involve a skipped exon [16]. Other generally recognized alternative splicing modes are the use of a different 3' or 5' splice junction site (15 %) by the splicing complex or the intron may be simply retained (4 %).

Of the more than 700 *FUM* gene ESTs, we found 87 alternative splice forms (ASFs) that corresponded to 8 of the 16 *FUM* genes (Fig. 2) [8, 9]. The percent ASFs had a bimodal distribution with *FUM11*, *FUM16*, *FUM18*, and *FUM21* with 47 %, 51 %, 44 %, and 67 % based on 19, 90, 9, and 16 ESTs, respectively, and the percent ASFs for *FUM8*, *FUM2*, and *FUM14* with 3 %, 2 %, and 11 % based on 65, 85, and 134 ESTs, respectively. In contrast to what is observed in higher eukaryotes in which a different protein is encoded by the ASF, almost all of the fungal ASFs result in a truncated protein due to the introduction of a stop codon. The lone exception for the *FUM* genes ASFs was the retention of the

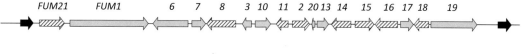

FUM21   FUM1   6   7   8   3 10 11 2 20 13 14   15   16 17 18   19

FUM genes with alternative splice forms

Fig. 2 Fumonisin genes with alternative splice forms in *Fusarium verticillioides*

third intron in *FUM15* ASFs which was in frame and did not include a stop codon.

The observation that the ASFs were differentially expressed over time suggest that they may serve a function [8, 9]. Over all, there were 4.4-fold more ASFs present in the 96-h culture than in the 48/72-h culture. The appearance of more ASFs in the older culture did not appear to be related to the age of the culture as we identified 29 other genes with ESTs present at 24, 48/72, and 96 h of which only 3 exhibited a similar pattern of expression to the *FUM* genes. Microarray analysis of four *FUM21* introns (introns 2, 3, 4, and 7) found that transcripts retaining the second intron decreased over time while transcripts retaining the seventh intron increased over time [9]. In order to test the hypothesis that truncated variants of the *FUM21* protein may serve a function, we created variants of the *FUM21* gene with stop codons in place of the 3′ intron border sequence. Transformants containing the different variant *FUM21* genes, driven by a constitutive promoter, exhibited wild-type levels of fumonisin production. Although we did verify expression of the variant genes in transformants, the failure to affect fumonisin synthesis could be due to a translation failure as we were unable to determine whether any recombinant protein was present.

5 Birth, Death, and Horizontal Transfer of the *FUM* Gene Cluster

Advances in sequencing technology also contributed to studies exploring the evolution of the fumonisin gene cluster. Early work examining fumonisin production and *FUM* gene presence using PCR and Southern analysis of species of the *Fusarium fujikuroi* species complex (FFSC) and related species found that the ability to synthesize fumonisins and the presence of *FUM* genes was restricted to a limited number of species of the FFSC and one species of *F. oxysporum* [17]. These findings indicate that *FUM* genes are discontinuously distributed in the FFSC complex and match the ability of different *Fusarium* species to synthesize fumonisins. Over all, phylogenetic analysis of *FUM* genes and primary metabolism genes found that their evolutionary history was not consistent.

Further studies of the evolutionary relationships between *FUM* clusters in *Fusarium* and the genomic context of the cluster suggest that the evolutionary processes culminating in the current

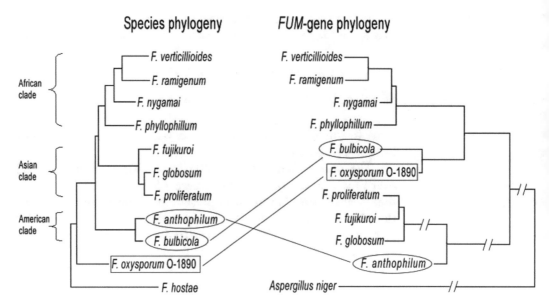

Fig. 3 FFSC species and *FUM*-gene phylogenies providing evidence for horizontal gene transfer of *FUM* cluster between *Fusarium oxysporum* and *Fusarium bulbicola* or closely related member of the FFSC. Adapted from Proctor et al. (2013)

fumonisin biosynthetic capacity across the FFSC resulted from a variety of processes including horizontal gene transfer (HGT) of the cluster and cluster duplication, sorting and loss [18]. In the case of HGT, species phylogeny based on 12 primary metabolism genes and *FUM* gene phylogeny based on 9 *FUM* genes provide strong evidence for horizontal transfer of the *FUM* cluster between *F. oxysporum* and *F. bulbicola* or closely related member of the FFSC [18]. Species trees based on primary metabolism genes resolved members of the FFSC into three well supported clades and *F. oxysporum* as distinct from the FFSC (Fig. 3) as previously described [19]. In contrast, in species trees based on *FUM* genes, *F. oxysporum* nested within the FFSC as a sister species to *F. bulbicola*. Further, the divergence of *FUM* genes was significantly less than the primary metabolism when comparing *F. oxysporum* and *F. bulbicola* [18].

6   Genome-Enabled Discovery

With the explosive growth of fungal genome sequences, the catalog of secondary metabolite genes greatly expanded. Phylogenetic analysis continues to be one of the best ways to characterize these genes. The first large phylogenetic analysis of fungal polyketide synthases took advantage of several genomes that included *Cochliobolus heterostrophus*, *Fusarium verticillioides*, *Fusarium graminearum*, *Neurospora crassa*, and *Botrytis cinerea* [20]. From this study, a polyketide synthase gene in the genome of *C. heterostrophus* with high similarity to that encoding the fumonisin polyketide synthase from *Fusarium verticillioides* was found. In addition to the fumonisin polyketide synthase, a *C. heterostrophus* gene cluster encoding orthologs of genes in the *F. verticillioides* fumonisin cluster was identified. Moreover, the genomes of two other *Cochliobolus* species, *Cochliobolus carbonum*, and *Cochliobolus sativus* appear to encode fumonisin clusters, although gaps in the assembled genomes make it more difficult to assess the structures of the associated gene clusters [21]. To date, biochemical and structural characterization of the predicted *C. heterostrophus* fumonisin cluster has not been performed. However, a related Dothideomycetes fungus, *Alternaria alternata* f. sp. *lycopersici* is known to produce AAL toxin, which like fumonisin is a sphingolipid analog mycotoxin [22]. It is tempting to speculate that the *C. heterostrophus* fumonisin cluster produces a fumonisin or something structurally related to fumonisin or AAL toxin.

7 Fumonisin in *Aspergillus niger*

Aspergillus niger is an industrial workhorse fungus that is commonly used as a production host for enzymes and organic acids, most notably citric acid. In addition, *A. niger* has GRAS (Generally Regarded as Safe) status. Because of its significant economic footprint, high quality genome sequences for two strains of *A. niger*, CBS513.88 and ATCC 1015, were generated [23, 24]. As in *Cochliobolus*, genome analysis of *A. niger* identified gene clusters predicted to encode the fumonisin biosynthetic gene cluster (Fig. 4) [24, 25].

These predicted fumonisin clusters drove the analysis of the extralites of these strains, leading initially to the discovery that *A. niger* does produce fumonisin B_2 [26]. Subsequent chemical isolation and analysis by NMR showed that *A. niger* produced fumonisin B_2, B_4 and a novel fumonisin referred to as B_6 [27]. Following the initial characterization of *A. niger* fumonisins, it was shown

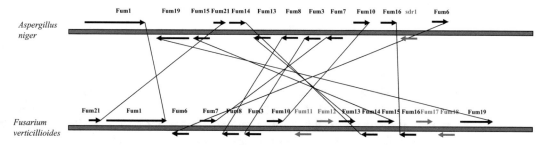

Fig. 4 A comparison of the fumonisin biosynthetic clusters of *Aspergillus niger* and *Fusarium verticillioides* shows conservation of gene content but not spatial organization

that the environmental and nutritional conditions needed for fumonisin biosynthesis were very different between *Aspergillus* and *Fusarium*, indicating that while the gene biosynthetic pathway may be conserved, regulation was most likely not conserved [28]. A significant number of *A. niger* strains have been tested for fumonisin production with over 80% testing positive in one study of wild-type and industrial strains [29]. There is currently not a consensus with regard to the ability of or inability of *A. niger* to produce fumonisins B_1 and B_3 [30–33].

In addition to its role in industrial microbiology and biotechnology, *A. niger* is an important member of microbial communities associated with grapes and other foods. Once the potential for fumonisin production by *A. niger* was demonstrated, strains isolated from these foods were isolated and tested. These studies indicated that *A. niger* associated with grapes maize, coffee, and peanuts have the ability to produce fumonisin [28, 29, 34–36]. These studies and others show the value of genome analysis in the study of mycotoxin production. *Aspergillus niger* is not the only *Aspergillus* section *Nigri* species to be isolated from food-associated microbial communities. Interestingly, studies indicate that in non-fumonisin production strains including *Aspergillus tubingensis*, *Aspergillus welwitschiae*, *Aspergillus luchuensis*, and *Aspergillus brasiliensis* there is evidence for loss of multiple genes from the fumonisin cluster as compared to *A. niger* [29].

8 Future Prospects

The explosion in genome sequencing for fungi has opened a new avenue for discovery in mycotoxin research. As more genomes are sequenced, more secondary metabolite biosynthetic pathways will be identified and products from these pathways elucidated at the structural level. As more secondary metabolites are correlated with biosynthetic pathways, genome sequencing will be able to rapidly point to the species that need to be monitored for their mycotoxigenic potential.

The acquisition of genome sequence data has highlighted a critical bottleneck in fungal research: gene function studies. Currently for most filamentous fungi, a single gene or multiple flanking genes is targeted for mutation analysis using a process that can take up to 2 weeks. Another limitation is the paucity of available selectable markers effectively limiting the "stacking" of multiple, non-linked gene mutations in a single strain. An exciting possible solution underdevelopment in a number of labs seeks to adapt CRISPR (Clustered Regular Interspaced Short Palindromic Repeats) to filamentous fungi. In bacteria, CRISPR serves as an immune system protecting the bacteria from invading viruses and plasmids. A modified version has been engineered that allows the introduction of mutations at multiple targeted locations in the genomes of eukaryotic organisms, including animals, plants, and yeasts [37, 38].

Acknowledgements

Mention of trade names or commercial products in this chapter is solely for the purpose of providing specific information and does not imply recommendation or endorsement by the US Department of Agriculture. USDA is an equal opportunity provider and employer.

References

1. Windels CE (2000) Economic and social impacts of Fusarium head blight: changing farms and rural communities in the northern great plains. Phytopathology 90:17–21

2. Amaike S, Keller NP (2011) *Aspergillus flavus*. Annu Rev Phytopathol 49:107–133

3. Bennett JW, Christensen SB (1983) New perspectives on aflatoxin biosynthesis. Adv Appl Microbiol 29:53–92

4. Chang PK, Skory CD, Linz JE (1992) Cloning of a gene associated with aflatoxin B1 biosynthesis in *Aspergillus parasiticus*. Curr Genet 21:231–233

5. Brown DW, Yu J-H, Kelkar HS, Fernandes M, Nesbitt TC, Keller NP, Adams TH, Leonard TJ (1996) Twenty-five co-regulated transcripts define a sterigmatocystin gene cluster in *Aspergillus nidulans*. Proc Natl Acad Sci U S A 93:1418–1422

6. Payne GA, Nierman WC, Wortman JR, Pritchard BL, Brown D, Dean RA, Bhatnagar D, Cleveland TE, Machida M, Yu J (2006) Whole genome comparison of *Aspergillus flavus* and A. oryzae. Med Mycol 44:9–11

7. Chang KY, Georgianna DR, Heber S, Payne GA, Muddiman DC (2010) Detection of alternative splice variants at the proteome level in *Aspergillus flavus*. J Proteome Res 9:1209–1217

8. Brown DW, Cheung F, Proctor RH, Butchko RA, Zheng L, Lee Y, Utterback T, Smith S, Feldblyum T, Glenn AE, Plattner RD, Kendra DF, Town CD, Whitelaw CA (2005) Comparative analysis of 87,000 expressed sequence tags from the fumonisin-producing fungus *Fusarium verticillioides*. Fungal Genet Biol 42:848–861

9. Brown DW, Butchko RA, Busman M, Proctor RH (2007) The *Fusarium verticillioides FUM* gene cluster encodes a Zn(II)2Cys6 protein that affects *FUM* gene expression and fumonisin production. Eukaryot Cell 6:1210–1218

10. Marasas WF, Riley RT, Hendricks KA, Stevens VL, Sadler TW, Gelineau-Van Waes J, Missmer SA, Cabrera J, Torres O, Gelderblom WC, Allegood J, Martinez C, Maddox J, Miller JD, Starr L, Sullards MC, Roman AV, Voss KA, Wang E, Merrill AH Jr (2004) Fumonisins disrupt sphingolipid metabolism, folate transport, and neural tube development in embryo culture and in vivo: a potential risk factor for human neural tube defects among populations consuming fumonisin-contaminated maize. J Nutr 134:711–716

11. Voss KA, Smith GW, Haschek WM (2007) Fumonisins: toxicokinetics, mechanism of action and toxicity. Anim Feed Sci Tech 137:299–325

12. Sydenham EW, Thiel PG, Marasas WFO, Shephard GS, Schalkwyk DJ, Koch KR (1990) Natural occurrence of some *Fusarium* mycotoxins in corn from low and high esophageal cancer prevalence areas of the Transkei, Southern Africa. J Agric Food Chem 38:1900–1903

13. Wild CP, Gong YY (2010) Mycotoxins and human disease: a largely ignored global health issue. Carcinogenesis 31:71–82

14. Hammond BG, Campbell KW, Pilcher CD, Degooyer TA, Robinson AE, McMillen BL, Spangler SM, Riordan SG, Rice LG, Richard JL (2004) Lower fumonisin mycotoxin levels in the grain of Bt corn grown in the United States in 2000–2002. J Agric Food Chem 52:1390–1397

15. Proctor RH, Brown DW, Plattner RD, Desjardins AE (2003) Co-expression of 15 contiguous genes delineates a fumonisin biosynthetic gene cluster in *Gibberella moniliformis*. Fungal Genet Biol 38:237–249

16. Sammeth M, Foissac S, Guigo R (2008) A general definition and nomenclature for alter-

native splicing events. PLoS Comput Biol 4:e1000147

17. Proctor RH, Plattner RD, Brown DW, Seo JA, Lee YW (2004) Discontinuous distribution of fumonisin biosynthetic genes in the *Gibberella fujikuroi* species complex. Mycol Res 108:815–822

18. Proctor RH, Van Hove F, Susca A, Stea G, Busman M, van der Lee T, Waalwijk C, Moretti A, Ward TJ (2013) Birth, death and horizontal transfer of the fumonisin biosynthetic gene cluster during the evolutionary diversification of *Fusarium*. Mol Microbiol 90:290–306

19. O'Donnell K, Nirenberg HI, Aoki T, Cigelnik E (2000) A multigene phylogeny of the *Gibberella fujikuroi* species complex: detection of additional phylogenetically distinct species. Mycosciences 41:61–68

20. Kroken S, Glass NL, Taylor JW, Yoder OC, Turgeon BG (2003) Phylogenomic analysis of type I polyketide synthase genes in pathogenic and saprobic ascomycetes. Proc Natl Acad Sci 100:15670–15675

21. Condon BJ, Leng Y, Wu D, Bushley KE, Ohm RA, Otillar R, Martin J, Schackwitz W, Grimwood J, MohdZainudin N, Xue C, Wang R, Manning VA, Dhillon B, Tu ZJ, Steffenson BJ, Salamov A, Sun H, Lowry S, LaButti K, Han J, Copeland A, Lindquist E, Barry K, Schmutz J, Baker SE, Ciuffetti LM, Grigoriev IV, Zhong S, Turgeon BG (2013) Comparative genome structure, secondary metabolite, and effector coding capacity across *Cochliobolus* pathogens. PLoS Genet 9:e1003233

22. Bottini AT, Gilchrist DG (1981) 1-aminodimethylheptadecapentol from *Alternaria alternata* f. sp. *lycopersici*. Tett Let 22:2719–2722

23. Andersen MR, Salazar MP, Schaap PJ, van de Vondervoort PJ, Culley D, Thykaer J, Frisvad JC, Nielsen KF, Albang R, Albermann K, Berka RM, Braus GH, Braus-Stromeyer SA, Corrochano LM, Dai Z, van Dijck PW, Hofmann G, Lasure LL, Magnuson JK, Menke H, Meijer M, Meijer SL, Nielsen JB, Nielsen ML, van Ooyen AJ, Pel HJ, Poulsen L, Samson RA, Stam H, Tsang A, van den Brink JM, Atkins A, Aerts A, Shapiro H, Pangilinan J, Salamov A, Lou Y, Lindquist E, Lucas S, Grimwood J, Grigoriev IV, Kubicek CP, Martinez D, van Peij NN, Roubos JA, Nielsen J, Baker SE (2011) Comparative genomics of citric-acid-producing *Aspergillus niger* ATCC 1015 versus enzyme-producing CBS 513.88. Genome Res 21:885–897

24. Pel HJ, de Winde JH, Archer DB, Dyer PS, Hofmann G, Schaap PJ, Turner G, de Vries RP, Albang R, Albermann K, Andersen MR, Bendtsen JD, Benen JA, van den Berg M, Breestraat S, Caddick MX, Contreras R, Cornell M, Coutinho PM, Danchin EG, Debets AJ, Dekker P, van Dijck PW, van Dijk A, Dijkhuizen L, Driessen AJ, d'Enfert C, Geysens S, Goosen C, Groot GS, de Groot PW, Guillemette T, Henrissat B, Herweijer M, van den Hombergh JP, van den Hondel CA, van der Heijden RT, van der Kaaij RM, Klis FM, Kools HJ, Kubicek CP, van Kuyk PA, Lauber J, Lu X, van der Maarel MJ, Meulenberg R, Menke H, Mortimer MA, Nielsen J, Oliver SG, Olsthoorn M, Pal K, van Peij NN, Ram AF, Rinas U, Roubos JA, Sagt CM, Schmoll M, Sun J, Ussery D, Varga J, Vervecken W, van de Vondervoort PJ, Wedler H, Wosten HA, Zeng AP, van Ooyen AJ, Visser J, Stam H (2007) Genome sequencing and analysis of the versatile cell factory *Aspergillus niger* CBS 513.88. Nat Biotechnol 25:221–231

25. Baker SE, Kroken S, Inderbitzin P, Asvarak T, Li BY, Shi L, Yoder OC, Turgeon BG (2006) Two polyketide synthase-encoding genes are required for biosynthesis of the polyketide virulence factor, T-toxin, by *Cochliobolus heterostrophus*. Mol Plant-Microbe Interact 19:139–149

26. Frisvad JC, Smedsgaard J, Samson RA, Larsen TO, Thrane U (2007) Fumonisin B2 production by *Aspergillus niger*. J Agric Food Chem 55:9727–9732

27. Mansson M, Klejnstrup ML, Phipps RK, Nielsen KF, Frisvad JC, Gotfredsen CH, Larsen TO (2010) Isolation and NMR characterization of fumonisin B2 and a new fumonisin B6 from *Aspergillus niger*. J Agric Food Chem 58:949–953

28. Mogensen JM, Nielsen KF, Samson RA, Frisvad JC, Thrane U (2009) Effect of temperature and water activity on the production of fumonisins by *Aspergillus niger* and different *Fusarium* species. BMC Microbiol 9:281

29. Frisvad JC, Larsen TO, Thrane U, Meijer M, Varga J, Samson RA, Nielsen KF (2011) Fumonisin and ochratoxin production in industrial *Aspergillus niger* strains. PLoS One 6:e23496

30. Palencia ER, Mitchell TR, Snook ME, Glenn AE, Gold S, Hinton DM, Riley RT, Bacon CW (2014) Analyses of black *Aspergillus* species of peanut and maize for ochratoxins and fumonisins. J Food Prot 77:805–813

31. Susca A, Proctor RH, Butchko RA, Haidukowski M, Stea G, Logrieco A, Moretti A (2014) Variation in the fumonisin biosynthetic gene cluster in fumonisin-producing and nonproducing black aspergilli. Fungal Genet Biol 73:39–52

32. Nielsen KF, Frisvad JC, Logrieco A (2015) "Analyses of black Aspergillus species of pea-

nut and maize for ochratoxins and fumonisins," A Comment on: J. Food Prot. 77 (5): 805–813 (2014). J Food Prot 78:6–8

33. Palencia ER, Mitchell TR, Bacon CW (2015) "Analyses of black *Aspergillus* species of peanut and maize for ochratoxins and fumonisins," A response to a comment on: J. Food Prot. 77 (5): 805–813 (2014). J Food Prot 78:8–12

34. Noonim P, Mahakarnchanakul W, Nielsen KF, Frisvad JC, Samson RA (2009) Fumonisin B2 production by *Aspergillus niger* in Thai coffee beans. Food Addit Contam Part A 26:94–100

35. Mogensen JM, Larsen TO, Nielsen KF (2010) Widespread occurrence of the mycotoxin fumonisin B2 in wine. J Agric Food Chem 58:4853–4857

36. Susca A, Proctor RH, Mule G, Stea G, Ritieni A, Logrieco A, Moretti A (2010) Correlation of mycotoxin fumonisin B2 production and presence of the fumonisin biosynthetic gene *fum8* in *Aspergillus niger* from grape. J Agric Food Chem 58:9266–9272

37. DiCarlo JE, Norville JE, Mali P, Rios X, Aach J, Church GM (2013) Genome engineering in *Saccharomyces cerevisiae* using CRISPR-Cas systems. Nucleic Acids Res 41:4336–4343

38. Sander JD, Joung JK (2014) CRISPR-Cas systems for editing, regulating and targeting genomes. Nat Biotechnol 32:347–355

INDEX

Antonio Moretti and Antonia Susca (eds.), *Mycotoxigenic Fungi: Methods and Protocols*, Methods in Molecular Biology, vol. 1542, DOI 10.1007/978-1-4939-6707-0, © Springer Science+Business Media LLC 2017

Printed in the United States
By Bookmasters